ROUTLEDGE LIBRARY EDITIONS: GEOLOGY

T0200551

Volume 11

GEOLOGICAL EXPLORATIONS IN CENTRAL BORNEO (1893–94)

GEOLOGICAL EXPLORATIONS IN CENTRAL BORNEO (1893–94)

G.A.F. MOLENGRAAFF

Routledge
Taylor & Francis Group

LONDON AND NEW YORK

This edition first published in 2020
by Routledge
2 Park Square, Milton Park, Abingdon, Oxon OX14 4RN

and by Routledge
52 Vanderbilt Avenue, New York, NY 10017

Routledge is an imprint of the Taylor & Francis Group, an informa business

First published in 1902 by Kegan Paul, Trench, Trübner & Co Ltd.

British Library Cataloguing in Publication Data
A catalogue record for this book is available from the British Library

ISBN: 978-0-367-18559-6 (Set)
ISBN: 978-0-429-19681-2 (Set) (ebk)
ISBN: 978-0-367-46445-5 (Volume 11) (hbk)
ISBN: 978-0-367-46451-6 (Volume 11) (pbk)
ISBN: 978-1-00-302881-9 (Volume 11) (ebk)

Publisher's Note
The publisher has gone to great lengths to ensure the quality of this reprint but points out that some imperfections in the original copies may be apparent.

Disclaimer
The publisher has made every effort to trace copyright holders and would welcome correspondence from those they have been unable to trace.

BORNEO-EXPEDITION

GEOLOGICAL EXPLORATIONS

IN

CENTRAL BORNEO

(1893—94)

BY

Dr. G. A. F. MOLENGRAAFF

FORMERLY PROFESSOR AT THE UNIVERSITY OF AMSTERDAM AND GOVERNMENT GEOLOGIST OF THE SOUTH AFRICAN REPUBLIC.

With 89 illustrations in the text, 36 Plates, 3 Maps and an Atlas of 22 Geological Maps.

ENGLISH REVISED EDITION

with an Appendix on fossil Radiolaria of Central-Borneo

BY

Dr. G. J. HINDE.

Published by the Society for the Promotion of the Scientific
Exploration of the Dutch Colonies

E. J. BRILL
LEYDEN

H. GERLINGS
AMSTERDAM

1902

CONTENTS.

PREFACE TO THE FIRST DUTCH EDITION.

The virtual promoter of the Borneo-Expedition of which this work is one of the results, was the Resident of the „Wester-afdeeling van Borneo" Mr. S. W. TROMP.

As early as 1891 Mr. TROMP considered the feasibility of a scientific Expedition to Central Borneo, in conjunction with the topographical survey, begun in 1886 and at that time in full action. While TROMP was considering his plans chance led Professor HUBRECHT to visit Pontianak. Prof. HUBRECHT was at that time president of the scientific Committee which acts as adviser to the Society for the Promotion of the scientific exploration of the Dutch Colonies, which then had the ex-minister Baron v. GOLTSTEIN as its president.

On Prof. HUBRECHT's return to Holland TROMP's projects were made known in the mother-country, and led to the inauguration of the Borneo expedition under the auspices of the Society mentioned above.

The expedition was chiefly organized by the Resident of West Borneo, and by the Indian Committee of the same Society. The financial question was happily solved by the direct and indirect support of the Dutch Colonial Government, who not only gave the entire yearly subvention for the promotion of Scientific exploration in the Dutch Indies, during the years 1893, 1894 and 1895 to the Society for this purpose, but also allowed the members of the expedition the free use of the Government means of transport in Borneo, which caused, as this privilege was extended to their luggage, a great saving.

A large share in the cost of the Expedition was borne by
the Society itself, while a considerable sum was obtained by the
donations of private individuals and Societies (Dutch Scientific
and Medical Congress, Society of Arts and Sciences of the province of Utrecht, etc.).

It was originally planned that the members of the expedition
should conjointly cross Borneo from west to east by ascending the
Kapoewas and descending the Mahakkam. After consultation with
the Government this plan was abandoned [1]), and preference was
given to a scientific exploration of Central Borneo, more especially
of the sources of the Kapoewas and its principal tributaries.

The participants in this expedition, which occupied the closing
months of 1893 and the whole of 1894, were: zoologist: Dr.
Büttikofer, at that time Keeper of the Royal Museum of Natural
History at Leiden; botanist: Dr. Hallier then assistant at the
Herbarium of the Botanic Gardens in Buitenzorg; anthropologist
and ethnographer: Dr. Nieuwenhuis Dutch Colonial Army-surgeon,
while the geological exploration was intrusted to myself.

This work embodies the geological results obtained. The honour
of assisting to obtain these results is shared by many. I feel
it my duty to name some specially; in the first place S. W.
Tromp, the efficient and energetic Resident of West Borneo whose
untimely death is much lamented. From beginning to end he
promoted with the greatest devotion the objects of the expedition,
and he may, without exaggeration, be called the moving force
of the undertaking.

I am further greatly indebted to the lieutenant-colonels H. Bosboom and J. J. K. Enthoven and their staff of the Military topographical survey of Borneo, who liberally placed at my disposition
the results obtained by them, including those as yet unpublished. It is owing to their courtesy that I was able to compare
the sketches I made during the journey with the very elaborate
and exact maps of the Topographical survey, which enabled me

---

1) It has since been fully realized by Dr. Nieuwenhuis in 1898.

to give a geological edition of the road and river maps of this survey, for the territory explored by me.

I have furthermore occasion to remember with great gratitude, Dr. and Mrs. VAN DER STOK of Batavia, the Assistent Resident of Sintang Mr. J. C. C. SNELLEBRAND, Dr. GOEDHUIS of Sintang, and Prof. Dr. TREUB of Buitenzorg, all of whom showed me much hospitality, and contributed greatly to my keeping a much cherished remembrance of my Indian voyage.

I greatly appreciate the help, support and hospitality rendered and shown to me by the "controller" of Tajan, the "aspirant-controller" of Nangah Badau, the Army Officers at Sintang and the Catholic missionaries at Sedjiram.

My heartfelt thanks are rendered to Dr. G. JENNINGS HINDE, who investigated the Radiolaria, to Messrs C. SCHLUMBERGER and R. BULLEN NEWTON who treated of the Nummulites, to Dr. P. G. KRAUSE and Prof. Dr. K. MARTIN who examined the fossil shells of my collection, and to Dr. J. L. C. SCHROEDER VAN DER KOLK, who assisted me in the microscopical investigation of a great many of the rocks collected.

Finally I wish openly to thank the Executive Committee of the Society for the Promotion of scientific exploration in the Dutch Indies for the liberal way in which they made the publication of my results possible, and the publishers, who by my departure to South Africa worked under difficult conditions.

Pretoria, October 1st 1899.   MOLENGRAAFF.

PREFACE TO THE ENGLISH EDITION.

The English edition is a faithful translation of the original Dutch text, in which nothing essential has been altered, and only a few small inaccuracies have been corrected. In preparing the English edition no account has been taken of the literature published after the Dutch text was completed (1899) [1].

To Dr. G. JENNINGS HINDE, who has kindly read all my proofs and revises and has given me much valuable advice on this occasion, I wish to tender my special thanks for the arduous labour involved in this task.

Hilversum, March 25th 1902.   MOLENGRAAFF.

---

[1] Among recent publications, treating of the geology of West Borneo mention may be made of:

N. WING EASTON. Voorloopige mededeeling over de geologie van het stroomgebied der Kapoewas-Rivier in de Wester-afdeeling van Borneo. Tijdschr. v. h. Kon. Ned. Aardrijksk. Genootschap 1899.

INTRODUCTION.

In the first place I wish to state my objects in writing this book, and to explain why it is written in the way it is now presented to the public.

The book aims at giving a true and vivid picture of the region explored by me, and at nothing more. Those who might expect much new information on ethnographical questions will be disappointed. My numerous notes on this subject remain unpublished for the very good reason that I understood the language of the different Dyak tribes whose countries I visited either imperfectly or not at all, while I could not obtain the services of an interpreter. For this reason I had frequently to guess at the significance of many of those objects, whose real import would ethnographically have been of the greatest interest. As I did not wish to augment the large number of conjectures of an ethnographical nature, I have thought it advisable to leave these subjects, with a few exceptions, undiscussed.

I have therefore limited my ambition to giving a description of the character of the landscape and the general geological structure of that part of Central Borneo which it was my good fortune to visit.

I have aimed, before everything else, at making the description useful to future explorers, and for this reason everything has been mentioned which can aid in the verification and identification of the geological and geographical observations recorded. I have therefore indicated with the greatest precision on the maps of the roads and streams, the localities where geological specimens were collected, while both in the text and on the maps the original numbers of the collections, which are now to be found in the museum of the University of Utrecht, are quoted. This exact information will doubtless enable future investigators to decide in any particular case respecting the material and the localities indicated in my work. The determination of all the rocks mentioned on the maps or described in the text, is based on microscopical investigation; elaborate petrographical descriptions, however, are not embodied in the work.

For the same reason the original form of a diary has been retained, although in so doing I run a certain risk of damaging the style of my story. The precise hours of arrival and departure of every day and every trip are indicated, while here and there the wages of carriers, rowers and other labourers are mentioned with the view of giving a basis for the calculation of the expenses of future travellers.

I wish here to make a remark in reference to the second and third parts of the twelfth Chapter, entitled "The Tectonics of Central Borneo" and the "Historical Geology of Central Borneo". Their hypothetical character deviates from the more descriptive one of the other chapters, and their right of existence must here be defended to a certain extent on the ground that the facts obtained by me are in reality insufficient for a complete insight into the subjects with which these chapters deal. This becomes painfully evident by a glance at the map inserted at the end of the index which shows the route followed by me. I was obliged to form for myself an idea of the geological structure of the whole of Central Borneo from the investigation of a limited number of sections. The way in which I tried to combine the facts in hand, and the results obtained from this combination are found in the parts of the chapters mentioned. Every future exploration and every new series of facts thus obtained will probably modify my interpretation. But, even if finally very little or nothing will have remained of my interpretation of the facts, it will yet, I think, have done good service in stimulating the comparative and critical investigation of future explorers.

Finally, the spelling of all local names is the same as in the recently published Dutch topographical map of Western Borneo on the scale of 1:200.000, and in the road and stream maps belonging to it on the scale of 1:50.000, being the result of the survey during the years 1886—1895. The illustrations in the text and atlas are prepared from photographs or sketches made by me on the spot. Many of the photographs and all the sketches have been improved by the able pencil of Mr. G. B. HOOIJER.

LIST OF THE MAPS, PLATES, AND TEXT-ILLUSTRATIONS.

— — — ·

MAPS.

PLATES.

TEXT-ILLUSTRATIONS. (*Abbreviated titles*).

Pl. I.

House in the Kampong Těbang.

CHAPTER I.

The River Kapoewas below Sĕmitau.

The name of „Mudland" by which Borneo and especially West-Borneo is known, has little in it to attract the geologist, but so far, I had not been able to judge for myself whether it deserved to be so called. As, however, on the 8th of February 1894, we approached the mouth of the Kapoewas, I realized whence the name must have originated. Mud greets the new-comer far out into the sea. At a distance of 50 kilometres from the land, the clear seawater becomes discoloured by the mud carried out by the Kapoewas, and only very slowly and gradually it mixes with the salt water. The line which separates the salt from the fresh water·is very clearly marked, there is a slight ripple on the water and an accumulated mass of vege-table matter and scum. When the water level is high in the rivers, which for the Kapoewas is usually during the months of November, December and January, this line of demarcation extends out as far as Poelau¹) Datoe, fully 62 kilometres beyond the mouth of the Kapoewas Kĕtjil.

The real coastline had not yet come into view, but looking north I could see the hills of the Chinese districts, like an archipelago of rocky islands, rising out of the morning mist. Right in front of us was the Goenoeng²) Lontjek (124 metres), the most westerly outpost of the hills east of the marshy delta

---

1) Poelau = island.
2) Goenoeng = mount.

of the Kapoewas, forming a capital landmark. Southeastward,
well within the delta, rises the Ambawang (450 metres) an
isolated mountain surrounded by flat marshland, once no doubt
surrounded by the sea, before the alluvial deposits of the
Kapoewas had annexed it.

We were now fast approaching the reputed bar[1]), which is
said to be most dangerous to cross at about 4 kilometres from
the shore, because of the sandiness of the bottom there. Further

Fig. 1. THE KAPOEWAS NEAR MOUNT KĔRAMAT.

out the bar is higher, but consists entirely of mud. A boat
of small size can work its way through the mud even if the

---

1) At the mouth of almost all rivers of Borneo, there are shallows or bars, running
parallel with the coast lines. They very materially deteriorate the value of the great rivers
as lines of communication. From what I know of the large rivers of Dutch Borneo, the
Sampit is the only one which is not impeded by a mud bar, while the bar of the
Sambas because of its comparatively low level — at high tide it is about 4 metres deep —
offers no serious obstacle to navigation.

depth of water is a foot or two short of the draught of the vessel, but when the mud is sandy this becomes impossible. As we entered, there were 9 feet of water on the bar, so we could glide over it without much difficulty. When the bar is passed, the bed of the river becomes very deep, and ships of any size might pass over it. From its mouth up to the town of Pontianak the Kapoewas Kĕtjil[1]) flows on smoothly between banks covered with nipah palms, and here and there along the shore may be seen the apparently floating houses of the Malay fishermen. At Batoe[2]) Lajang, the monotony of this ever verdant scenery is somewhat relieved by the narrowing of the stream and the appearance of a small island. On the right side several masses of rock stand out of the water, and the only navigable channel is on the left side of the stream.

A little higher up, the river curves slightly to the southeast, and the beautiful picture of the Malay portion of Pontianak unfolds itself. The Sultan's residence, the missigit[3]) and the greater part of the Malay Kampong[4]) are situated on the tongue of land at the confluence of the Kapoewas and the Landak. The remainder of Malay Pontianak is on the right shore of the combined rivers. On this same shore a few small sailing-vessels and bandongs[5]), lie at anchor. Presently the landing-place of the European portion of Pontianak comes in view, and behind it, a little further upstream and on the same shore, a dense mass of bandongs, some moored to the bank, others lying at anchor in the river secured by long rattan ropes, a sure sign that the Chinese passir[6]) is there. At 8 o'clock we reached the landing-place and after a few formalities I went to the badly managed hotel, close

---

1) The Kapoewas Kĕtjil is the most northerly of the many branches in which the Kapoewas is divided in this gigantic delta-district; it is estimated that this branch carries down to the sea a sixth part of the total mass of water of the river.

2) Batoe = stone, rock.

3) Missigit = Mohammedan place of worship.

4) Kampong = settlement, village.

5) Bandong = large roofed boat for the transport of merchandise.

6) Passir = market-place, also part of a settlement, where the native or Chinese shops are found.

by. That same day I took a sampan[1]), and rowed up to the Batoë
Lajang already mentioned. It is a small hill about 7 metres

Fig. 2. BATOE LAJANG.

high composed of coarse amphibole-biotite-granite, large boulders of which lie on the surface, (see fig. 2). This granite hill
partly forms the foundation of a decayed Malay fortress, where
still a few cannons remain in position. Downstream, close to Batoë
Lajang, are a missigit and the graves of the sultans of Pontianak.
The granite rock slightly extends into the river and stretches
across to the other side as a submerged bank which forms a
small island close to the right shore. I was told that on the left
bank rocks of the same granite may be found, evidently the
continuation of the hill in that direction. I did not visit the
place because a strong westwind blowing straight from the sea,
the precursor of heavy rain, ruffled the water to such an extent
that it was not safe to venture any further in our small sampan.
So we returned to Pontianak with the greatest possible speed.

Pontianak itself lies low, and the ground is swampy. At high
tide it becomes flooded daily, where not artificially protected, where-

---

1) Sampan = small, open boat or canoe.

as at low tide the river leaves an unsightly mass of mud behind.
To these tidal occurrences the drainage of the town is intrusted.

As Pontianak had no charms for me, I took the first opportunity which presented itself to go further upstream. I left on
February 10ᵗʰ at half past eight in the morning in the Ban Tik,
a small Chinese boat which had to pull up three bandongs.
The Kapoewas Kĕtjil is at first fairly thickly set with houses
on either side, but after the kampong Rasan is past there is
a decided falling off in their numbers. The shores reveal a terrace
vegetation with brushwood from 3 to 4 metres high bordering
the river and behind this the real forest appears, with trees

Fig. 3. Mount Bĕloengai.

ranging from 20 to 30 metres in height. The shores of the Kapoewas are so monotonous that I got tired of them before the first
day was over; this was my first experience of one of the darkest
sides of travel in Borneo, viz: the oppressive, leaden monotony
of its rivers. Soeka Lanting is a beautiful spot; here the river
divides into two branches, the Poengoer Bĕsar and the Kapoewas Kĕtjil. I was delighted to find the river at this point opening out into a large clear sheet of water a kilometer in width.

Shortly after passing Soeka Lanting, the Kapoewas divides
into several branches, which meet again soon afterwards and
enclose small islands in their embrace.

We spent the night on the deck of our vessel and drifted past
Djamboe Island, but we got so entangled in the bushes which

skirt the banks that in the morning we found ourselves not much beyond that place. At some little distance a few solitary mountains rise in bold relief from the low lying land; the most striking are Mount Běloengai (770 metres) (fig. 3) E. N. E. and Mount Kédikit (429 metres) (fig. 4) E. to S. On the south-side of Djamboe Island another branch of the river tends southward, and by afternoon I reached the upper part of the

Fig. 4. MOUNT KĚDIKIT, SEEN FROM THE KAPOEWAS NEAR DJAMBOE ISLAND.

delta, above Sěparoh Island, and then saw the entire width of the Kapoewas river before me. Straight in front the prospect was cut off by Mount Sěbajan and other hill-tops of the ranges situated between the Lower Tajan, and the Landak river. At Tjempěde Island near the mouth of the river of that name the characteristic sharply pointed cone of Mount Tijoeng Kandang (887 metres) (fig. 5) first comes in view. This mountain is held sacred in the neighbourhood, and the Dyaks moreover believe it to be the abode of the departed; its summit can be reached in two days from Tajan. On the first day one travels as far as Batang Tarang and the second suffices for the ascent. We reached Běloengai Island at 4 p. m. The river is here at its widest, being from 1400 to 1600 metres in width. At 6 p. m. I landed on the island Tajan, close to the house of the con-troller[1]) Westenenk, who shortly after welcomed me most heartily.

1) Controller, i. e. the highest civil official of a district.

His house stands on the highest extremity of the island,
and the fast fading daylight only just enabled me to enjoy for
a few moments the glorious view of the majestic stream, from

Fig. 5. MOUNT TIJOENG KANDANG, SEEN FROM THE KAPOEWAS NEAR TJÉMPEDE.

the spot where a neat block of masonry indicates the position
of the astronomical station Tajan ¹).

In the evening the controller and myself made our plans for
the next two days at my disposal, before the Kwantan ²) would
be ready to take me on to Sintang. We decided to make an excursion in the hilly district in the vicinity of the kampong Tĕbang.

At sunrise next morning I went about ³/₄ kilometre down-
stream to examine some granite blocks exposed on the right-
shore of the Kapoewas. These blocks are not boulders, but form
the outcrop of a granite boss, the rock is in many places much
weathered and altered into yellowish brown laterite. One of these
rocks, the *batoe bĕlang* ³) or „spotted rock", is held in great awe
by the natives. It is almost black with a coating of lichens but
in some places the lighter weathered surface of the granite may
be seen. The undecomposed rock both here and lower down the
stream is a light coloured amphibole-biotite-granite.

At 8.43 a. m. we left Tajan, in a long open sampan, here called
sampan djaloer, to go up the river Tajan. The Malay kampong
is built on the two shores close to where the river joins the

---

1) The astronomical stations, mentioned in several places in this book, are beacons,
indicated by pillars of masonry, whose astronomical position was accurately defined during
the topographical survey of this colony.

2) A small Chinese steamer doing regular fortnightly service between Pontianak and Sintang.

3) The batoe bĕlang is „pantang" for the natives. They believe that evil would come upon
them if the ground round about were disturbed, trees felled, or the rock itself in any
way meddled with. The probable reason for this, besides the curious marks produced on
the rock itself by the lichens, is that many years ago an execution had taken place there.

Kapoewas. The Chinese settlement is on the island Tajan which is Government property.

Along the shores of the Tajan, a number of ladangs ¹) have been laid out, and all the wood is cleared with the exception of some tall *rĕnggas*-trees. This tree has a very stout white trunk, and the bark, which is always pealing off, hangs down in long red scaly festoons. The rĕnggas is feared both by Malays and Dyaks, because of the acrid milky juice which oozes from the bark when the tree is felled, and they therefore, generally, leave it untouched.

A long pintas ²), here called *bĕntassan*, considerably shortened our passage upstream, and we entered a rantau ³) which brought us up to a tributary ⁴), on the right side of the Tĕbang, which we followed. In its lower course, this stream runs through low lying land and, as soon as it rises above its normal height through a high state of the water in the Tajan, it immediately overflows its banks and a 'danau' or lake is formed. A 'danau' of this description is not, of course, a lake in the ordinary meaning of the word, i. e. an open sheet of water. It is most often merely submerged woodland, which with the fall of the river becomes dry ground again. We happened to get there when the land was deep under water, and I was struck with the accuracy with which the pilot steered the boat through the channel of the river, which as far as I could make out, was in no way distinguishable from the rest of the lake. Thus proceeding we soon found ourselves in the very heart of a tropical virgin forest. How I enjoyed the sight of it. There stood the mighty forest giants, their trunks covered with climbing and parasitical plants, supporting like so many columns the leafy dome overhead while, below, the foot of each trunk, covered

---

1) Ladang = a dry, non-irrigated rice-field, in opposition to a sawah, i. e. a wet or irrigated rice-field, which in the common type of cultivated ground on the island of Java.

2) Pintas = short waterway connecting two parts of a river by cutting off one or more bends.

3) Rantau = long and straight or hardly perceptibly winding part of a river.

4) Malays and Dyaks determine the position of a tributary in exactly the opposite way to what we do. They reckon from the mouth, and call a leftside tributary „*kanan moedik*", i. e. to the right when going upstream, or simply „*kanan*", i. e. right.

Pl. II.

A SACRED TREE NEAR THE KAMPONG TÉBANG.

with plants and flowers, formed a little garden in itself. It is only a first impression that can give us such exquisite enjoy-ment. Often afterwards I have been disappointed with myself for growing so soon accustomed to these novel sights; never again does that first enthusiastic rapture return, nay, worse, after having roamed through the most glorious virgin forests for months together and experienced the many obstacles and difficulties they offer, I began to loathe the very sight of these endless, incomparably beautiful, silent forests, and positively longed for space and motion and air and sunshine!

The banks were now fast becoming higher, and the stream perceptibly narrower. In the afternoon we took a short rest close to some very picturesque Dyak graves, hidden in the forest, not far from the river side. The ornamental designs on

Fig. 6. DYAK GRAVE ON THE TÉBANG.

these graves reminded me of those in vogue amongst the Bataks of Sumatra, (fig. 6 and 7). One of them (fig. 7) was the grave of the wife of a Dyak Chief; underneath was the kat-jang-matting, and by the side of the grave were ranged all the

Fig. 7. DYAK GRAVE ON THE TĔBANG.

belongings of the deceased, from personal ornaments to domestic articles, used by her during her lifetime; there was also food, ready for use, to support her in the "hereafter".

At this point navigation became much impeded by numbers of trunks of trees lying in all directions in and across the stream. Sometimes they had to be cut through, sometimes we were able, by partially unloading the boat, to drag it underneath or over the trees. The shores continued low and sandy until about 200 metres below Tjempĕda, where the solid rock is exposed for the first time on the right shore. It is a diabase, deeply weathered into brown clay. Arrived at the kampong Tjempéda we found it impossible to work the large sampan any further upstream, we therefore left the waterway and proceeded on foot along the Dyak footpath as far as Tébang, accompanied by the Toemenggoeng (a title claimed by the majority of Dyak Chiefs) of Tjempĕda. This was my first acquaintance with the Dyak footpaths, which, as I learned by experience, have the same peculiarities everywhere. They are, so to speak, the only lines of communication by land in Borneo, and are therefore well worthy of a brief description. The Dyak path is invariably so narrow that no two persons can go abreast along it. It is made through the forest by cutting down all minor obstructions and avoiding the larger ones. Any tree that does not easily yield to a few smart strokes of the parang¹), is left standing. As a rule the Dyak path makes straight for its destination, regardless of hills and mountains and taking still less account of brooks, marshes or small rivers. Marshes are avoided only if too deep to ford; often in very swampy places a sort of bridge is made of thin trees resting lengthways one upon another. Small rivers and brooks are forded; if they are too deep, a trunk of a tree is laid across by way of bridge. In crossing great heights the path retains generally its straight course, and is therefore as a rule very steep. The Dyaks never use zig-zag paths when scaling high mountains, and the strain upon the European is very considerable on such occasions, and constant attention is necessary, for both head and feet are always exposed to danger. As the forest soil in Borneo consists for the greater part of rich clay, the paths are often very slippery, which makes mountain-climbing very tiring. Unless one has learned the trick of hoisting oneself by the lesser trunks and branches of trees, or leans with all one's might on a long stick, it is almost impossible for a booted European to accomplish the ascent. Greatly to be recommended for the purpose are mountain shoes, the soles of which are furnished with several rows of brass nails. The natives do not experience the least inconvenience not even when carrying heavy loads; they somehow dig

---

1) Parang — long hatchet, a kind of sword.

their toes deep into the ground, and are thus saved from all danger of slipping.

The choice of the direction of the Dyak footpath depends on experience. Many accounts of travel state, as a peculiar characteristic of primitive nations, including amongst them the Dyaks, that intuitively they seem to know the right direction, even when in a dense forest, but my experience with the Dyaks was different. They possess, alike with all primitive nations, a keen sense of observation, and a good memory for locality, and a way, once gone over, they can easily identify by slight landmarks. Moreover they cross forests which are practically impenetrable to Europeans, with the greatest ease and swiftness. I always noticed that they are not guided by the direction of the wind or the position of the sun, and if they do lose their way, they do not instinctively recover it again. On such occasions they try different directions one after the other until they find the right one, and here their excellent memory, and their keen observation of small details are of great assistance. When crossing the Madi plateau we were on several occasions lost in the forest, and at last I took the lead, compass in hand, to avoid the endless delay of trying all the different tracks made by the gĕtah-seekers [1]).

At last after a tramp of an hour and a quarter we arrived at Tébang (Pl. II) at 6 in the evening. Tébang is a Dyak kampong, consisting of six houses (Pl. I) inhabited by 21 families. The four principal houses of the settlement are situated on a tongue of land, enclosed by a bend of the Tébang river. The solid rock, consisting of diabase-porphyrite, crops out through the muddy soil near the waters edge on both sides of this strip of land.

On Febr. 13 we resumed our journey at an early hour, and at first we went though a wood rich in arèn palms. Arèn sugar is made and sold here in large quantities; this sugar is so famous that in Borneo the 'goela arèn' is generally 'called 'goela Tajan'. We crossed the Témbawang Oedjoeng ridge

1) Gĕtah i. e. the native name of the milky juice of a tree, which after preparation is known as gutta percha.

(30 metres high) and made for Mount Kétoejoe, a hill, east of
Tébang. Its western slope is almost entirely laid out as a ladang,
belonging to Galak, the Pasirah (sub-chief) of Tébang. On this
ladang, about half-way up the mountain is a shed whence one
has a most glorious view over the valley of the River Tébang
and the mountains surrounding it. Opposite to us, and almost
due west, was Mount Intingin, the most northwesterly peak of a
range of hills, enclosing the Tébang on the west, and continuing
southward as far as the Kapoewas near Tajan.

To the north of this mountain the eye roams over a low
hilly district, above which in the far distance to the west-south-west
rises Mount Ambawang. Still further away above the limitless
lowland to the west (the Kapoewas delta), the sharply outlined
Mount Pajoeng is just visible. The most conspicuous feature in
the landscape is Mount Tijoeng Kandang (fig. 8), a cone-shaped
mountain, deeply furrowed by erosion, with a prominent and very
steep top, resting on a sloping pedestal.

On the southwestern slope of Mount Kétoejoe porphyrite
appears in several places. (I, 164—66).

We continued our journey southward and soon reached an
amphibole-biotite-granite area; the first time I met with this
rock was at Parong Pétandan, where the sources of the River
Sémattan, a leftside tributary of the Tébang, arise. There in
the solid granite rock a dyke of mica-porphyrite occurs, and scattered in the bed of the brooks numerous pieces of white vein-quartz
are found. In one of these brooklets, near Lobang Batoe, a
beautiful granite landscape has been formed, where in the hollows under the roughly piled up granite blocks, often as large
as a house, colonies of bats have taken up their abode. In several
of these hollows the ground is covered with a thick layer of bat-
dung. Up to Tjempéda, granite continues to be the solid rock.

At one p. m. we left Tjempéda in the sampan and shooting
downstream very rapidly, we reached Tajan at 5.30 that evening.
At 10 p. m. the Kwantan steamboat also arrived there and I
resumed my journey up stream.

The hilly district of Tajan was now soon left behind, and my little excursion had convinced me that these hills are built up principally of granite, occasionally alternating with porphyrite,

Fig. 8. MOUNT TIJOENG KANDANG.

which in many places has forced its way through the granite. There can be no doubt that this granite boss is a continuation of the granite of the Chinese district with which we may also reckon that of the Batoe Lajang near Pontianak, and these again are probably connected with the granite territory of the Riou archipelago and Malacca by means of the islands of the Chinese-sea.

The Kwantan had only to tow up a few barges, and went pretty fast so that at daybreak we had come as far as the

Malay kampong Sěmarangkai from where one has a beautiful view over the flat topped hills Mt. Ganting (215 metres), Mt. Koenta (252 metres), Mt. Matjan (255 metres) and Mt. Kě-ramat (260 metres, fig. 1). They give the impression of being the remnants of a formerly united mountain plateau.

According to EVERWIJN (*11 pp. 41 etc.*) the country rock here is all along composed of nearly horizontal strata of sandstone and claystone, amongst which occasional layers of lignite may be found. These rocks which are said to occur along the Kapoe-was from Tajan as far as the mouth of the River Sěpauk, he considers to be of miocene age, and to form part of the great tertiary basin of the Kapoewas, the oldest division of which is of eocene age, and comes to the surface above Sěpauk.

As the domain of my investigations lies much further upstream, beyond Sintang, I was not able to judge of the accurateness of these observations which were made chiefly along the banks of the Kapoewas. At the only place however, where I had the chance of a cursory examination, I found the rocks spoken of by EVERWIJN. This was at Kajoe Toenoeh, where the Kwantan stopped for an hour to take in wood for the boilers. This gave me time to row across to the other side where, a little lower down, the shore is steep and rocky. The following strata could here be seen in succession ranging from bottom to top, i. e. from the water surface to the vegetable earth:

1°. Sandstone (3 metres) in layers of 10—30 centimetres.

2°. Sandy claystone, thinly laminated, with round, flattened, harder concretions.

3°. Claysandstone in distinct somewhat thicker layers.

All these strata dip 5° to the northeast.

After a short delay at Sanggau (Pl. III) at the house of the con-troller we reached Sěkadau that same day, and the next morning at sunrise I saw in a bend of the river above Bělitoeng, the grotesquely shaped peak of Mount Koedjau, rising before me to the south. This mountain is the twin brother of Mount Kělam, and like two gigantic citadels they command the lower part of the Mĕlawi valley. I was not able to visit Mt. Koedjau, it can be reached from the mouth of the Sĕpauk in $4^1/_2$ days, and from the mouth of the Tĕmpoenak in three days.

About midday we came in sight of Sintang (Pl. IV) after working our way through the terribly tortuous windings of the river beyond Bĕlitoeng. To the left of the plate (i. e. on the right shore) is seen the Malay kampong. It is thickly overgrown with palm trees, and boasts of a missigit and the house of the panembahan[1]), a very lofty building, otherwise it is similar to all other Malay kampongs; on the left shore lies the straggling, well-built Chinese encampment, with several bandongs moored in front of it, some laden, others empty. Before us rises the Dutch fort, surrounded by black palisades. It is built on the extreme point of the strip of land where the Kapoewas and the Mĕlawi converge. Beyond this, on the left shore of the Kapoewas, comes the European part of Sintang with an avenue of high canary-trees by the waterside. In the background, towering above the trees, rises Mount Kĕlam, its mighty crest shrouded in clouds which hang over it like a transparent veil. (Pl. IV, compare also page 21 of the atlas, where this part of the Kapoewas is shown from the opposite direction).

Soon after we were moored to the tottering pier, and on landing I was welcomed by the assistant-resident S. J. J. SNELLE-BRAND, and introduced to one of my fellow members of the Borneo expedition, Dr. A. HALLIER, who had just come back to Sintang from his botanical investigations on Boekit Kĕlam. Sintang is the key to the interior of West Borneo, on account of its situation at the confluence of the Kapoewas and the Mĕlawi. There is a good deal of traffic and trade, which is however chiefly confined to the Chinese quarter. Sintang is the residence of an assistant-resident, and it has a small European garrison. From the European quarter one gets an exceptionally beautiful view of the river Kapoewas of which fig. 10

1) Panembahan = Malay prince.

Pl. III.

SANGGAU ON THE KAPOEWAS-RIVER.

will give some faint idea. The commodious house of the assistant-resident (Fig. 9) where I was hospitably entertained on this and subsequent occasions, is entirely built of ironwood, and it is a

Fig. 9. House of the Assistant-Resident at Sintang.

striking example of the beauty and exceptional durability of this wood, and of its suitability for building purposes. In spite of over 25 years standing, there is not a flaw nor any appearance of decay in the wooden flooring of the gallery, nor indeed anywhere inside the house.

About half a kilometre upstream beyond this house, not far from the right shore of the Kapoewas, there are alluvial gold-diggings, which occasionally are still worked by Chinamen. Down to a certain level not much above the average water level of the river the low gravel hills have been entirely excavated, and the pits remaining are surrounded by the perpendicular walls of the as yet un-excavated material. A conduit runs through the low lying ground which is always partially submerged, and carries the water and the mud down to the river. The manner of working the diggings is quite simple. A furrow, a very simple kind of sluice, is constructed all along the foot of the perpendicular wall which is subsequently undermined, so that the sand and gravel fall together into the furrow. As long as the digging is going on, the water, collected in a higher lying reservoir is kept back by a small dam, and as soon as a sufficient quantity of earth has been heaped up in and about the sluice, this water is allowed to run down; the mud and sand are thus carried away, and only the gold, with other heavy minerals and pebbles of quartz are left behind in the sluice. This operation is repeated until a sufficient quantity of gold has accumulated in the sluice; it is then carefully cleaned and the gold collected. In one place where the wall of the digging was very high, I observed the following section:

Top. 2 Metres clay and iron-stained sand.

4 Metres layers of gravel, alternating with sand. The upper layers are composed of very fine gravel, the pebbles increase in size downwards.

$9^1/_2$ Metres more compact gravel. The larger pebbles have the size of an egg or a fist.

Rich sandy clay in thin layers differing from each Bottom. other in the proportion of iron they contain.

The gravel, which is left after the washing is completed, is exclusively composed of pebbles of crumbling quartz. Not unfrequently the quartz is intergrown with tourmaline. A careful examination of the perpendicular walls shows that, besides quartz, the gravel contains several felspar-bearing rocks, amongst which tourmaline-granite particularly attracted my attention. These rocks are quite decomposed and altered into soft clay, but the kaolinized felspars remain visible in the dark clay as small angular white spots. On seeing such immense gravel deposits, one asks involuntarily:

how did they come here? Their situation about 10 metres above
the present average water level in the river, is easily explained,
for one can well imagine that the Kapoewas or the Mélawi, or
both, formerly deposited the material at this height, and afterwards
cut their beds deeper down. This fact alone however, not to speak
of the existence of coarse gravel at a place where the said rivers
no longer possess sufficient transporting power to carry gravel away
and can at best only form fine sand or mud deposits, would
in itself necessitate us to admit of the theory that these rivers
possessed a much greater carrying power during the time that
the gravel beds were being formed.

When we come upon deposits of this kind anywhere in
Europe, the explanation is always easy, having only to remember
the greater carrying power of the rivers during the ice age.
In Borneo however this explanation is of no avail, for there
is nothing to give us the right to suppose that at one time the
rainfall was perceptibly greater there than it is now. The only
conclusion therefore we can come to, is that at one time the
fall of the rivers near Sintang was greater than it is now.
The gravel, as mentioned above, contains a good percentage
of pebbles of tourmaline-granite, and tourmaline-fels, rocks which
do not occur *in situ* in the neighbourhood but which, as I
afterwards found, play an important part in a portion of the
territory which is now almost entirely occupied by the Upper
Kapoewas plain. This fact leads one to conclude that, at a time
when the waters of Kapoewas were forming the gravel deposits
here, the difference in altitude between the granite mountains
of the Upper Kapoewas and the environs of Sintang, must have
been considerably greater than it is now. At the same time we
must allow for the possibility of the Mélawi having brought
down a part of the gravel, and the origin of the tourmaline-
bearing quartz may therefore very possibly be traced back to the
mountains of the Upper Mélawi, say to the Schwaner Mountains.
Tourmaline is known to occur in these mountains, but, considering their great distance the explanation is not made easier.

Later on it will be shown, that there are further reasons to justify our coming to the conclusion that the Upper Kapoewas plain obtained its peculiarly low position after the direction of the Kapoewas stream had become mainly the same as it is now.

Thanks to the kind help of the assistant-resident, I engaged that day a Chinaman called Bon Toeng Diam, to break the large stones with a sledge-hammer and to assist in the field-work. I make special mention of the fact, because this Chinaman was with me during the greater part of my travels and distinguished himself by his great zeal, skilfulness, reliability, and imperturbable good nature, even under trying circumstances, and thereby contributed in no small degree to the success of my expedition. He also was sometimes useful as interpreter, as he was often able to understand the defective Malay language mixed with Dyak words, which is spoken by some of the chiefs of the Dyak kampongs of the interior, and to translate it into the limited number of pure Malay words which I had at my command. He was perfectly acquainted with some of the Dyak dialects, such as that of the Batang-Loepars and the Kantoeks, but he did not understand the dialects of the tribes of the interior such as the Oeloe-Ajers (Ot-Danoms) and the Kajans.

On February 18th at 6 a. m., I left Sintang for Sémitau, in the Poenan, the steamboat of the controller of Sémitau, and reached my destination at 6 o'clock the next morning.

CHAPTER II.

— — — —

Sĕmitau and Neighbourhood.

Sĕmitau is a small settlement on the left shore of the Ka
poewas, inhabited by Malay and Chinese. The houses stand ir
a row along the road which separates them from the river.

Fig. 10. MALAY SINTANG.

Close to the shore, opposite each house lies a small bathing
raft floating on beams. Huge trunks of trees are the means ol

communication across the marshy parts of the shore between the houses and the floating baths (Pl. V). The Chinese camp, recognisable at a great distance by the large bandongs in the river in front of it, is situated in the lower part of the settlement; the Malay houses are higher upstream, and the highest of all is the house of the controller, then occupied by Mr. W. A. van VELTHUYZEN.

As late as 1894, Sĕmitau was the highest place on the Kapoewas where a Dutch official was stationed. It was made the residence of a controller in 1885 during the time of the disturbances of the Batang-Loepars, and the choice was doubtless a good one for the emergency. There was a military station at Nanga [1]) Badau besides the koeboes [2]) armed with pradjoerits, at Poelau Madjang, Pangkalan Pĕsaja, Genting Doeryan and Lantjak.

Poelau Madjang can be reached from Sĕmitau by steamboat in 4 or 5 hours, and Sĕmitau was therefore a capital starting-point for any operations in the Batang Loepar lands.

The natural position of Sĕmitau makes it so to speak, the gate of the higher districts on the Kapoewas. Further upstream, as far as the Upper-Kapoewas mountains, the bed of the river is generally wide, and at high water the land on both sides becomes submerged for a considerable distance, but at Sĕmitau the bed contracts to within its proper limits, as the river here breaks through a low range of hills which keep it within its borders, no matter how high the water may rise. Above Sĕmitau therefore, communication may be carried on with rowing boats, through the many canals and lakelets over the submerged land, without it being perceived from the river itself, but below Sĕmitau this is not the case. No matter how high the state of the water, the entire volume must of necessity pass through this one valley with the exception of a few known anastomoses.

The movements by water therefore of the natives of the interior could always be controlled with certainty from this station. Since however ten years ago peace was restored in

1) Nanga = mouth. Nanga Badau = settlement at the mouth of the Badau.
2) Koeboe, local name for small fortifications, armed by native policemen, called pradjoerits.

the Batang Loepar lands and through the influence of the
Dutch Government the head-hunting raids have been put a
stop to, the military station at N. Badau has been relinquished,
and the place left in charge of an aspirant-controller with a
few pradjoerits under him, — Sĕmitau has lost much of its
former importance. To this may be added the fact that under
the skilful management of the Resident TROMP, the influence of
the Dutch Government has so considerably increased in the
district of the Upper Kapoewas that the want is being felt for the
establishment of a Dutch settlement more in the vicinity of the
sources of the Kapoewas. Boenoet seems to be the most likely
place for such a station, being situated just at the point where
the fairly well populated Boenoet River flows into the Kapoewas.
A little distance above Boenoet the Embaloeh discharges into
the Kapoewas, and 10 hours higher upstream the Mandai
also. Boenoet is the centre of trade on the Upper Kapoewas,
and may well be called the key to that district, in the same way
that Sintang may claim that title for both the Mĕlawi and
Kapoewas. Boenoet is moreover the highest place upstream,
which even at low water can be reached by steamboats of no
more than three feet draught. At a high or even an average
state of the water in the Kapoewas, the river is navigable much
higher up for small vessels, sometimes as far as the mouth of
the Mĕndalam River above Poetoes Sibau, an actual length of
902 kilometres from the sea.

At Sĕmitau, on a little hill not far from the house of the
controller, a good sized building had been erected, which did
duty as the Central-station of the Borneo-expedition. It was
built on high posts and consisted of two parts, with a fair
space between them, and it was entirely surrounded by a broad
verandah. The one part was used as storeroom and night quarters
for the domestics, the other was divided into three small rooms
for the members of the expedition. The roof hung well over

on all sides and overshadowed a spacious gallery where many goods could be stored leaving plenty of room for the perfor-

Fig. 11. HEADQUARTERS OF THE BORNEO EXPEDITION AT SĔMITAU.

mance of the ordinary domestic functions. This house, built entirely in native fashion, fully answered to our requirements, and its elevated position rendered it cool enough for unbroken study even on the very hottest days.

When I arrived, there was much bustle and activity at the station. Messrs BÜTTIKOFER and NIEUWENHUIS had already been there some little time. BÜTTIKOFER was putting the finishing touches to a large zoological collection, made in the neighbourhood of Mt. Kĕnĕpai. The first day I devoted to the arranging and packing of my goods, and to see to the victualling for my further travels. Through the care of the Resident, there was a plen-

Sintang and the Bokit Kĕlam.

tiful supply of excellent provisions [1]) stored here. I also began
to plan more in detail my intended examination of the geolo-

Fig. 12. MY FAITHFUL COMPANIONS.

gical structure of Central Borneo, more especially of the Upper
Kapoewas and its principal tributaries. My chief object was to
cross in several places the borders of the Kapoewas-Mĕlawi
territory in order to obtain geological sections in different directions throughout the area under examination. The way in which
I accomplished this object will not be related in actual historical order but I have given the account of my different excursions so
as to show the geological connection to the best advantage.

---

1) Supplied by Messrs Tieleman and Dros of Leiden.

THE GEOLOGICAL STRUCTURE OF THE NEIGHBOURHOOD
OF SĔMITAU.

The environs of Sĕmitau consist of low hilly ground, with smooth, gentle slopes, showing only occasionally sharper defined mountain ridges · such as the Mount Djoewaran (200 metres). Viewed from any point close to the river, the aspect of this territory differs widely according to the state of the water. If the river is low the valleys between the hills lie several metres above the level of the water in the Kapoewas, and the many brooklets which arise on these ranges flow down with great rapidity towards the valley of the main river. At high water it is altogether different. The Kapoewas water then freely penetrates all the connecting valleys, surrounds the nearest groups of hills and causes a totally new system of drainage. To illustrate my meaning: just behind the hill on which stands the house of the Borneo-expedition, runs a clear rippling brook, the rivulet Sĕmitau, which joins the Kapoewas a little higher up. When the water is high the Kapoewas entirely surrounds this hill and in the valley behind there is no sign of the clear little streamlet. All that one sees, is an unsightly mass of muddy riverwater, running in an opposite direction towards the higher portion of the valley, across the water-parting and round by the hill, to rejoin the Kapoewas some distance lower down.

The hills in the neighbourhood of Sĕmitau are the remnants of a deeply eroded mountain chain, consisting of a system of highly folded strata the trend of the strata being on an average W.-E. with a steep dip to the south, sometimes they stand absolutely vertical. Amphibolite, chlorite-slate, quarzitic-slate and silicified clay-slate alternate constantly in this complex system of rocks. The quartzitic-slate contains very thin layers of haematite which close to the surface are generally completely altered into limonite. It is moreover intersected in all directions by quartz veins of varying dimensions, the thicker, (as far as could be ascertained) generally following the strike and the dip of the country rock, while the smaller ones intersect each other in all directions. The weathered jagged pieces of these quartz veins remain strewn on the surface and form a brilliantly white but barren soil [1]), on which after the wood has been cleared away, hardly anything grows but a kind of fern very like the *Pteris aquilina*. The amphibolite on the contrary weathers into a brown clayey soil rich in iron which is very fertile even without any further cultivation. The brooks generally spring from the valleys on the borders between the hard amphibolite and the porous quartz-veined silicified slate. In this sloping hilly country near Sěmitau in which no deep ravines are to be found, it is only possible to come upon rock *in situ* in the dry season, and even then it is found to be weathered to a considerable depth. If the water be very low, rocks may be seen above the surface on the right bank of the Kapoewas, just above Sěmitau; they belong to this same system. The same thing occurs further up-stream near the left side, at the lower end or boentoet [2]) of Měrassak island, where at low water, navigation is often endangered by the rocks in the riverbed. They consist of strata of amphibolite, alternating with quartzite, whose planes of partition are coated with numerous flakes of muscovite. Above Měrassak island, no solid rock is exposed along the Kapoewas for several day's journey, till some distance beyond Poetoes Sibau.

I gleaned these particulars with regard to the geological structure of the environs of Sěmitau on an excursion to Mount Djoewaran (200 metres) on March 30th. This hill lies about 6 kilometres in the direction E 4° S. from Sěmitau, and can be reached from there within 5 hours, by first rowing up the Kěněbah for

---

1) The soil of the hill, on which our house was built, consisted of the same kind of rock, and this explains the presence of the many angular pieces of pure white quartz scattered about, which attracted my attention from the first. When digging the holes for the posts on which the house rests, these angular pieces of quartz had been brought to the surface.

2) Kapala poelau = upper end (lit. head) of the island. Boentoet poelau = lower end (lit. tail) of the island.

Sĕraboen-Sĕtoenggoel range which rise, as seen from this side, like a steep wall, their trend being about east to west. In the far distance in the direction S. 5 E. rises the dome-shaped Mt. Bang, (1080 metres). It lies about 51 kilometres away. In several places to the west and north, the Kapoewas water may be seen sparkling amongst the hills, which keeping their uniform character, stretch away westward on the other side of the Kapoewas. Directly north, lies the wide expanse of the great lake district, one unbroken gloomy forest scene, the crowns of the trees, as seen from this distance, forming a completely horizontal plane, and from this standpoint it is also distinctly seen that the hills of Sĕmitau are the southern boundary line of the low plain of the great lake district. In westerly and north-westerly directions, on the other side of the Kapoewas, the evenness of the landscape is rapidly interrupted by some isolated mountain-groups, ranged in a line trending from S. E. to N. W. These are the Sĕkadau, the Kĕnĕpai and the Toetoop.

CHAPTER III.

The Mount Kĕnĕpai [1]).

Whenever I spent a few days in the House of the Expedition at Sĕmitau, upon the conclusion of any of my excursions, in order to arrange the collected materials and to prepare for the next campaign, I never omitted at sunrise and sunset to visit the top of a little hill not far from the house. It was a constant delight to me to view from there the wide expanse of the Kapoewas plain and the lake districts with the grotesque mountains which surround it [2]). The centre of attraction, the glory of this panorama, is Mount Kĕnĕpai; I never tired of studying its exquisitely pencilled outlines. Sometimes, especially on hot days its contours were fine and delicate, and the entire rock-mass seemed to melt away in the quivering atmosphere. At other times its outlines were clean and sharply defined, and the mountain stood forth quite oppressively near in its sombre threatening attitude. Most beautiful of all is the aspect of the mountain when shrouded in a thundercloud, the little detached cloudlets frolicking round, encircling now this, and then another portion of its mighty crest. Often I have seen the highest cone, standing out far above the clouds, illuminated by a mystic reflection of light, while the lower part of the mountain remained wrapped in vapour.

My first excursion was to this glorious mountain, which has been my beacon during a great part of my travels, and as

1) For this Chapter see Maps III and IV of Atlas.
2) This landscape is reproduced in Plate VI with perfectly accurate outlines from a photograph.

such has been of untold service to me. On the day after my
arrival at Sĕmitau, on February 18ᵗʰ I started at 7.45 a. m.,
accompanied by NIEUWENHUIS, and with three sampans manned
by Dyaks from Soehaid. The Kapoewas was high and the
current strong, so that it was necessary to keep very close to
the shore. After paddling laboriously for 3¹/₄ hours we reached
the mouth of the Kĕnĕpai River, at 5¹/₂ kilometres distance from
Sĕmitau. The high river banks both there and further upstream,
are of a beautiful green, and seen from a distance they call
to one's mind the soft tender verdure of the slopes of the
riverbanks at home but this illusion vanishes when one touches
the shore, for the grass is stiff and hard and difficult to walk
over. Moreover the bank is only about 10 metres wide, beyond
which it falls and gradually passes into submerged marshy and
for the greater part impassable forest land. We entered the
Kĕnĕpai. The left bank vanishes almost directly and the waters
spread over limitless expanses of submerged wood land. The right
shore remains at first fairly high; it is formed by the most
northerly spurs of the Sĕmitau hills. Presently however the
river leaves the hills and only a wide streak of water through
the submerged forest indicates the direction of the river-bed. We
had reached the "danau" (see p. 8); there is no perceptible current
and only the natives know, as it were by intuition, which is the bed
of the river among all the many waterways through the wood.

Shortly after 10 oclock, the danau, lay behind us, and
muddy banks began here and there to show us the actual
limits of the river again. But again and again the stream diverges
from its original course, divides and encloses small islands, thickly
wooded with different kinds of high standing graceful rattan- and
pandanus palms.

At eleven o'clock we saw the masses of water contract once
more within a distinct bed in which the current rapidly increased. Here we got a foretaste of the usual obstacles so
prevalent in the small rivers of Borneo viz. the trunks of trees. It
soon appeared that the Kĕnĕpai was quite low, and that many

trunks stood out above the water, which under ordinary circumstances do not reach up to the surface, but allow of a small boat passing over them without difficulty. In consequence of the great delay caused hereby, we could not reach the pangkalan [1]) Kĕnĕpai or place where the path commences to Roemah Mĕnoewal and Mt. Kĕnĕpai, and although it rained in torrents we had to make up our minds to spend the night on the bank. Our little pondok [2]) was soon erected and so comparatively well sheltered, we watched with interest the welcome rising of the water.

At 7 a. m. next day we continued our journey, and as the condition of the water was most favorable (about 1 metre higher than the previous day) we were able to reach the pangkalan by 8.30 a. m. The path at first crosses the river several times, and then it continues for a considerable distance through a marshy forest-land, on which grows a variety of *Nepenthes'* in such abundance, as I have not seen anywhere else in Borneo. The ground gradually rose and the path became easier. Taking it as a whole, I should reckon this among the very good Dyak footpaths. The pure white sandy soil at once betrays the nature of the solid rock underneath, which is a soft yellowish white sandstone, but only at one place was it well exposed. Generally speaking, the path followed the left bank of the River Kĕnĕpai and higher up the Entjik, and in the bed of this rivulet I found the first boulders: silicified slate, sandstone, porphyrite and a quartz-tourmaline rock.

Until close to the house Mĕnoewal the Kĕnĕpai remained hidden from sight by the thick foliage, but when we reached the ladangs belonging to that house, we suddenly found ourselves face to face with the mountain giant, its precipitous top shrouded in vaporous clouds. (see sketch pag. 33).

---

1) Pangkalan = place where a river ceases to be navigable i. e. the place where the path over land begins.

2) Pondok, i. e. hut or shed, built of thin straight pieces of wood, fastened together by strings of rattan and covered by large leaves. The forest affords suitable material in abundance and the Dyaks erect such temporary abodes in less than half an hours time. For this reason it is a useless burden for travellers in the interior of Borneo to carry tents with them.

PL. V.

Sᵐ Kapoewas near Sěmitau.

At 3 p. m. we reached the Dyak house Mĕnoewal, inhabited by Kantoek Dyaks.

I often came across members of this Dyak tribe which in consequence of the repeated head-hunting raids of the Batang-Loepars had become much scattered. Of all the Dyaks, these are to my mind the most obliging, the pleasantest and the

Fig. 14. Top of Mount Kĕnĕpai.

gentlest. In point of ability they also rank high, quite on a level with the Batang-Loepars. Their weaving is particularly beautiful.

Early next morning we set out in the direction of Mt. Kĕnĕpai. For a short distance the path led over hilly ground through ladangs, but we soon came to one of the spurs of the mountain. The undulating ground, just traversed, was composed of sandstone, slate and hornfels, but at the edge of the mountain we entered upon a new formation, viz. uralitised diabase-porphyrite, large pieces of which lay scattered about on the ground.

3

The bed of a small brooklet descending from the mountain is strewn with rounded pieces of tonalite, tourmaline-granite, tonalite-porphyrite, diabase-porphyrite and augite-porphyrite.

The path goes straight up the mountain, and being very steep and slippery, the climb is a most difficult one for a booted European.

The south slope of this southern spur of the Kĕnĕpai mountains is composed of augite-tonalite-porphyrite (I, 109), uralitised diabase-porphyrite (I, 110), and quartz-labrador-porphyrite (I, 122).

A little before nine o'clock we reached the top of the spur, and proceeded through a most beautiful virgin forest, along the ridge which connects the spur with the principal cone of the Kĕnĕpai range. The rock *in situ* is augite-tonalite (I, 115). At 9.30 we came to a point where a little side path leads down into a valley in which a short while ago BÜTTIKOFER and HALLIER had built a pondok on the brink of a running stream. Some of the coolies were at once despatched to get the place ready for us, and to await us there in the evening. We continued our way along the ridge which rose and fell alternately. The lowest point at no great distance from the principal cone is ± 1000 metres high. The cone itself which we subsequently ascended has an incline of 42°, (in some places) locally increasing to 55°. With great difficulty we pulled ourselves up by tree trunks, lianas and roots. The ground was still covered with high standing forest, occasionally interrupted by a precipitous mass of rock but at an altitude of about 1100 metres the character of the vegetation suddenly changed; the forest is superseded by a scrub vegetation, chiefly composed of Rhododendrons and Nepenthes', which covers the summit entirely. The height is 1156 metres. The top of the highest cone or poentjak of the Kĕnĕpai range consists of uralitised diabase-porphyrite (I, 122), while the southern slope is composed of augite-tonalite-porphyrite (I, 117) which near the foot of the mountain passes into augite tonalite. On the slope of the cone facing West, labrador-porphyrite is found (I, 121).

The top of the Kĕnĕpai is flat, in a N. W.-S. E. direction only
a few metres, and at a right angle to this about 30 metres
in breadth. Owing to the low scrubby vegetation the view
from the top is free in all directions. Then however the mountain was enveloped in clouds, and only an occasional break
allowed us a glimpse of the clear sky above, with some slowly
moving cirri high up in the air. Sometimes the cloud roof rent
in one direction, sometimes in another, revealing piecemeal the
grand panorama which this summit offers. To the east and
northeast stretches an almost immeasurable wooded plain, in
which far in the distance the lakes Sĕryang, Loewar and
others may be seen to sparkle. The extreme limits of this
plain are formed by the Sĕmbĕroewang mountain-group, and more
to the north by the boundary mountains of Sarawak. In the
centre of the plain and far down below the place where we
stood stretches a range of hills, the most easterly of which is
the Lĕmpai (250 metres) opposite the island Madjang. This range
increases in height towards the northwest, and reaches its cul-
minating point in Mt. Bĕsar (890 metres). In the foreground of
Mt. Bĕsar is Mt. Sĕmĕlawi (360 metres) on the opposite side of
the valley of the river Empanang. The upper part of this valley is
inclosed in a ring of beautiful and singularly shaped mountains, amongst which besides Mt. Bĕsar, Mt. Toegak (1020 metres)
and Mt. Toetoop (1220 metres) attract special attention.

Heavy thunderclouds were rising from the Chinese sea in the
northwest, silhouetting in a most striking manner the mountains
in the northern portion of the Kapoewas plain, lying somewhat
isolated from the boundary mountains of Sarawak. Far down
below us in the foreground rose the Paoeng (370 metres) and
the Limau (418 metres); behind these on the other side of the
valley of the Mĕrakai river, the elongated forms of the Pĕgĕdang and
Entawak (401 metres) and immediately behind these again Mt.
Bangkit (810 metres) which has every appearance of being a volcano.
A little more to the west, already partially wrapped in clouds,
we can just distinguish the outlines of the Kĕhoema 1210 metres

high, round which the Kĕtoengau river describes a wide curve. To the west and southwest stretches the gigantic Kĕtoengau valley: one enormous dark wooded plain, with occasional light patches, old or new ladangs edging the streams. The southeastern boundary of the Kĕtoengau plain was lost to view by the gathering rain clouds.

At 3.30 p. m. we hurriedly left the top, with the thunderstorm close upon us. With quickened pace, not always voluntary, we descended the steep incline, and no sooner had we reached the wood, than the rain came down in torrents on the thick green canopy above our heads, but it was a short-lived shelter, and presently the rain broke through and came streaming down upon us. At 5 o'clock we reached the deserted bivouac of Büttikofer and Hallier, and a pondok was soon erected for us. Directly after sunset commenced that many voiced concert of the insects which at that time still possessed the charm of novelty for me. Our bivouac was on the southeastern incline of the Kĕnĕpai, sheltered by a beautiful lofty wood in a wide ravine through which a little brooklet ran its course. Large blocks of rock lay scattered about, some being porphyrites which had fallen from the top, some pieces of the solid rock. The principal rock is tonalite (I, 128) which is intersected by dykes of mica-porphyrite (I, 131). Close to the pondok lies a massive piece of porphyrite apparently stratified by numerous more or less parallel veins of pegmatite which are very rich in quartz and stand out from the weathered surface like raised framework. In the tonalite strings and nests may be found, containing much tourmaline. There are moreover a large number of pieces of a beautiful blue-black rock (I, 134) with round white spots, in the centre of which there is generally a little dark speck. This is a fine sugar-grained quartzite so intergrown with tourmaline that it has got a blue-black colour, the parts free from tourmaline appearing like white spots, and the dark speck in their centres being again formed by a spherulitic aggregation of tourmaline needles. In some places this rock passes into a typical tourmaline fels. It is moreover streaked in all directions by thin veins which owe their dark hue to the closely packed tourmaline needles. Besides these there are pieces of finely-grained sandstone and of arkose (I, 135) which contain a smaller proportion of tourmaline more equally divided, but like the other rock is intersected by fine stringlets of tourmaline. In this latter variety also partly decomposed flakes of biotite are found. All these rocks are sandstones metamorphosed and impregnated with tourmaline, through contact with the tonalite, while we find tourmaline as a mineral component also in the tonalite itself in proximity with the contact zone.

At 9 a. m. we left the bivouac on our return to Sěmitau. A steep path brought us in 15 minutes to the ridge which unites the hill behind the house Měnoewal, with the central part of Mt. Kěněpai, and there we regained the path by which we had travelled the previous day. We followed the ridge to its terminus and then descended in a S. S. E. direction towards the house Měnoewal, of which we caught occasional glimpses deep down through the foliage. At an altitude of 200 metres above the Dyak house I found pieces of metamorphosed quartzitic clay slate (I, 144). At 11.15 we emerged from the forest and entered the ladangs of the house Měnoewal. The bottom rock, as shown by the many loose angular pieces of stone strewn about, consists of quartzitic sandstone (I, 151) and metamorphosed claystone (hornfels) (I, 152). I could find no opening which would enable me to examine the solid rock, so that the trend and the dip of this formation remain unknown to me. The claystone corresponds with the fragments which I found on the slope of the mountain at an altitude of 200 metres.

At noon we reached the Dyak house, where we only made a short stay and took leave of the kind-hearted inhabitants in order, if possible, to reach the pangkalan that same day.

Not far from Roemah-Měnoewal the path leading to the pangkalan traverses some ravines about 12—15 metres in depth, intersected by brooklets running in a southern direction, which arise on the Kĕnepai mountains, and send their waters to feed the Rekadjau, which again discharges into the Kĕtoengau. Here I had an opportunity of more closely examining the different rock exposures.

In the first valley (Fig. 15 A) I found a beautiful coarse-

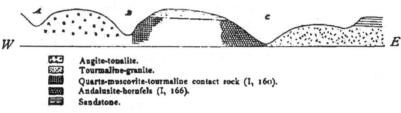

Augite-tonalite.
Tourmaline-granite.
Quartz-muscovite-tourmaline contact rock (I, 160).
Andalusite-hornfels (I, 166).
Sandstone.

Fig. 15. A. B. C. RAVINES INTERSECTED BY BROOKLETS, ARISING ON THE KENEPAI MOUNTAIN AND FLOWING IN A SOUTHERN DIRECTION TOWARDS ONE OF THE FEEDERS OF THE REKADJAU.

grained augite-tonalite, which continues to be the predominating rock on the ridge between the first and second ravine. This second ravine (Fig. 15, B) cuts through the zone of contact between the augite-tonalite on the west slope, and a quartz-muscovite-tourmaline rock, being a sandstone altered by contact metamorphosis, on the east slope. The narrow ridge separating the second and third ravine is composed of crumbling sandstone with a feeble south-eastern incline, lying uncomfortably on the contact rock. It contains some rounded fragments of tourmaline. On the west slope of the third ravine (Fig. 15, C) typical andalusite-hornfels underlies the sandstone, being in contact with and very probably metamorphosed by the tourmaline granite, which is exposed in the ravine, and forms the next ridge. This is overlain about 250 metres further on by thick beds of the previously mentioned more recent sandstone. This latter formation closely resembles the sandstone (I, 171*) which forms the sandy soil in the forest on the way to the pangkalan, and is in all probability part of the sandstone formation which is exposed on the northern shore of the Kĕnĕpai river, more to the east, where it expands into a lake; coal-seams are also found there, which are still being worked at intervals.

Shortly after, we unconsciously once more crossed the water-parting between the Kĕnĕpai and the Rĕkadjau; with slight rise and fall our path now took us through high forest and over sandy soil, generally keeping close to the waterside. In some few places the ground was rather more hilly. We reached the highest point at 4.30. A brilliantly white sandy soil at once betrayed the underlying sandstone, which fact was further confirmed by the presence of scattered pieces of white crumbling sandstone. I had delayed long in the ravines with the beautiful contact rocks, and had lost so much time that in order not to be overtaken by the darkness in the middle of the forest I had to hurry along almost at a trot over the Dyak footpath with its many obstructions, roots, morasses, etc. The twilight however once deceived me, and close to the pangkalan I fell from a treetrunk which, by way of bridge lay across the river Kĕnĕpai, and suddenly found myself in the water, stuck fast in the mud. There was nothing for it but to wait patiently till one of the carriers came up who managed to extricate me out of my awkward position. After spending the night at the pangkalan, we started again at 7 a. m. and further descended the Kĕnĕpai.

The shores consist almost entirely of white sandy soil, with here and there hard lumps of clay wedged in, which through erosion of the sand stick out and might be mistaken for boulders.

At 11 we reached the danau of the Kĕnĕpai and as the water was low, the many treetrunks in the river much impeded our progress; at 2 p. m. we disembarked at Sémitau.

The Kĕnĕpai is a rock massive, trending from N. W. to S. E. and appears to be entirely composed of granite, porphyrite having forced its way through in different places. The upper steep cone of the mountain, the poentjak as it is called by the natives, consists of porphyrite, the broader base of granite. At one time the granite must have been surrounded by slate and sandstone, which have been entirely removed by erosion except at the very foot of the mountain, and there the sandstone and slate have become so intensely metamorphosed in contact with the granite that the slate has passed into andalusite-hornfels, and the sandstone has become silicified or impregnated with tourmaline and muscovite.

Resting uncomformably on this formation of metamorphosed strata are found almost horizontal banks of a more recent sandstone. Probably the coal-beds on the Kĕnĕpai river belong to this sandstone formation, and we may conclude that it is of tertiary age.

The combined Kĕnĕpai and Oelak form part of a mountain ridge trending from N. N. W. to S. S. E. of which the Toetoop, Kĕnĕpai and Sĕkadau are the most conspicuous heights.

On account of their situation and general form, it is fair to suppose that these mountains, and especially the Toetoop, partake of the same geological structure.

CHAPTER IV.

THE MANDAI RIVER AND THE MÜLLER-MOUNTAINS [1]).

During the early morning hours of February 26th 1894, the landing-stage in front of the house of the controller at Sĕmitau was the scene of unwonted bustle and activity, caused by the organisation of our flotilla for the expedition to the Upper Mandai. All the members of the Borneo Expedition intended to undertake this trip together, and the controller was going to accompany us as far as N⁴ Raoen, the first object of our journey. The Poenan, the government steam-boat would tow our bidars [2]) and sampans as far as the mouth of the Mandai-river and from there we should have to proceed in the smaller boats. Some of our goods were stored in the Poenan, and the remainder were divided among the smaller vessels. At 11 a. m. all was ready and half an hour later we steamed away. The Poenan was towing 5 bidars and 2 sampans, fastened to the steamboat with strong rattan ropes. It was a heavy pull, aggravated by the addition, at Soehaid, of another large native cargo boat, as the bidars proved to be too heavily laden. Slowly and vigorously the Poenan steamed against the strong current in the muddy waters of the Kapoewas.

Each time we approached the mouth of one of the little side streams the water of the Kapoewas had a strangely spotted look, dark patches seemed to rise up from the yellowish grey depths. This was caused by the fact that the brownish coloured

---

1) Consult map VI of Atlas.
2) Bidar — small, roofed rowing boat, see Pl. VIII.

but clear water of the side rivers could not quickly amalgamate with the muddy waters of the Kapoewas. This phenomenon is always a sure sign of the existence of lakes within the river basin of the tributaries. For in these lakes all the particles of clay which the water held suspended sink to the bottom, whereas from the floating or submerged vegetable matter, which abounds in these flooded lakes, certain ingredients become dissolved which dye the water a light yellow or brownish colour. We see the same thing occurring in Europe in those regions, where marshy peat occurs. At the mouth of all the smaller tributaries of the Kapoewas between Sémitau and Poetoes Sïbau, i. e. in the whole of the Kapoewas valley this phenomenon can be observed.

At 2 a. m. we passed Djongkong where the Embau or Embahoe discharges into the Kapoewas. The banks continue low and the landscape extremely monotonous; the only break being an occasional glimpse of the picturesque mountaingroup of Měnjoekoeng (630 metres) which, as the river here describes such an exceedingly meandering course, we saw in the distance now to the right, then to the left of us.

At noon we reached Boenoet, a lively pretty little settlement, on the mouth of the river Boenoet. There seemed to be a good deal of trade and business in the roomy Chinese passir. Boenoet is the highest point on the Kapoewas, that can be reached by small steamboats of slight draught, whether the water be high or low, and where travellers can supply themselves with provisions and other necessaries in case of emergency.

Above Boenoet the same dreary monotony prevails. The river is flanked on either side by narrow strips of land which on an average lie $1^1/_2$—2 metres above the waterlevel, and behind them again is lower lying, often inundated, marshy forest land. The land near the river has been brought under cultivation, and a few small houses are scattered about.

Next day at 9 a. m. we arrived at N° Mandai. The loading of our sampans took a good deal of time; a very important tem was the luggage of our zoologist. The water was low, and mudbanks of a dirty grey colour $1^1/_2$—2 metres high, lined the river on either side.

Here and there we noticed deep cracks in the mudbanks, through which the water from the surrounding lakes spurted with great force into the Mandai. We bivouacked that night on a sandbank on the right shore, just below the mouth of the Kadjoetan.

At 8 a. m. we were en route again. The weather was fine March 1. and the water low, leaving the steep banks well exposed. Strata of sand and loam alternate with thin layers of tightly pressed peat resembling brown-coal. At 11 a. m. we reached Nanga Kalis, the highest Malay settlement on the Mandai. The kampong N⁴ Kalis consists of nine houses, and like the general run of Malay kampongs, has nothing specially interest-ing about it; it is built on a wide strip of land, separating the Kalis river from the marshy forest land stretching away between the Kalis and the Kapoewas. Behind the Malay kampong are several small danaus, deserted beds of the Kalis. Both above and below N⁴ Kalis these danaus communicate with the Mandai, and thus, as a matter of fact, the kampong lies on an island called Poelau Entaban. Surrounded by high trees, small projecting rattan bushes and Pandanus-groups, these danaus are perfect types of a picturesque tropical landscape (see Plate VII). We delayed a few hours here to make the necessary arrangements with Abang Beak, the wakil (representative) of the Dutch Government in the Mandai district. Like the Upper Kapoewas district and the Batang-Loepar lands, the Mandai district is not governed by a native prince, but is under the direct control of the colonial Government.

Standing on the bank near the kampong and looking south, one has a beautiful view of the mountains which separate the Mandai valley from the Soeroek or Boenoet valley. The most character-istic peaks are, from east to west, the Liang Soenan, the Sagoe, the Liang Paoeh, the Liang Pappan, and the Mensassak or Sassak.

Above Nª Kalis, on the convex side of the bends of the river, the sand becomes considerably coarser, and about $7^1/_2$ kilometres further up (see map) the first bank of boulders appears on the right shore. Its upper part (*kapala*) consists of gravel, the centre of coarse sand, and the lower part of fine sand and mud. The gravel is composed of debris of sandstone, quartzitic sandstone, coal, agate, chalcedony, and basaltic and andesitic rocks. On washing the sand I found traces of gold in it.

At 6 p. m. we passed the mouth of the S. Taman Tapah, and camped that night on a mudbank just beyond, on the right shore of the Mandai. Close by was a rapid, where for the first time our sampans had to be pulled upstream by strong rattan ropes, the men who held them walking along the edge of the stream.

March 2.   In the first bend above our bivouac the left shore exposes two interesting sections (see map VI Section A and B). In both instances coal-beds are intercalated between the strata of sand and clay. These are in part largely mixed with clay or sand, finely laminated and but little coherent, much like the deposits of vegetable remains in the clay at Kapala Toewan, but in part they are purer, darker, more coherent and with high lustre, particularly in the lowest bed. The dip is slight, only o—12°, varying in different directions but always inclining towards the river. The coal is evidently recent and its mode of formation was probably the same as that which we see even now constantly in progress along the uncontrolled streams of Borneo. With each flood the bed of the stream and the flooded district surrounding it, is covered by large quantities of vegetable remains upon which again deposits of sand and clay are laid down. This process repeats itself constantly and the result is a succession of alternate layers of sand or clay and vegetable matter. Pressure and decay under insufficient access of air change these vegetable sediments into laminated coal-seams; and where the deposits happen to be thick, they pass into true coal-seams. The layers of coal, here under consideration, possess cleavages generally in two directions both at right angles to the plane of stratification and they in consequence break up easily into parallelopipèd fragments.

At 9.15 we passed the kampong N⁰ Penyoeng consisting ot one large and two small houses. The shores remain the same for some distance, but above N⁰ Penyoeng a few bends further upstream, gravel gets more plentiful along the banks, and near the large boulderbank R. R. where we took our afternoon rest, the shores, now about 2 metres high, are entirely composed of gravel cemented into a loose conglomerate by tough clay and clayey sand.

On this boulderbank and in the conglomerate I found pebbles of silicified wood, silicified slate, quartzitic sandstone, sandstone, basalt, dolerite, quartz and chalcedony. The transparent pieces of agate and chalcedony are evidently the filling of gas pores in igneous rocks.

The pebbles are the size of an egg or of one's fist. The water which is muddy in the lower course of the Mandai, here begins to run clear; we were getting near to the upper course of the river. Boulderbanks and islands become more numerous and these, together with the trees thrown into the water, when the grounds along the river were cleared from bush for the preparation of ladangs, hindered us from making rapid progress. We spent the night on a boulder island, beautifully situated and thickly lined with high standing virgin forest. The principal components of this boulder island are sandstone, quartzite, silicified wood and basalt, besides which one finds occasionally pieces of coal, andesite, and porphyritic varieties of amphibole-granite.

We continued to be favoured with most delightful weather, March 3. the only drawback being that the water was getting gradually lower, and we had some difficulty in passing certain boulder-banks which at a high water level are hardly noticeable, except for a slightly stronger current and an increased turbulency in the stream.

The landscape increased in beauty as we approached the mountains, and more especially the picturesquely outlined Tiloeng group. The Tiloeng is a typical table mountain with horizontal terraces, and I felt so certain that this mountain and those in the neighbourhood of the same shape must be composed of sandstone, that I was greatly perplexed as to the origin of the pebbles of volcanic rocks, which began to be very numerous in the boulderbanks. The river in its meanderings swerved now to the south then to the north, but no matter what our direction, we could always see the Tiloeng mountains either before or behind us forming an exquisite picture in a setting of lovely river scenery.

Fig. 16. THE TILOENG AND DAROEWEN, SEEN FROM THE WEST.

Fig. 16 is a sketch made from a photograph taken at a point marked on the map[1]). From this point one can distinctly recognise three principal terraces, formed by massive horizontal beds, overlying each other. The third or highest terrace is in reality part of a large plateau which extends over the entire top of the mountain on which however rises another narrow ridge, the last remaining vestige of a fourth terrace.

---

[1]) On the map it is erroneously marked fig. 22 instead of fig. 16.

Mt. Tiloeng is held very sacred by the Dyaks and the natives of the Kapoewas and Mandai districts, who believe it to be the abode of the souls of their departed. I was told that the small excrescence on the top was taken for their real home. Mt. Tiloeng is 1112 metres high, and the ascent is fraught with so many great difficulties, that neither BÜTTIKOFER nor NIEUWEN-HUIS, who both spent over two months in the neighbourhood of of N° Raoen and attempted the ascent from different directions, ever succeeded in reaching the highest plateau or top of the mountain.

Most Dyak tribes hold the belief that the souls of their departed adjourn to some high mountain, this being always one of the highest or most conspicuous peaks within the area of their habitation; such a mountain they hold sacred, and it continues to be held in reverence by those Dyaks who are converted to Islam, and even the Malays bestow a certain amount of veneration on it. In the districts of the Lower Landak and the Tajan river the Mt. Tijoeng Kandang (887 metres) is the sacred mountain; in the territory of the Lower Mělawi it is Mt. Saran (1758 metres); on the Sěběroewang Mt. Bélimbing[1]; in the Upper Kapoewas district, the Upper Mahakkam and possibly also on the Upper Boeloengan and the Upper Baram it is Mt. Tiloeng; on the Upper Mělawi and in the river basins of the Kahajan and the Katingan it is Mt. Raja (2278 metres); for the entire Barito district and the east coast of Borneo south of the Mahakkam it is Mt. Loemboet or Loemoet where the Kias river, a leftside tributary of the Teweh, takes its rise.

It must strike one as strange[2] that in so many cases the sacred mountain lies at so great a distance from the present home of the tribes who venerate it, that it is not even visible to them. The natural inference of course is that the tribes used to live closer to their sacred mount, and the sphere of vene-

1) J. C. E. TROMP. 59, p. 114.
2) Compare S. W. TROMP. 60, pp. 737 and 759.

ration allotted to each mountain might possibly give us a clue to the migrations of the different tribes.

A little higher up the river the first solid rock is exposed, it is a fine-grained, clayey sandstone fairly firm and rich in small flakes of muscovite. The sandstone beds are each about 40 centimetres thick. Higher up this same formation of feebly inclined sandstone rises to a greater height, sometimes even forms rock groups along the shore. The pure sandstone beds alternate with strata containing much clay and sometimes so many particles of coal that the sandy shale passes into coal-shale (see Section C.).

Where solid rock is absent the shores either slope and are wooded down to the water's edge, or they are underwashed and broken down a metre or two above the waterlevel, exposing compact gravel overlain by sand, clay and vegetable mould. Just below N^a Sahoei a good specimen may be seen of such a river-bank; it is on the right side, where at the point D the following section is exposed (see map VI):

Section D.
{ Humus, 1 metre.
Sand, peat and vegetable remains distinctly stratified 4 metres.
Solid gravel, 1.50 metre.

We did not stop at N^a Sahoei, a house inhabited by Poenan Dyaks, but took our afternoon rest on a boulderbank on the left shore looking aslant upon a well-exposed section three metres high on the opposite shore (see map, Section E). Here the sandstone rests on thinly laminated layers of sandy claystone in such a manner that the lower strata of the sandstone and the banks of claystone encroach upon each other revealing clearly the origin of this deposit as formed by running water.

A little higher upstream we met in the afternoon some Oeloe-Ajer Dyaks from N^a Raoen in a sampan. I well remember how at this first meeting I was particularly impressed by the shyness of their demeanour, and the extraordinary length of their earlaps, stretched out of all proportion by the weight of the wooden ornamented discs, which are forced into their perforations.

Pl. VII.

Danau kear Nanga Kalis, S^a Mandai.

On the right shore, above the prettily wooded island Daän, I found to my surprise, basalt as solid rock, built up in reclining columns dipping feebly towards the water. The columns are on an average 1.50 metres long and 0.50 metre wide. They are distinctly foliated, the planes of cleavage standing vertical and parallel to the longitudinal axes of the columns. The rock is anamesite rich in olivine, whose groundmass contains very small flakes of biotite. It was impossible to ascertain whether this was a dyke or a flow, the manner however in which the parting into columns occurs points to a dyke, but not with absolute certainty. The rock was only exposed over a distance of about 10 metres, and on either side stretched the inevitable marshy plain. Most probably the basalt is continued in the high, wooded part of Daän island. The river which a little higher up flows from N. 35 E. to S. 35 W., when reaching the basalt describes a sudden curve of about 90° to the west; the basalt is evidently the obstacle which compels the stream to alter its course, and it is most certainly a redeeming feature in the monotonous character of the river-banks all the way up. Clay and sand with layers of peat predominate in the lower parts, and sandstone occurs where the banks are higher.

All this time we had a beautiful view of Mt. Tiloeng on our right. It was quite near to us now, and seen from our standpoint bore a striking likeness to the Table Mountain near Capetown. Even the so called table-cloth, the familiar cloudcap of the latter mountain was not missing when on my return I again passed this spot. The photograph reproduced on Plate VIII was taken just below Daän island. One can there only distinguish three terraces, as the small one, the highest of all, lies more to the south and is hidden from view. Almost all the mountains in this area are of the same type as Mt. Tiloeng, except that not far off in a south-easterly direction a few mountains of totally different outlines attract the attention. They are either rounded or else they form long ranges with fairly sharp crests.

· One bend higher upstream the water had run so low, that

4

near the right shore cliffs of a loose sandstone rich in coaly particles and containing numerous stringlets of coal become exposed. They are full of pot-holes varying in their dimensions from 10

Fig. 17. Pot-holes in the bed of the Mandai.

to 70 centimetres. The biggest of them average 60 centimetres in depth, and even the smaller ones never measure less than 25 or 30 centimetres. The small ones are generally round, the larger often oval and elongated in the direction of the current. These latter are often funnelled at the top, the funnel leading to a round almost vertical hole, the pot-hole proper, which lies with regard to the current at the lower end of the hole. Sometimes the pot-hole is bottle-shaped so that the diameter at the bottom is larger than at the top. The pebbles in these pot-holes are not globate but rather discoidal. Both the bottle-shape of the pot-holes and the imperfect rounding of the pebbles are explained by the fact that the rock — a crumbling sand-

stone — is much softer and less capable of resistance than the
pebbles, principally of basalt and andesite, which, moved by the
current, bore the holes in it.

At a bend just above this spot, we came to long rapids which
we passed after sunset. The rushing, splashing, foaming water
combined with the unaccustomed sensation of the trailing of the
boat over the big boulders, made a deep impression upon me,
and in my diary I noted this riam[1]) as decidedly awkward. Afterwards, however, when I had gained more experience of the rivers
of Borneo and their rapids, I came to the conclusion that the
riams in the Mandai below N⁰ Raoen are by comparison insignificant and quite harmless.

We encamped on a boulderbank a little higher up on the left March 4.
shore. Except silicified wood and a few pieces of sandstone I
found exclusively igneous rocks, viz., several types of basaltic rocks,
such as anamesite, dense basalt, basaltwacke, also hypersthene-
andesite, quartz-biotite-andesite and finally pieces of tuff and tuff-
breccia. I knew that this tuff ought to appear somewhere close by
in situ, as transport by water pounds it up directly, and for the
first time I began to doubt whether after all the Tiloeng did consist
of sandstone like its twin-brother at Capetown. Along the shores
the rock continued to be a crumbling white or yellowish sandstone, interspersed with particles of coal. We had now passed
round the Tiloeng leaving it to the west, and to the south and
south-east arose a many sided mountainous district, with grotesquely
shaped table mountains, built up like the Tiloeng in horizontal
terraces. The one nearest to us was Mt. Liang-Agang which seen
from here is pointed and conical in shape. At 8.30 we passed the
mouth of the Raoen and shortly after the deserted old kampong
N⁰ Raoen. At 9 we arrived at New-Nanga-Raoen, a kampong
consisting of two houses at the foot of the Liang-Agang.

At N⁰ Raoen the Mandai is still a fairly large river, about
60 metres wide flowing in a valley, about 4 metres deep. On
either side stretches a strip of level ground which on the left

1) Riam = rapids.

soon begins to rise and merges into the slopes of Mt.- Liang-Agang. On the right there are no hills of any importance so close to the river, and there the plain passes slowly into undulating ground. At a distance of about 30 metres from the left shore is a large Dyak house standing with its greatest length parallel to the stream. It is 155 metres long by 10 wide and is built on 568 poles placed in 8 rows of 71 poles each.

The style of architecture is much the same as that in vogue amongst the Batang Loepars and the Kantoeks, the type that I first became acquainted with in the house at the foot of Mt. Kěněpai. The principal difference between the two is that the notched tree-trunks which do duty as stairs are here placed right underneath the house in the centre, while in the Batang Loepar house they are at the sides. The space underneath the house at Nᵃ Raoen is higher than usual, quite 4¹/₂ metres.

It is new and all its dimensions are on a larger scale than the ordinary run. It has 39 pintoes[1]). The general room and the gallery opening out of it face the river, the family rooms face landward. By the side of the large house is a smaller one. On the space between this latter and the river we built our pondok, and Dr. NIEUWENHUIS lived there till some time in May. The inhabitants, belonging to the great tribe of the Oeloe-Ajer or Ot-Danom Dyaks, received us with indifference. They were neither hostile nor friendly, and did not evince the slightest interest in our doings. They certainly had their curiosity well under control, and all the time we were there we had never once to complain of intrusion on their part. The men are tattooed; they all have the tribal mark of the Oeloe-Ajers on the calves of their legs, viz. a large indigo-coloured patch. On the arms a perfect network of diagonal lines, on the chest more or less elaborate designs, the simplest being a star on the inner side and two vertical lines on the outer side of the nipples. The women are not tattooed.

1) Pintoe means literally translated door, but it is also used to indicate the abodes of each family in a Dyak house, which are separated from the general apartment by a door. Thus in this instance a hous of 39 pintoes means a house, in which 39 families live and possessing therefore 39 separate private rooms.

We spent the night in our bidars; I was kept awake a long time by a heavy thunderstorm with torrents of rain, which however brought us much desired coolness. No sooner had I settled down to sleep, when I was roused by the most infernal screams and howls I ever heard in my life. By degrees I became conscious that this awful concert proceeded from the house N' Raoen; the dogs were holding an ovation; some, a little longer winded then the rest howled out their solo fortissimo after the general uproar had died away. Zoologists maintain that the Dyak dog is a separate variety, he is short haired, usually of a yellow-brown colour and medium size, thin, with curly tail and ears standing erect. The dog is a necessary appendix to the Dyak house. He lives principally on refuse, and may be considered the chief sanitary officer of the place. He is neither petted nor ill-treated. The mutual relationship of the dogs in a Dyak settlement is not of the friendliest, and the defects on their ears, legs and tails, testify to many a fierce battle. But in their musical effusions they are always of one mind, and their perseverance is truly wonderful.

I was up early next morning and completed my preparations March 5. for an excursion into the mountains. My scant knowledge of the Malay tongue caused me a good deal of difficulty in organizing my escort, but by 8 o'clock I started full of anticipation of what the mountains might hold in store for me. My company consisted of Bon Toeng Diam, Sidin, five Dyak carriers from Soehaid and a local guide. This latter could not speak a word of Malay, and our intercourse was limited to gestures, but I was quite satisfied with his attentions to me, a novice in the difficulties of a Borneo forest, and the careful way in which he drew my attention to the many snares which beset the forest mountain path. There are two things against which a European has to be particularly on his guard, when traversing a tropical forest; first the creeping roots and tendrils which grow across the path a few inches from the ground, and are generally tightly held on either side. In low marshy forests these roots are often bent and form loops, and if one has the misfortune to catch one's foot in one of them a fall is inevitable. It is therefore advisable in walking to lift the foot very high, and before taking the next step forward, to swing the leg well backward. When after a long weary march my attention began to flag, I have had to pay for it by many a tumble. In the second place I would mention the lianas, or worse still the thorny rattan-shoots which hang down suspended from the trees. The Dyak heading the procession cuts them down as he goes along, but only just above his own head, so that a European of medium height is in constant danger of knocking his head against them. I found that this nuisance is best avoided by going bare-headed through the forest. With no brim to intercept the sight, one can more easily avoid the contact with these sharp customers. A hat is bound somehow to come into collision or to be thrown off, especially when going uphill, and in trying to save it, one's face gets often terribly scratched.

We passed underneath the Dyak house and took our way across old ladangs, and over the hilly ground formed by the

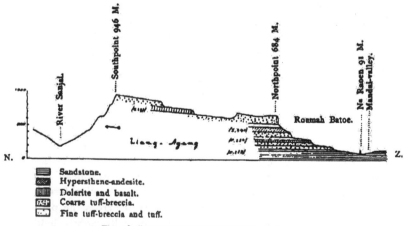

Sandstone.
Hypersthene-andesite.
Dolerite and basalt.
Coarse tuff-breccia.
Fine tuff-breccia and tuff.

Fig. 18. SECTION THROUGH MOUNT LIANG-AGANG.

spurs of the Liang-Agang. The solid rock is again the same crumbling sandstone, familiar to us from the Mandai-valley, but the pebbles in the brooks are almost exclusively of igneous rocks.

Fig. 18 gives the position of the rocks as observed by me when I ascended Liang-Agang on its north side. At an altitude of 175 metres an unexpected revelation awaited me. The sandstone is covered by volcanic tuff and tuff-breccia overlaying each other in thick beds, forming terraces, and, with some intercalated lava-flows, building up the entire upper part of the mountain. Here was the origin of the igneous rocks, the principal component in the boulderbanks on the Mandai!

At 11.30 I came to a high, steep tuff-wall, reaching up to the top of the hill. This wall is bare up to a certain height, and is conspicuous from the valley below as a white band encircling the mountain. The lower part projects, and underneath are several spacious grottos. The rock-wall consists of very coarse volcanic tuff, but near the bottom, layers of fine tuff and ashes predominate. These latter are eroded both by wind and water, to such an extent that grottos have been formed. The first grotto we reached was very commodious and in it a little rill of pure water came trickling down from the upper plateau. I named it Roemah Batoe, and chose it for our bivouac. It lies on the north side of the mountain. In another grotto close to ours, we found a Poenan encampment; the occupants had evidently fled at our approach. Their fire was still alight, and the bamboo-tubes and cocoanut shells were full of water.

Fig. 19. Mt. Tigin and Mt. Miran as seen from the northern top of Mt. Liang-Agang.

In order to gain the ascent we had to make a long round by the vertical tuff-wall (Plate XI) which forms the upper terrace of Mt. Liang-Agang, but at last we succeeded in finding a crevice by which we could climb up. We found evident traces of the Poenans all round, they must have been there quite recently. The Dyaks from Soehaid became frightened, and the guide from N⁴ Raoen did nothing to set their minds at rest, and did not even speak a word to them. With my very limited vocabulary, I had a great deal of trouble to persuade them to go on, however I succeeded in the end. I continued my way with the guide, and they followed in the rear. At 12.45 we reached the crest of Mt. Liang-Agang. It is thickly wooded but on the north side we managed to make a little clearing, no easy task because of the overhanging rocks. I was rewarded however by an extensive view, reaching as far as the Oeloe ¹) Kapoewas territory. The hills in the foreground in a north-eastern and eastern direction are of the same terrace-type with perpendicular rockfaces as Mt. Liang-Agang. Close to us, with only the Mandai-valley intervening, rose Mt. Tĕnaboen, Mt. Tigin, N 65 E¡ and Bt. Miran N 85 E. They are all without any doubt tuff-mountains. The saddle between Mt. Tĕnaboen and Mt. Tigin, separates the valley of the Kĕryau in the foreground from that of the Mĕnoeki, which discharges into the Kapoewas, and behind this we could distinguish the table-mountains, forming the boundary line of the Kĕryau valley on the south side. The prospect is in that direction intercepted by Mt. Hariwoeng, a volcano-shaped mountain, far away on the other side of the river Kĕryau. The extreme top of Mt. Liang-Agang is covered with a thick layer of vegetable earth, in which I found scattered large pieces of pyroxene-andesite (I 243), derived from the underlying tuff.

At 4 p. m. I was back again at Roemah Batoe. This grotto is 8 metres deep and 12 wide; its roof is formed of coarse volcanic tuff-breccia. The rock fragments in this tuff are composed of different varieties of andesite and basalt, the pieces varying very much in

¹) Oeloe = upper part of a river near its source.

Pl. VIII.

BOEKIT TILOENG AS SEEN FROM THE NORTH.

size, some have a diameter of 1—1.50 metres. The space now forming the grotto [1]) had been previously occupied by seams of fine tuff, i. e. solidified ash, which were intercalated between the strata of coarser material as is very clearly shown in the sidewalls of the grotto. The tuffbeds are generally horizontal, but they wedge slightly into one another. Thus the beds of fine tuff which form the side walls of Roemah Batoe, rapidly decrease in thickness just beyond there, and strata of coarser material take their place. With the exception of a few places where the water trickles down through the roof, the ground is dry and covered with volcanic ash — disaggregated fine tuff — mixed here and there with large pieces of rock, fallen down from the roof overhead. I followed the terrace in yet another direction and soon came upon a chasm, severing the mountain into two halves, from the top down to this terrace and opening a passage to the north-west- and west-side of the mountain. On that side the structure of the terraces was in principle exactly the same. When in the twilight I returned to Roemah Batoe, our grotto presented a most picturesque scene. On one side was my camp-bed and on the other the coolies had made themselves comfortable round two or three small fires, and were cooking their rice. Quite away from the others, at the very furthest end of the cave, was my guide, forming a grotesque picture in the weird fire-light; with his long hair, his weathered frame, only covered with a tjawat [2]) of bark, and his general rough wild appearance, he seemed to me the very personification of the original cave-dweller, who frightened by the sudden intrusion, had retired to the furthest corner of his den.

I spent the greater part of the night in making barometrical observations, but apart from that there was not much rest for anybody in our camp that night. A severe thunder-storm and the

March 6.

---

1) In the grotto (see Pl. X) Dr. NIEUWENHUIS is seated in the foreground. He visited me there on March 12.
2) Tjawat = Along, narrow piece of cloth or bark wound round the loins and between the legs.

consciousness of the nearness of the Poenans, added to the intense and unexpected cold, drove all sleep away from our quarter, and before sunrise I was about again enjoying the unspeakable glory of the early morning scene. As I was not equipped for a longer stay we began the descent at 9 a. m. and at noon we were back at N° Raoen. They had got on so well with the pondok that I could take possession at once. One part was covered with bark (koelit kajoe) which makes a capital protection and is often used by the Dyaks for roofing their houses; the rest was protected by kadjang-mats, but these, unless they are in a very slanting position, do not keep out the water effectually. During a severe thunder-storm which raged that evening we found that it rained almost as heavily inside our pondok as out, and frequently during the days which followed we had to sit in our temporary dwelling with an umbrella over our heads.

March 7.   Accompanied by Nieuwenhuis I climbed the slope of Mt. Liang-Agang in order further to determine the boundary between the sandstone of the Mandai valley and the volcanic formations overlaying it. On my first ascent I had crossed a little brook which flows into the Liang-Agang and forms a deep ravine which at the time seemed to me a promising spot for observation. I now went to this same place and followed the brook upstream through the ravine, and I succeeded in finding a good section which reads as given in Fig. 20.

The crumbling sandstone (a) is of a light grey colour, and consists of small, perfectly rounded quartz grains, and numerous minute flakes of muscovite, cemented together by clay which does not contain any lime. There is not a trace of volcanic material in this sandstone which is overlain by clayshale and sandy clayshale (b and d), containing claystone-concretions and vegetable impressions. Intercalated is a coal-seam (c). This coal is rich in fragments of carbonized tree trunks sometimes partially silicified. Following the section of the brook further upstream, sandstone (e) again covers the sandy clayshale in strata from 20 to 40 centimetres thick. Next comes a perpendicular rock-wall, over

which the water comes down in a superb fall about 15 metres
high. The bottom stratum of this wall is composed of fine sand-

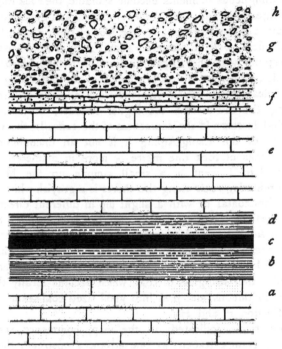

Fig. 20. Section on the Northern slope of Mt. Liang-Agang,
at an altitude of 160 Metres. Scale: 1 : 100.

stone (*f*) in which the quartzgrains are largely mixed with
particles of ash. Then follows a conglomerate (*g* and *h*) composed of volcanic material, chiefly andesite; besides this it contains quartzpebbles and pieces of quartz-sandstone, occasionally
holding quartz-strings. As a rule the pebbles are small, seldom
larger than one's fist and distinctly rounded and flattened. The
larger flattened side of these pebbles generally lies horizontal,
and gives the bottom part of the wall (*g*) a stratified appearance.
The pebbles in this conglomerate are cemented together by quartz-

sand, mixed with fragments of igneous origin. Towards the top the conglomerate gradually passes into a true volcanic tuff (*h*) in which the non-igneous elements become sporadic, the rock fragments are no longer flattened, they are less rounded, and the cement does not contain any quartzgrains. At the foot of the tuff-wall large pieces of silicified wood lie scattered about; they are not water-worn but correspond with the fragments found in the boulderbanks of the Mandai, which led me to presume that the latter originated from the volcanic tuff, which opinion was afterwards confirmed. Comparing the lower with the higher terraces, the latter show a decided increase in the coarseness of the tuffs. In the upper most terrace of the Liang-Agang it is not an uncommon thing to find blocks of rock one metre or more in diameter.

March 8. At 8.30 I went in a small sampan up the Mandai. The predominating rock along the shores is a greyish-white sandstone, lying horizontal or with a slight dip in different directions. Now and then the strata are locally disturbed (see Map VI, Section G). In the sandstone and more frequently still in the layers of sandy claystone which alternate with it, one often meets with impressions of leaves and other vegetable remains. Occasionally also it contains thin stringlets of coal, which become locally thicker and more continuous and pass into good sized coal-seams. (See Map VI, Section F and H).

The riams indicated on the map, commence shortly after passing Nanga Raoen; as one advances upstream they become more and more awkward, and it would be impossible, with a bidar, to get much beyond N° Raoen, for there the river-bed is full of little sandstone islands, and projecting sandstone rocks narrow the channel considerably. Still higher up where the river flows from S 30 E, islands of tuff rise above the surface, being in reality huge blocks of rock which have been hurled into the stream from the table-mountains close by, perhaps more especially originating from the bleak heights of Bt. Aoe. Riam Djala is one of the most interesting spots, as the river is there almost entirely obstructed by huge blocks of tuff-breccia. The view from there over the Upper Mandai ranks certainly amongst the most picturesque river views I have seen in Borneo. The fantastically shaped rock-masses in the river obstruct its course incessantly and create a series of rapids; the waters flicker and shine in the dazzling sunshine in weird contrast with the dark solemn trees on either side, above which the flat-crested tuff-mountains are sharply outlined against the clear blue canopy of heaven.

About one kilometre above riam Djala solid tuff begins to form the banks of the river. At riam Tĕmahapan the channel becomes so intensely narrowed and divided by the masses of tuff in the stream, that a series of cataracts is formed (known collectively as the goeroeng Bĕroewang); these are impassable, but there is a footpath on the left shore by which the luggage has to be carried, while the empty sampans are drawn up through the goeroeng. Owing to the solid andesite- or tuff-banks and the huge rocks in the goeroeng, the river-bed is in many places narrowed to a few metres in width.

On the left shore a large flow of enstatite-andesite built up of coarse columns is exposed. It is about 400 metres wide and extends to close to the house Goeroeng Bĕroewang. This house, situated just above the rapids is now deserted. The inhabitants, Oeloe-Ajer-Dyaks, have removed to a new house a couple of tandjongs[1]) higher up the river. Büttikofer visited this settlement later on.

The large boulders in the riverbed just above Goeroeng Bĕroewang are chiefly composed of igneous rocks, amongst which, dense or porous varieties of pyroxene-andesite and basalt largely predominate; but there is a fair proportion of hard sandstone containing thin layers of clay with carbonized vegetable remains, which show that the sandstone formation which constitutes the bottom rock at Nᵃ Raoen and further downstream in the Mandai valley, also extends higher up the valley.

When, in the evening, I returned to Nᵃ Raoen I found that Büttikofer had also come back from his expedition to Mt. Tiloeng after fruitless efforts to reach the highest plateau of the moun-

---

1) Tandjong or tandjoeng = bend or loop of a river.

tain. NIEUWENHUIS, who afterwards made the attempt from another direction, was not more successful in reaching the top. Probably

Fig. 21. MOUNT TILOENG SEEN FROM NANGAH RAOEN.

the natives of N° Raoen know quite well the direction from whence an ascent could be made, but they pretend ignorance for fear lest misfortune should befall them if the „sacred mountain" were scaled. From what NIEUWENHUIS and BÜTTIKOFER told me, I gathered that the terraces of Mt. Tiloeng, like those of Liang-Agang, are composed of volcanic tuff resting on horizontal beds of sandstone.

March 9. On the preceeding day I had noticed on my way to the goeroeng Béroewang that the Liang-Agang is a long ridge trending from North to South, terminating southward in a very steep rock-wall, and therefore promising in that direction an unimpeded view of the southern mountains. I was most anxious to secure a sight of the mountains south of the Liang-Agang, and determined to make the ascent once more, and to try, starting

from my old bivouac, to follow the ridge to its southern termination.

I started at 9 a. m. and reached Roemah Batoe in about two
hours. First I went through the chasm mentioned on page 57, and
inspected the west side of the mountain. As on the south side
the rock-wall which terminates the upper terrace here juts out
considerably so that the water which trickles down from the
edge, falls on an average at about 5 metres distance from the
foot of the mountain-wall. The intervening strip of dry ground
and some of the deeper grottos in the rock, had been converted
by HALLIER [1]) into a station for botanical investigations. It is
not quite so cheerful a situation as that of Roemah Batoe, but
it has the great advantage of a much better water supply. In the
afternoon HALLIER and myself climbed to the top, and after
reconnoitreing the neighbourhood, we both decided that it would
be quite possible to reach the southern limit by following the
highest ridge of the mountain.

Early next morning we proceeded to the south side. Sometimes March 12.
we followed the base of the upper terrace, sometimes we went
right across the ridge. It appeared that the Liang-Agang forms
part of a plateau with a feeble northern incline, and terminating
on all sides in almost perpendicular rock-walls. The distance
between the eastern and western extremities varies slightly, but is
never very great, and this gives the plateau the character of a
long narrow ridge trending from north to south. This ridge
although flat is by no means easy to traverse, for the regularity
of its course is constantly interrupted by fissures and queerly
shaped hillocks with very steep sides. In places where the ridge
is not quite so narrow and the drainage imperfect, marshy jungle
appears, and thick tangles of strong rattans considerably obstruct
the way. We took a short afternoon rest on the east slope near
to a beautiful cataract 25 metres high which falls from the
highest plateau, leaving sufficient room to pass behind the watery
veil dryshod. From there, always continuing in a southern direc-

---

1) H. HALLIER. *16*, p. 440 and following.

tion our progress was slow and very difficult, and more than once some of our native companions tried to bring us to a standstill, by exclaiming in a very decided dictatorial way "di sini poentjak habis, toewan" [1]), whenever a turretted crest of the ridge compelled us to go downhill for some little way. But at last, about 1.30 p. m. they really were right, and we had practically reached the southern extremity of the ridge of Mt. Liang-Agang at an altitude of 946 metres. We found ourselves suddenly on the edge of a precipice, and far down, about 700 metres below us, stretched they valley of the Sanjai, a rightside tributary of the small river Raoen. The Liang-Agang proceeds in a south-eastern direction and with the northern slope of Mt. Pali it forms a narrow gateway through which we could look high up into the valley of the Upper Mandai (Plate XII). Horizontally grown trees and shrubs protrude a great way beyond the rock wall, so that the view is nowhere quite free and unimpeded. My clever Chinaman, however, soon constructed a balé-balé [2]) for me in the branches of one of the spurs of this hanging forest, and from there hovering over the abyss, I took some photographs which have served as the basis of the sketches, reproduced in Plates VI and XII [3]).

Right in front of us, on the opposite side of the valley of the Sanjai, rose the Liang-Koeboeng and to the right, with a deep valley between, the Liang-Pata whose pointed finger may be seen far and wide. In the far distance behind Liang-Koeboeng in a S 12 W direction rises another mountain evidently higher than the rest. It can clearly be seen that Liang-Koeboeng, Liang-Pata and Liang-Pali [4]) have been built up of the same

1) "Here the top ends, Sir".

2) Balé-balé = stretcher of rattan.

3) The photographs for Pl. XII were taken from a point on the westside of the upper terrace of Mt. Liang-Agang, at a distance of about ³/₄ kilometre N N W of the point from where Plate VI was taken.

4) The names are given as the natives pronounced them, but they are not guaranteed to be correct. On the right, south of Mt. Tiloeng, the topographical map happens to mention by name only a very few of the peaks. The mountain called Liang-Koeboeng on the map

Pl. IX.

House: Nanga Raven.

kind of tuff-terraces as the Liang-Agang and the Tiloeng, and
the surrounding mountains all show the same type, with the
exception of a few peaks arising far away in the south above
or next to the Liang-Koeboeng. Beautiful waterfalls rush down
with great force from the terraces on the other side of the
valley, especially from the Liang-Koeboeng. This view together
with the observations I made on the north summit of the Liang-
Agang and in the Mandai valley, quite convinced me that the
Liang-Agang forms part of an extensive volcanic mountain-land,
a hitherto unknown region of, geologically speaking, recent
volcanic activity. I decided to call the whole of this district the
"Müller Mountains" in memory of GEORG MÜLLER, the first
European, who it is believed entered this volcanic region near
Pĕnaneh, much further east, a few days before he was murdered.

Returned to my bivouac I spent the evening and the next March 13.
morning in exploring a grotto about 250 metres S. E. of Roemah
Batoe. In this grotto was a small pool of water constantly supplied by droppings from the roof. Some deer came to drink there
every evening and left their marks in the soft soil. We therefore
named it the Deergrotto. The section of the strata exposed in the
grotto is given in Fig. 22. At the bottom, lies thinly stratified
tuff (1), corresponding in thickness with the height of the cavern;
this is overlain by finer tuff, the cleavage fissures of which are
filled up with fibrous gypsum, the fibres standing vertical or nearly
so on the cleavage planes. In this tuff are leaf-impressions which
become more numerous in the somewhat coarser upper layers [1]).

Then follows another layer of tuff-breccia similar to that found
in Roemah Batoe. It contains here silicified pieces of branches
and tree-trunks in horizontal position, besides silicified erect tree-
trunks which partially project from the roof of the grotto and

---

seems to be the curiously shaped peak which the Dyaks call Liang-Pala. BÜTTIKOFER on
the other hand heard the natives give the name of Liang-Koeboeng to the mountain which
on Pl. XII and also on the topographical map is called Amai-Ambit.

1) In the fine brown tufflayers at Roemah Koetoe I also found leaf-impressions. They are
very brittle and I only managed to secure a few perfect ones.

hang down as straight columns about ³/₄ metre in length. This
phenomenon is easily explained. Originally the trees were entirely

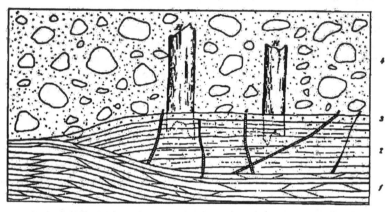

Fig. 22. SECTION IN THE DEER GROTTO AT THE FOOT OF THE HIGHEST TERRACE ON
THE NORTHEAST SIDE OF MT. LIANG-AGANG.

1. Chocolate-brown fine tuff; thinly stratified with irregular cleavage planes.
2. Fine brownish-grey tuff with leaf-impressions, in thicker less sharply defined strata,
than (1), they split in so many directions, that it is almost impossible to obtain perfect impressions. The wider fissures (S) are filled with gypsum.
3. The same tuff as in 2, only coarser; the leaf-impressions are more distinct.
4. Coarse tuff-breccia, in which are silicified treetrunks (II) in an erect position; probably
they were originally rooted in the fine tuff. From the roof of the grotto, where these
trunks protrude, the tuff-breccia has crumbled away considerably, and pieces of the
trunks have also fallen down, so that it is now impossible to indicate the exact place
where they once were rooted in the soil.

enclosed in the tuff-breccia, but pieces of this breccia successively
broke off and fell down through the roof of the grotto, as is
proved by the many loose stones at the bottom, originating from
the tuff; the silicified trunks did not fall because they penetrated
deeper in the breccia and were still firmly held by it. I could
therefore ascertain to my entire satisfaction that the trees really
stand upright in the breccia. No doubt they were once rooted
in the underlying layers of tuff, and were buried by the tuff-
deposits, shown in the section, and then partly destroyed, partly
enclosed without being overthrown.

I returned that same day to Nˢ Raoen, where I received a
visit from some Poenans belonging to the family, which our
arrival on the Liang-Agang a few days previously, had frightened
away from their home. They had learned since that we had
not come with any hostile intentions, and they now wanted to
satisfy their curiosity. The trio, man, woman and little child,
made a pleasant impression upon me. They were of a stronger
and broader build than the Oeloe-Ajer Dyaks, and the young
wife had regular features and beautiful eyes.

As a farewell greeting the mountains of the Upper Mandai
treated us to most magnificent fireworks, in the shape of a
terrific thunderstorm, and the peals of thunder resounded through
the mountains till far into the night.

The next day I took leave of my fellow travellers. Henceforth March 14.
I was to travel almost entirely alone. BÜTTIKOFER intended to
stay for some time yet at his mountain-station on the Amai-
Ambit, HALLIER could not tear himself away from the rich flora
of the Liang-Agang, and NIEUWENHUIS had settled himself at the
pondok we had just built, to continue his studies amongst the
Oeloe-Ajer Dyaks.

The heavy rain which fell during the night had swelled the
Mandai considerably; my two little boats glided swiftly down
stream and in a few minutes Nˢ Raoen was lost to sight.

We were hardly aware of the existence of the riams now,
as most of the boulder banks and islands were under water;
only here and there a more rapid current betrayed the whereabouts of a riam. I carefully examined the boulders in the
neighbourhood of the mouth of the Aroeng Bĕsar, because,
when going upstream I had noticed there some non-volcanic
rocks, and now these banks proved to be rich in biotite-granite,
amphibole-granite, augite-granite, granite-porphyry, tourmaline-
granite, tonalite, tonalite-porphyry, felsite-porphyrite, quartz-porphyry and quartz-porphyrite with exceptionally beautiful micro-
pegmatite-structure, hornfels, and silicified slate, none of which
rocks I had come across higher up in the valley. At this point

the Mandai approaches a hilly and mountainous region, situated
a short distance to the northeast, in every way different in structure
to the type of the table-mountains. Most likely these mountains lie
outside the volcanic area, and the non-volcanic rocks in the river-
bed of the Mandai come from there. They would be carried by
the Aroeng Bĕsar and other smaller streams which rise in these
mountains and flow into the Mandai a little higher up. We
continued rowing all day and at 9 p. m. reached Nᵃ Kalis in a heavy
rain-storm. From the representative of the Dutch Government
there, the Malay Abang Beak, I gathered some information about
the mountains in the south which had attracted my curiosity on
my former visit, and I now learned that in order to reach them
one must follow the Kalis river for some distance and then branch
off by one of the paths which lead straight up into the mountains.

In the space adjoining the outer gallery some Malays were
squatting round two open Korans, and gave vent to their reli-
gious emotions, by shrieking out their prayers and lamentations
in the most execrable vocal and nasal sounds. They stopped
every few minutes to recover their strength by chewing some
sirih[1]). Abang Beak invited me to spend the night in his outer
gallery, but added that these devotional exercises, intended as
an act of homage to "the only true God and to His Prophet",
on the occasion of the Poeassa (Great Fast), would probably be
continued till 3 a. m. I therefore preferred to retire within my bidar.

March 15. Owing to the heavy rains, I could not leave till 9.30 a. m.
next day, and started in a small open sampan which I had
hired there. We rowed up the Kalis, which discharges into the
Mandai, just above the Kampong Nᵃ Kalis. The lower part of
the river offers little variety; both shores are flanked near the
mouth by some Malay houses with their floating bathing rafts in
front, followed by the ordinary river scenery of deserted ladangs,
with a few scattered Dyak houses. The banks are steep and
composed of rich yellow clay.

---

1) Sirih (Mal.) ▬ A mixture used for chewing by the natives in the greater part of the
East Indian archipelago. One of the chief components are the leaves of the sirih-plant.

As we proceeded upstream the current visibly increased with the rising flood and we were obliged to row quite close to the land making very little progress. Large tree-trunks swept past us carried away on the current, sometimes they veered right round in their mad career, and it behoved us to keep well out of their way, — or else woe betide us.

With great difficulty we managed to row past the pintas Toeboeh. This pintas is evidently of quite recent formation, and is not yet marked on the topographical map. The water coursed with boisterous fury through the gaping hole which was partly blocked up with tree-trunks and in which remnants of the original bank could yet be seen. Considering its recent formation this pintas is very deep, and in all probability it will be a permanent one.

About 1 o'clock we reached the Dyak kampong Boentoet Toeboeh, consisting of two houses. Here the rowers took their siesta. As by 2.30 the second boat had not arrived I sent a small Dyak sampan with two men to its assistance. About 3 o'clock both sampans made their appearance. A terrible bandjir[1] was raging; during the two hours that I had been waiting here the water had risen 90 centimetres. One could see it rise. Wave after wave dashed over the existing water-level and swelled the waters with incredible rapidity. A Dyak staircase which led from the high shore to a boulder-bank in the river stood under water up to its highest rung.

At 3.15 we went on again and with tremendous exertion we reached the pangkalan at 4 p.m. From there a road led to the Mt. Sassak. I was told that the two rivers which we had to cross to reach the mountain were so swelled by the heavy rainfall that they had become quite impassable. But I took no notice of the warning, remembering the „di sini poentjak habis", which I had heard so often on the Liang-Agang.

We reached the first stream after half an hour's march. I ordered a tree to be felled on each side of the water, both of which reached beyond the middle of the stream, and so I

1) Bandjir = flood.

managed to get across. A small chest with tools and a few
other things which we did not absolutely need, I left there,
perfectly unguarded. I mention this as a proof of the absolute
honesty of the Dyaks; although this path was used every day,
the chest could be left with perfect safety by the roadside,
covered with a piece of matting, without the slightest fear of it
being taken away or in any way interfered with. Another good
hour's tramp across fallen trees, roots and mud, brought me to
the second stream, a fierce mountain torrent; which we had to
ford. The water came up to my shoulders, and I had great difficulty in retaining my footing. We made our bivouac on the
opposite shore under a large tree encircled by a Ficus.

The boulders in the mountain stream were entirely of volcanic
origin, chiefly andesite, basalt and tuff-breccia, which fact led me
to suppose that the Sassak would in structure be similar to the
Liang-Agang.

March 16. At 7.30 a. m. I resumed my journey; the path led in a south-
westerly direction, through several brooks, feeders of the Sépangin,
which discharges into the Kalis below the Toeboeh. Between
two of these mountain streams is a small ridge, entirely covered
with blocks of doleritic basalt rich in olivine (I. 343).

About 9 o'clock we reached the Dyak house at the foot of
Mt. Sassak. There are several other houses more to the
west, on the northern slope of the mountain, all surrounded by
extensive ladangs. They reach as far as the sources of the Méri-
poeng, which flows into the Soeroek, a tributary of the Boenoet.
For one florin I hired a guide from the Dyak house to lead us
up the mountain, and we immediately commenced the ascent
through the ladangs on the steep southern slope. At an altitude
of 200 metres we turned to the right, i. e. to the west, and
scrambled up and down the naked, sloping and slippery tuff-walls,
without getting much nearer to the top, until we reached a little
stream where we took our midday rest.

The coolies proposed to make our night quarters here, but I
ordered them not to unpack anything for the present, but to

await my further instructions, and with a couple of men I continued the ascent. Again through very steep old ladangs we reached a height of 430 metres.

The prospect was glorious; I sighted Mt. Kĕnĕpai W. 5 N., Mt. Toetoop W. 13 N., Mt. Mĕnjoekoeng W. 20 N. and more in the foreground a small cone-shaped mountain (probably Mt. Sĕnarah near Djongkong) W. 8 N. A little higher up there was still good water to drink and I sent one coolie back to tell the others to make ready there for the night. Yet another steep ladang rose above this point, it was the highest in this part of the district. Beyond it came the primeval forest and I was particularly struck with the large number of beautiful *Begonias* and *Piperaceae*, amongst them some creeping plants with beautiful variegated leaves, 30 centimetres long. We were now ascending a north-westerly ridge of the Sassak, and proceeded therefore in a south-easterly direction. The almost perpendicular tuff-walls which we had to climb occasionally, gave us a good deal of diversion. At 3 p. m. I stood on the north-west summit, 690 metres above sea-level.

This summit consists of tuff and tuff-breccia, in fact the whole of Mt. Sassak is built up of mighty deposits of basaltic and andesitic tuff which in the upper part of the mountain are well exposed in the naked vertical rock-walls stretching S.W.—N.W. By climbing up a tree I managed to get a good view to the West, North and North-east.

I sighted the Kĕnĕpai W. 6 N, the Sĕkĕdau W. 1 S.

There were other mountains quite near, more or less connected with the Sassak, in fact the continuation of it, viz. Mt. Kĕtoengang W. 15 S, Mt. Pĕkaong W. 5 S, and Mt. Liang-Pahoe S. 15 W.

The view from Mt. Sassak is particularly instructive. To the west and north stretches a vast wooded plain, with here and there a shimmering patch, indicating a danau or a stretch of the Kapoewas river. To the far west this plain is bordered by the Toetoop, Kĕnĕpai and Sĕkadau ranges, adjoining which are the hills of Sĕmitau, and further on the hilly district of the Upper Soehaid, the Embahoe and the Boenoet. In a west-north-west direction the Sĕmberoewang range arising in the north-west, winds its way through the lake-district. The Mĕnjoekoeng is the highest south-eastern extremity of this range. To the north and north-east this region is bordered by the high mountains of the Oeloe Kapoewas district, where I sighted E. 49 N. a very high mountain, presumably Mt. Oendjoek Baloei (1670 metres), in the mountainous district between the Upper Kapoewas and the Mĕndalam. Turning to the east one overlooks the Müller mountains with all the spurs which trend towards the Kapoewas Plain (Fig. 23). The south-eastern boundary, above Boenoet, is formed by the sheer abrupt north-north-western fringe of a large table-land, the Mandai tuff-plateau, built up entirely of tuff-deposits. This ragged fringe is the result of erosion and the steep cliffs form long ridges trending S.S.E.—N.N.W. According to the topographical map the general trend of this south-eastern boundary of the Upper Kapoewas plain is E 20 N—W 20 S; the ridges of the Tiloeng and the Sassak protrude farthest into the plain. All these mountains terminate abruptly towards the north and north-west [1]), and the Mandai tributaries have all cut their beds very deep into the plateau, so that the rock or fringes present almost in all directions perpendicular terraces. The Sassak itself is one of these ridges trending N.N.W.—S.S.E. I followed it for about half an hour when I found myself face to face with a perpendicular rock-wall 150 metres high. I had no time to find out whether there was any chance of scaling it, and so this highest south-eastern summit of Mt. Sassak had to remain unexplored.

Returning by the same way we soon descended the mountain and I found my tent pitched in the place that I had selected for that purpose on the upper edge of the rock-wall facing north. In the middle of the night I was awakened by a weird muffled rumbling noise, and on opening the curtain door of my tent a

---

[1]) The strong contrast between the richly tinted mountains and the uniform greyness of the slightly clouded Upper Kapoewas plain, gave this latter the appearance of a sea in which the spurs of the Müller mountains stood forth as so many promontories. This impression has been reproduced in the sketch (Fig. 23).

Pl. X.

ROEMAH BATOE.

4. Gĕgĕri-chain.

3. Baki-chain.

2. Liang-Karang.

1. Soenan.

Plain of the Upper
Kapoewas.

Fig. 23. Northern spurs of the Müller Mountains, seen from the top of
Mount Mĕnsassak in an Eastern direction.

wonderful, mysterious scene displayed itself before my astonished gaze. A heavy thunderstorm was raging far down below me in the Kapoewas plain, and the flashes of lightning revealed at intervals the exceedingly great dimensions of the cloud-masses, which barricaded my view at a few kilometres distance. At my feet stood some half decayed trunks of trees, enveloped in an intense, blueish-white, smoky light caused by phosphorescent fungi. The light was so strong that I could see to read by it at several metres distance, and even after closing my tent I could see the mysterious blueish shimmer. This phenomenon is a common one in the forests of Borneo, but I have not seen it anywhere so strongly marked as here on that night.

March 17.   In the morning with a temperature of 22° C. we were shrouded in a thick fog, and as we descended it became evident to me that the bank of clouds which always envelops the mountains in the morning is perfectly level on the under-side, for when I reached the highest point of the lower lying ladang, I also reached the lower limit óf the cloud wrap, and one step forward brought me out of impenetrable fog into bright daylight. Before me arose a cloudless unlimited panorama; to the north my eyes roamed over the vast uninterrupted expanse of the Kapoewas plains; to the east rose the mountains of the Upper Mandai, whose crowns seemed to be cut off, so entirely were they enveloped in the clouds, at an altitude about on a level with the place where I stood.

I did not stop at the house, which possessed seven pintoes. The Dyak type is far from beautiful here. Many individuals, our guide among them, were suffering from Korap[1]). The young girls wore dark blue skirts edged with red, which had a very bright effect.

At 10.15 I regained our former night-camp, and at 11.30 we were back at the pangkalan on the Kalis.

The water had fallen three metres since I went up the river, and several boulder-islands and banks had become exposed; besides

— ·-·—— — —·-

1) Korap (Dy.) = A skin-disease from which the Dyaks in West-Borneo suffer heavily. It is caused by a fungus. Compare: A. W. NIEUWENHUIS. Ueber Tinea imbricata, Archiv für Dermatologie und Syphilis Bd. XLVI. Heft 2, 1898.

volcanic rocks I also found pebbles of granite which proved that also in the river-basin of the Kalis, the older non-volcanic formation comes to the surface.

We rowed on at a good rate and reached Nanga Kalis at 3 p. m. I only stopped to pay what I owed there and went on again. We rested for half an hour at 6 p. m., and rowed on as far as Kwala Mandai which I reached at midnight.

I had to spend two monotonous days after that in my sampan going down the Kapoewas river before reaching Sĕmitau where I arrived in the evening of March 19.

THE STRUCTURE AND ORIGIN OF THE MOUNTAINS OF THE UPPER MANDAI.

Before the volcanic eruptions took place which have been the cause of the formation of the Müller mountains, that region must have been occupied by a much denuded mountainous area, composed of granite, silicified slate, chert, hornstone, etc., probably corresponding in structure to the older formations, which are exposed occasionally in the lake-district and which I afterwards designated by the name of the Danau-formation. This old formation must have been entirely or partially overlain by more recent deposits of sandstone and claystone, in which coal seams occur. Below Goeroeng Bĕroewang these latter deposits are still exposed in various places along the banks in the Mandai valley and they also form the lower parts of the Liang-Agang and the Tiloeng. I consider these coalbeds to be stream-deposits formed by a process similar to that we now see in course of action in the flats within reach of the large rivers of Borneo, where at the time of floods, layers of vegetable debris become intercalated between beds of mud and sand. The banks of the Lower Mandai show many examples of similar vegetable beds lying between layers of riversand and mud, which geologically belong to the present era. The constant occurrence of false bedding in the sandstone and claystone confirms their deposition from running water.

The section of the northern slope of the Liang-Agang (page 59, fig. 20) shows that on the pure sandstone (e) containing almost exclusively quartz-grains, a little clay, particles of coal and minute flakes of muscovite, another sandstone is deposited (f) containing particles of ash and small fragments of tuff. The volcanic eruptions therefore must at that time have occurred at some distance, and material devided from them was carried along by the water and deposited here. That the volcanic deposits gradually gained ground, is proved by the increase of volcanic material in the sandstone and finally the beds of tuff (h) which compose the entire upper part of the Liang-Agang (above a height of 160 metres), themselves covered the sandstone deposits. The period of volcanic activity must have been of very long duration and was characterised locally by long intervals of rest. This is proved by the fact that large fragments of tree-trunks have been enclosed even in the uppermost layers of tuff. The tree-trunks standing erect in the tuff, such as those I found in the Deer grotto on the Liang-Agang (see fig. 22) prove moreover that during such a period of rest, a forest must have grown upon the then existing tuff-plateau, which afterwards with the renewal of volcanic activity was buried under a fresh deposit of tuff.

At first, no doubt, the area covered by the tuff formed one consecutive whole, a plateau-land in fact, at a considerably higher level than that occupied by the Kapoewas plain. The rivers which flowed down from this higher level, afterwards cut out for themselves deep valleys of erosion, and this is the reason that the character of the plateau especially on the side facing the Kapoewas plain, has been lost, and a grand and magnificent mountain region has been created. The mountains themselves are remnants of the original plateau, which so far have escaped erosion, and they stand forth as beacons of varied and often very curious shape between the deep valleys of erosion. As a rule, the trend of these valleys on the boundary between the area of the volcanic activity and the Kapoewas plain, is vertical, that is to say, the water has generally taken the shortest cut to the Kapoewas valley. In consequence of this the plateau has been divided into sections, elongated in a S.S.E.—N.N.W. direction, as may be clearly seen from the top of Mt. Mĕnsassak (see fig. 23). In the south they all unite with the central part of the Müller mountains, which so far has retained its uniformly coherent character, and seen from the Kapoewas plain they look like the fringe of this mountain-mass.

Every one of those sections is in its turn composed of a range of more or less coherent table-mountains or mountains with terrace-structure, all trending N.N.W.—S.S.E.

Going from west to east, the following ranges and valleys may be noted in this region:

1. On the borders of the Kapoewas plain near the sources of the Mĕripoeng, Mt. Sĕkaroeang (650 metres) and Mt. Toenggoel (699 metres).
2. Valley of the principal feeder of the Mĕripoeng.
3. Mt. Sassak or Mĕnsassak (780 metres), Mt. Liang-Papan (820 metres), and Mt. Liang-Paoeh (837 metres); the Sassak-range.
4. Valley of the Toeboeh.
5. Mt. Sagoe and Mt. Liang-Soenan (986 metres); the Soenan-range.
6. Valley of the Kaoek, a left side tributary of the Kalis.
7. Mt. Liang-Karang (664 metres).
8. Valley of the Kalis river.
9. Mt. Baki or Bake (613 metres), Mt. Liang-Satoe (670 metres); the Baki-range.
10. Valley of the Pĕnyoeng.
11. Mt. Toeryan (248 metres), the Gégari-massive (1390 metres).
12. Valley of the Sĕmĕran.
13. Mt. Tiloeng (1112 metres), Mt. Amai-Ambit (1081 metres) and Mt. Liang-Koeboeng (± 1350 metres).

West of the Tiloeng are two lower summits evidently connected with it, Mt. Mahakoer (722 metres) separated from the

Tiloeng by the Daän rivulet and Mt. Daroewen (564 metres)
which is connected with Mt. Mahakoer by means of a saddle
about 100 metres high; between these two latter mountains the
Patan flows towards the Mandai.

Between Mt. Tiloeng and Mt. Amai-Ambit lies the deep narrow
valley of the Raoen-kanan [1]).

14. Valley of the Mandai below Nᵃ Raoen, of the Raoen
    kira and of the Sanjai.
15. Mt. Liang-Agang (946 metres), Mt. Aoe (1051 metres)
    and Mt. Pali (1338 metres); the Agang-range.
16. Valley of the Mandai above Nᵃ Raoen.
17. Tuff-mountains east of the Mandai valley.

None of the tributaries of the Mandai drains so large a piece
of this part of the Müller mountains as the Kalis. The river
rises not far from the source of the Sanjai west of the Agang-
range. At first its course is very sinuous, running in a western
direction, on the south side of the Tiloeng-range and the Gěgari
massive, and only much further west does it turns in a north-north-
west direction towards the Mandai valley. Mt. Tiloeng and the
Gěgari massive are the two broadest remaining remnants of the
original plateau. The top of Mt. Tiloeng is nearly a flat, plain
covering about 3 square kilometres with a feeble northern incline.
The shape of this plain corresponds to a square, its sides pointing
to the four corners of the wind, only on the south side the regularity of its shape is interrupted by spurs, and on the same side
the Tiloeng-plain is also considerably higher as there rises on it
a ruin of a higher terrace. Its height there is 1112 metres, while
on the north side it is 872 metres. Larger and higher than Mt.
Tiloeng is the Gěgari massive; its spurs are much more numerous,
and it lacks the severe simplicity of outline which makes Mt.

1) *Kanan*, right, and *kira* or *kiba* (Dyak) left, are words used in the reverse sense to the one we attach to it. They count right and left facing upstream. Sometimes they add *moedik*, upstream or going upstream; thus what we should call a left side tributary, they call a *sintang (papan) kanan moedik*. The Raoen originates from the confluence of two mountain streams, the left of which is called the Raoen kanan, the right the Raoen kiba.

Tiloeng so striking an object. Its position, moreover, is less isolated
and therefore less conspicuous. Seen from the Mandai, at a point
in the vicinity of N⁑ Pényoeng, the Gĕgari bears a close resemblance to the Tiloeng. The best point from which to survey the
enormous massive of Mt. Gĕgari is from one of the summits west
of the Kalis, for instance from the Mĕnsassak or the Soenan.
The total area covered by the highplain on Mt. Gĕgari is certainly
not less than 2¹/₂ square kilometres ¹). This plateau also dips slightly
to the north and reaches its highest elevation on the southern edge
(1390 metres). To the east, the Müller mountains, with uniform
characters extend at least as far as the Middle and Upper
Kĕrijau, where later on I found tuff-mountains of the same structure
as those of the Mandai. Still further east in the Oeloe Boelit and
Oeloe Boengan, the Müller mountains although still composed of
volcanic rocks, are somewhat different in character. It is not yet
known how far south the region of the tuff-mountains, the volcanic
Müller mountains, does extend. An examination of the neighbourhood of the Mandai above Goeroeng Bĕroewang in the Upper
course of the Kalis, could certainly throw some light on this
subject. But the Kalis is only navigable up to just above
N⁑ Kaoek, and neither on the Mandai nor on the upper course
of the Kalis are there any houses, the only inhabitants being
the nomadic Poenans. They are very shy, and as they live in a
constant state of enmity with the neighbouring tribes they are
always on the alert for danger; and the very unusual sight of
a European might drive them to make use of their terrible
weapon, the blowpipe (soempitan) with poisoned arrows. The
only casualty during the topographical survey of West-Borneo
that I know of, occurred in the Upper Mandai district not far
from the Poenan settlement Kĕloeka which belongs already to
the river basin of the Kĕrijau. One of the coolies in the service

---

1) The high plateau (i. e. that part of it which forms one unbroken whole) is therefore
smaller than that of the Tiloeng, although the entire mountain is much larger; this is
accounted for by the fact that the Gĕgari massive is much more divided and eroded than
that of the Tiloeng.

of the surveyor Werbata, was walking close behind him, carrying one of the instruments, when on a thickly wooded track he was quickly and cleverly despatched by a Poenan. It was done so suddenly that no one even saw the perpetrator. Werbata had the courage to go on to the Poenan settlement, and there he learned that the Poenan had not really intended to kill the coolie but had done the deed merely in a frenzy of fright. Werbata eventually completed his survey of the entire Upper Mandai, without any further molestation, and the Poenans paid to the Dutch Government an indemnity for the murder committed, consisting of a certain number of gongs, etc. This is a telling instance of the influence of a cool, self-possessed, tactful man, over these crude, uneducated, but at bottom thoroughly honest and trustworthy people.

Until now, this territory had not been known as volcanic. It was surmised that Mt. Tiloeng consisted of granite, but I have not been able to find out on what grounds. TROMP for instance (*60* p. 759) calls Mt. Tiloeng a granite-mass. The mining engineer EVERWIJN [1]) in 1853 approached to within a short distance of these mountains during his search for coal-beds in the valley of the Kapoewas and its tributaries. From a hill on the Mĕntĕbah, which he called Miangouw, he could see the volcanic mountains at a short distance in front of him; „he called them a series of hill- and mountain-tops, sometimes of considerable height". He also ascended a low hill one kilometre inland from the right-bank of the Soeroek, not far below N° Mĕntĕbah, composed of a succession of sand, coal-slate, claystone and clay strata. Resting upon these he found alluvial coarse sandy soil, which has been worked for gold. Besides many quartz pebbles he observed here also much petrified wood, amongst which were pieces 1—2 metres long and from 0.5—0.8 metres in circumference. It seems to me quite reasonable to suppose, after my observations in the Mandai district, that these were derived from eroded tuff-beds. During the

1) R. EVERWIJN, *11* p. 21 and 22 and *10* p. 379 etc.

Pl. XI.

UPPER TUFF-TERRACE OF THE BT. LIANG AGANG.

time that these alluvial sand and gravel beds were deposited the tuff-formation must have extended much further down towards the valley, in this case to the west-north-west, and since then the entire Mandai district has become considerably smaller through erosion. This explains the presence of the large pieces of silicified wood, which refute the possibility of transport over any considerable distance. Besides which, EVERWIJN cannot have been more than 7 kilometres away from the nearest tuff-mountains, for instance Mt. Sékaroeng, and one can only be surprised that he could resist the temptation of visiting them. It is interesting to note that he found pieces of tourmaline rock in this same gravel formation. We may therefore confidently surmise that somewhere within the Soeroek valley outcrops of the older formation also exist upon which both the coal and volcanic deposits rest. If we take into account that in the boulderbanks of the Mandai tourmaline-granite also occurs, we may conclude that in this old formation tour-maline-granite played an important part.

CHAPTER V.

The Lake-district [1]).

April 9.   A freak of the moon caused one day's delay in my departure
for the great lakes, originally fixed for the 8th of April. The
New Year of the Malays, being the close of the Poassa (great
fast), falls on the day that the moon's crescent becomes visible
after the first new moon subsequent to the 21st of March. The
new moon fell on the 6th of April, the crescent therefore ought
to have been visible on the 7th and bring the fast to a close.
But the moon not showing itself on the 7th the fast was prolonged,
and it was not until the evening of the 8th that the festivities
commenced. This was the reason that I was not able to start
for the lakes until the morning of the 9th, when at 7 o'clock
I embarked in the steam-launch Poenan, Poelau [2]) Madjang being
my provisional point of destination.

In passing the little side rivers which above Sĕmitau frequently
serve to connect the Kapoewas with the lakes, I was more than
once the victim of a peculiar optical delusion which constantly
gives one an impression that the water of the Kapoewas is higher
than that of the side rivers and therefore that the principal river
discharges into them. The cause of this phenomenon is very
easily explained. In the lake district the Kapoewas is bordered
on both sides by a bank of clay and loam from 1 to 3 metres
higher than the country beyond. When one of the side rivers
breaks through this bank and we look through the gap from the

1) For this Chapter consult Maps I and III of Atlas.
2) Poelau (Mal.) = island.

side of the Kapoewas, we see the banks which in the immediate neighbourhood of the Kapoewas are still tolerably high, gradually becoming lower at a distance from 20 to 100 metres, and at a high state of the water, it flows over the banks there, forming danaus. This in itself would be sufficient to give the impression that the water of the Kapoewas flows into the side rivers, but the effect is still enhanced by the slight stir of the wind always noticeable on the Kapoewas, which causes the light of the sky to be very strongly reflected in it, while the silent water of the side rivers, screened by high trees looks very black and appears to recede. This makes it appear as if the water in the side rivers is on a lower level, and consequently as if the Kapoewas flows into them.

At half past nine we steamed up the Tawang river, and very soon the banks disappeared, and the river was bordered by submerged trees. The water though transparent, looks dark, being not at all muddy but of a brownish yellow colour. Ramifications of the stream now become more frequent, and they, like the Tawang itself, have no perceptible current. The submerged forest is not very high and has evidently not the same variety of trees as is found elsewhere in Borneo. The whole landscape is very monotonous, once only a pretty glimpse of the Séligi mountains, closing a long rantau of the Tawang, gave a little variety to the scene. Gradually the anastomoses become wider and more numerous in this marshland, and at last, shortly before noon, the water spread rapidly, and soon the Danau Sérijang lay before us, an immense stretch of water, lined by unsightly swampy underwood. When bending down, so as to exclude from my view the circle of beautifully outlined mountains which encompass the lake-district, I could fancy myself back again on one of the large fen-lakes of the mother country. At a distance the swampy underwood of the great danaus of Borneo bears a striking resemblance to the willow and alder groves which border these Dutch lakes. We crossed the lake almost in its greatest diameter, and at 12.30 we anchored off Poelau Madjang. There is a fort on this little island, a so-called Koeboe, a small square

building on high posts surrounded by a fairly strong ironwood[1])
palisade. In this Koeboe some few dozen of police soldiers
(pradjoerits) were garrisoned, under the command of a native
sergeant. I did not worry myself about the evidently very rudi-
mentary development of military discipline in this stronghold, but
entered fully into the delight of contemplating the glorious land
or rather waterscape which lay before me (see Plate XIII on
page 21 of the atlas) as I stood on the raised platform at the
entrance of the fort. The calm and peaceful waters reflected in
minute detail, the scrubwood and the clouds of the afternoon
sky heavy with thunder, as well as the delicately outlined mountains around. Right in front of me, on the opposite side, rose
Mt. Lĕmpai, behind it Mt. Sĕligi, then Mt. Patjoor and lastly
Mt. Bĕsar. They all turn their steep inclines towards the north
and slope gently towards the south. In the distance I recognized
my old friend Mt. Kĕnĕpai. The ground of Poelau Madjang is
flat, the soil a mixture of sand and clay.

We did not long delay at Poelau Madjang and at one o'clock
continued our journey to Nᵉ Badau in two small open sampans. At high water one can go from here in a steam-launch
some distance up the Sĕrijang which discharges into the north-
western corner of the Danau Sĕrijang; at low water which it
happened to be in my case, this is impossible. But as long as
it is not dead low water, one can quickly run from Poelau
Madjang into a pintas, which is a kind of winding waterway connecting one part of the river Sĕrijang, just above Kwala Bĕsar,
with Danau Sĕrijang. In spite of the already critically low state
of the water we chose this short cut, but shallows forced us to
make constant little détours and we did not reach the river Sĕrijang
till four o'clock. The submerged forest between Poelau Madjang
and this point, the Danau Loepa Loewar and its surroundings,
is composed for the greater part of medium sized, charred tree-

---

1) Ironwood, a well known timber in Borneo, it particularly resists the effects of moisture
and the ravages of ants.

trunks, all broken off at a certain height and in different stages of decay; a sprinkling of lighter live-wood may be seen amongst them. This extensive charred forest has called forth many strange theories as to its origin, the true explanation however is this. In times of prolonged drought (two dry months are sufficient) these lakes quite dry up except for the deeper channels of the rivers. The fish collect in great shoals in the few pools and channels which remain filled with water and can there be caught easily and in very large quantities. The Malays and Dyaks of the neighbourhood congregate there for that purpose, and prepare an enormous amount of ikan kĕring [1]) over large fires. These are times of abundance and merry-making; the fires are not carefully watched and soon the forest is in flames, which, favoured by the drought, may spread over a considerable portion of the lake-district until their progress is stopped by wider channels destitute of vegetation. The result is a forest of charred trunks. Then comes the rainy season, the Kapoewas fills the lakes and the marvel of the inundated charred forest is explained. It will easily be understood that under the existing unfavorable conditions the forest cannot quickly recover itself. It is quite evident from the description of this submerged forest given by IDA PFEIFFER [2]) in 1846, which corresponds exactly to its present condition, that these forest fires did not merely occur in 1871, as we might be led to suppose from the descriptions of GERLACH [3]) and CROCKER [4]).

The winding Pĕsaja rivulet (only from 4 to 6 metres wide) was soon reached, and, the water being very low, presented the usual difficulty of the half submerged fallen tree-trunks. By the time we had reached the pangkalan Monggo Pinang, close to the Dyak house of that name, darkness had set in and it became evident that the state of the water higher up would not allow my sampan to pass through. At high water the land route from the

---

1) Ikan kĕring (Mal.) = dried fish.
2) IDA PFEIFFER. 44, I, pp. 82—128.
3) L. W. C. GERLACH. 14, p. 293.
4) W. M. CROCKER. 6, p. 203.

lake-district to N⁴ Badau begins at pangkalan Pésaja, when somewhat lower at pangkalan Djaroep, after that at pangkalan Monggo Pinang, and at a still lower state of the river at Nanga Pésaja, until finally Poelau Madjang can be reached on foot. At Monggo Pinang the land route commenced this time, and we followed it for that day as far as pangkalan Pésaja, where there is a deserted koeboe. This land route proved to consist of a deep layer of soft mud over which a sort of roadway of tree-trunks had been laid a so-called "djalan batang". Now it is a well known fact that in all acrobatic performances, darkness is synonymous with failure; it was therefore not without serious misgivings that I started on this undertaking. Very soon however some Batang Loepar women from the wayside house Monggo Pinang came to our rescue; they carried some of the luggage and two of them took me under their special charge, and guided me safely across the slippery tree-trunks, warning me whenever there was a hole deeper than usual. These women were very loud and talkative, they questioned me upon all manner of things, but as I could not make out half of what they said, they had it all their own way, and chattered on. Thanks to this diversion, and thanks also to the moon which now and again lighted up the pathway, I reached pangkalan Pésaja at 7 p.m. and took possession of the empty koeboe there. I paid my stalwart carriers partly in money which was welcome, and partly in tobacco which was eagerly received, and was soon for the greater part turned into quids.

The next morning I started again at 7.30 and was agreeably surprised to find a well kept broad pathway, the military road from Poelau Madjang to N⁴ Badau. On the way to N⁴ Badau we passed fourteen small bridges, the only bridges I remember seeing in West Borneo outside Pontianak and Sintang. This road was made in 1880 when there was a military post stationed at Nanga Badau, and it has ever since been kept in repair. The country we now traversed is gently undulating with low hills. There are no high trees as all the ground has already more than once been utilised as ladangs, one can therefore command a fairly good prospect from almost all points of view, which is not generally the case in Borneo. The low lying parts of this district, the valleys between the hills are mostly marshland (bogs) over which a roadway of trunks of trees is laid. This roadway consists of two or three small trunks each about the thickness of a man's arm, fastened together at intervals with rattan and resting on branches placed crosswise, thus forming a sort of floating bridge over the swamp [1]).

The soil in most parts is a hard yellowish brown clay and although the absence of natural or artificial sections made observations on the substratum an extremely difficult matter, I was yet able to ascertain that there is only a thin cover of clay and soil over the solid rock. The soil does not appear to be productive, which could hardly be expected otherwise considering the nature of the underlying rocks. On the hills previously used as ladangs, there is now a scanty jungle of brushwood mingled with ferns, resembling the vegetation on the hills near Sĕmitau. The predominating rock for the first three kilometres from pangkalan Pĕsaja is a fractured chert and hornstone, the fissures of which are filled up with limonite and in some places moreover coated with small quartz crystals. The prevailing strike is, as at Sĕmitau, almost due east and west, with a considerable dip chiefly to the north, but occasionally to the south, frequently also the beds are vertical. Beds of Limonite are sometimes intercalated with the chert, with the same strike and dip. These Limonite-beds are to my mind the decomposed outcrop, the "gossan" so to speak, of intercalated beds of iron bearing ore. The ore has moreover filled up the wider as well as the narrower cracks and has occasionally angular particles of chert imbedded in it. In the hilly district of Sĕmitau with a somewhat similar geological

---

1) It already existed in 1881, but at that time the trees were put across the deepest places only, and not made into bridges, see GERLACH. *14*, p. 285. BUYS, who without a doubt is more of a narrator than a traveller has greatly exaggerated the difficulties of this territory and of the road between pangkalan Pĕsaja and Nanga Badau. See M. BUYS. *4*, p. 112 etc.

structure I noticed the same relation between the chert and the iron-ore in the strongly folded strata.

About midway between pangkalan Pĕsaja and Nᵃ Badau, we came upon another rock (I 706, 707) which is particularly well exposed in the gorge of a streamlet, a right hand tributary of the Boenoet. This rock, which has here a strike E. 5 S. with nearly vertical dip, plays an important part in the geology of the lake-district and the Upper Kapoewas; it is of a greenish gray colour, tough, though much fractured and indistinctly stratified. Lighter coloured patches give it a marbled appearance and frequently numerous veins of white quartz intersect it, which occasionally results in a brecciated structure. Crevices and cracks are generally filled up with serpentine, so that when breaking up a piece of it, one is almost sure to come across the characteristic fracture of serpentine rock. This peculiar rock is the result of the combined action of decomposition and dynamo-metamorphism on diabase-tuff partly also on diabase and diabase-porphyry. The diabase-tuff occasionally passes into tuff-breccia in which sometimes the fine tuff-cement, and sometimes the manifold and varied rock fragments predominate. I shall call this characteristic rock the Poelau-Mĕlaioe rock, because the small island Poelau-Mĕlaioe is entirely composed of it and therefore all confusion is avoided. Moreover large fresh pieces of it can easily be obtained there.

In the neighbourhood of Nᵃ Badau the limonite and the fractured chert again appear, and are well exposed in the valley of the Boenoet river, which is six metres deep. The prevailing strike is east and west with a steep dip to the north.

At 11 a. m. I reached Nanga Badau, where I was kindly and hospitably received by the assistant-controller[1]) Mr. Spaan. His dwelling is within the fort where from 1880 up to 1888 a small garrison was stationed, but which is now occupied by a few pradjoerits under the command of a civil official. Like all koeboes

---

1) In the Dutch civil service in the colonies the sequence of ranks is as follows: Resident, Assistant-Resident, 1ˢᵗ class Controller, 2ⁿᵈ class Controller, Assistant-Controller.

iper terrace of the Liang-
Agang from where the
view is taken.

Mandai valley in
the distance.

Liang Koeboeng ± 1350.
Valley of the Sanjai in the foreground.

Table mountains composed of stratifie
formations in the distance

S 40 E   S 30 E   S 20 E   S 10 E   S   S 10 W   S 20 W   S 30 W

MÜLLER-MOUNTAINS. VIEW TAKEN FROM THE SOUTHERN SUMMIT OF THE LIANG-AGANG. (946 M.)

in Dutch Borneo it is surrounded by a square wall of heavy palisades, at two of the extreme opposite corners of which there is a projection from which one can at a glance survey the entire outside of the palisade.

In the afternoon I visited Mt. Pérak, on the top of which (90 metres high) there is a stone pillar indicating the place of an astronomical station of the topographical survey. The entire hill is composed of the already mentioned fractured chert, here much weathered and for the greater part altered into pure white, amorphous silica. The cracks are filled up with limonite and where they are very numerous, as is the case on the eastern slope of the Pérak hill, the rock resembles a beautiful mosaic of white angular fragments, cemented together by veins of a brown colour.

Although the hill is not high, the prospect from it embraces the greater part of the surrounding hilly territory. From here one can distinguish even better than on the way from the pangkalan Pésaja to N° Dadau, that the greatest diameter of all the hills is invariably from east to west and that they are grouped in the same direction in more or less clearly defined ranges separated by elongated valleys. But as these lines of hills vary from a fairly good height to but slightly above the level of the valleys around, a cursory observation gives the impression that they are irregularly dotted over a plain, while as a matter of fact we are actually standing on one of the ridges of a very strongly denuded mountain range. The most conspicuous mountain group is that of the Bésar, Pangoer Doelang, Patjoor, etc., which separates the Batang Loepar district from the valley of the Empénang. To the north and far into Sarawak there is much hilly ground but no really high mountains. In a north-eastern direction lies the higher boundary range between the Batang[1]) Loepar and the Lakes, called the Tintin Kédang; the far western outpost, the steep Mt. Pan (480 metres high), being the most conspicuous of this group. An east to west trend in this mountain group is easily discernable. To the east, at least in so far as the

---

1) Batang, from batang ajer (Mal.) = riverbranch, river.

view is not intercepted by the Badau (138 metres high) standing very close to us, our eyes roam over undulating hilly ground until they rest on the mountains of the Sémbéroewang range, also called the Lantja or Lantjak mountains, to the east of Danau Loewar. In the south-east we see the glimmer of the lakes, and to the south we look with pleasure upon the beautiful group of the Lémpai, Séligi, etc., which now turn their steep inclines towards us, and have, as seen from here, the appearance of belonging to the table mountain-type.

Returning to the koeboe from my point of observation, I found the controller's house beset by a number of Batang Loepars of both sexes. Their purpose in coming was of a peaceable nature, principally to see the toewan doctor batoe[1]), to ask him a hundred and one questions, and to offer him their goods for sale. The end of it was that the front gallery was filled with quite an assortment of swords, garments, musical instruments, etc. As we could not quickly agree about the prices — the market having been spoiled by the recent visit of Count Teleki — I placed by the side of every article that I wished to buy, the money that I was prepared to pay for it, after which I quietly went indoors and sat down to my work. In about half an hour I came out to see the result. Nearly all the Dyaks had left, but in every instance either the article itself or the money had been left, according as to whether the money offered seemed to them sufficient or not. No irregular dealings need ever be feared with these thoroughly honest people.

April 11.   On the 11th of April 1894, the mist at sunrise was not quite so thick as is usually the case in the wet forest-land of Borneo. The early morning temperature, rising as a rule to 22° and 24° Celsius registered at 6 a. m., was on this particular day as high as 25.5° Celsius.

At 8 a. m. I set out with a few carriers for Mt. Pan, our terminus for that day. The winding path led principally in an eastern direction right through the hilly district. We left Mt. Badau close by, to our left, to the north therefore of the path we were

1) Toewan doctor batoe (Mal.) = The doctor who knows the rocks.

following. The path is fairly good, although a good deal of delay was caused by the marshes in the deeper gorges between the hills which we had to wade through. This country has also been cultivated by the Dyaks, the prospect from the hills is consequently open and extensive. Highly cleaved chert still predominates, the strike here being E. 5 N. with an average dip of 75° to the north. The colour of the chert varies considerably; sometimes it is blueish; sometimes gray, white, or red. As a rule the chert here is also decomposed into white amorphous silica which is cleaved into small angular fragments and re-cemented with broad streaks of limonite. It presents the appearance of a dark brown breccia, with angular clear white flakes; in some places the iron-ore lies in purer broader layers, which however always contain angular fragments of decomposed chert.

Nowhere in the whole of the lake-district have I come upon the outcrops of lodes or beds the surface indications of which were promising enough to be considered worth prospecting to a greater depth; nevertheless it is quite possible that a more detailed investigation in this formation may lead to the disclosure of bodies of ore of some considerable importance. The chert also here alternates with rocks of the Poelau Melaioe type, as is clearly shown in a section along a brooklet close to the place where the path traverses the Gĕrendjang river.

Gradually the path verged to the north-east and we now cut through a most interesting complex of strata, well exposed at the point where the path traverses the bed of the river Pĕsaja in the immediate vicinity of Roemah [1]) Siengka. Beds of red jasper, chert and hornstone overlie the variegated cherts with a similar strike and dip. Microscopical examination shows that these hornstones are for the greater part composed of siliceous tests of Radiolaria. The researches of Dr. G. JENNINGS HINDE (see the Appendix to this work) have shown that these Radiolaria are probably of pre-Cretaceous age.

These remarkable rocks which occur over a vast extent in

1) Roemah (Mal. and Dyaks) = house.

West Borneo, in many respects bear a likeness to the recent deep-sea deposits which under the name of Red Clay and Radio-larian-ooze have obtained universal renown through the researches of MURRAY and RÉNARD. I found them in the lake-district in loose pieces in the bed of the Télijan near Génting Doerijan, as solid rock on Mt. Badau, which seems for the greater part composed of them; more westward in different places in Sarawak, and lastly at the point where the path from N° Badau to Roemah Antjeh traverses the Pésaja river, and in a few more places in that neighbourhood. Again I came upon the same rocks in larger quantity and greater variety in the basin of the Upper Kapoewas river, and its tributaries the Kérijau and the Boengan, where I could trace them till close up to the borders of East and West Borneo, and finally I got from the natives a piece picked up far beyond the water-parting in the river-bed of the Upper Ma-hakkam[1]). Wherever found these rocks always form part of a system of strongly folded strata, the strike being almost due east and west just as in the Batang Loepar lands. In the Upper Kapoewas this system of strata is, as a matter of fact, merely the eastern continuation of the system occurring in the Batang Loepar lands. Most probably in both places the Radiolarian chert occurs in the same horizon in this system. In the Upper Kapoewas district in the neighbourhood of the Goeroeng Délapan, I found a younger complex of strata, consisting of coarse sandstones with *Orbitolina concava* Lam. belonging to the middle of the Cretaceous period, which overlie and are folded in the system to which the cherts with Radiolaria belong, and so by two different ways we come to a confirmatory result as to the age of these presumable deep-sea deposits.

After wading through the Pésaja we left Roemah Siengka to our right and struck into a path in a north-north-easterly direction towards the Sérijang. The direction of this path being aslant to

---

1) By means of correspondence I learned from NIEUWENHUIS, who travelled along the Mahakkam in 1896, that the rocks in the great rapids of that river consist of milk-white hornstone. No doubt we have to deal here with the same Radiolarian rock.

the trend of the ranges of hills we were forced to climb up and down the hills much oftener than at first, when we could frequently follow the direction of the longitudinal valleys for some considerable distance. The predominating rocks in these hills are in some places chert, in other places hard fine-grained quartzitic sandstone. All traces of forest had long since disappeared and in a temperature of 34° Cel. I quickly regained my deep respect for the power of the tropical sun, which I had sometimes been on the verge of losing when traversing the primeval forest. The object of our journey, Mt. Pan, lay now to the N.N.W. of us; a heavy thunderstorm breaking over it that afternoon, partly hid it from our view.

At 1 p. m. we reached the river Sĕrijang near Roemah Antjeh, a house with 16 pintoes. Our way now led over some hilly ground through ladangs, at some distance from the left bank of the Sĕrijang. It was only in the little streams which flowed south-westward into the Sĕrijang that I caught occasional glimpses of the rocks, which proved to consist of quartzites similar to those found in the neighbourhood of Nᵃ Badau. Shortly after 2 p. m., just past Roemah Inoeh, the clouds brought us shade and rain. At 3 p. m. we again struck the principal tributary of the Sĕrijang near Roemah Patti Banang; it runs eastward past the southern foot of Mt. Pan, fed by a multitude of little streams descending from the mountain. Our surmise that there had been heavy rain in the mountains here proved to be correct, for the Sĕrijang, which the Dyaks here called Mălabau, had overflowed its banks and inundated the surrounding ladangs. The tree-trunk forming the bridge over the river stood about 1 metre deep under water, and the bamboo ropes, one finger thick, which did duty as a kind of balustrade, just touched the water, while in mid-stream they were washed away altogether. There was no time to lose, I therefore determined to risk the crossing, but took the precaution to have a rope stretched across the river, which precaution proved not to be superfluous for in mid-stream it was all I could do to keep my footing on the tree-trunk.

Arrived at the opposite side we had to wade through an inundated ladang and at 3.30 we reached Roemah Haloe, a house
with 5 pintoes, situated on one of the promontories of Mt. Pan.
As my coolies needed refreshment, I went for a few minutes
into the house. Immediately a few tikars[1]) were spread out in the
general part of the dwelling before the pintoe of the head of
the house, the kapala or toewa[2]); a rice-block[3]) was placed on the
tikars wrong side up, and covered with a piece of home-made
cloth. I was invited to seat myself upon this and was thereupon
welcomed by the toewa and his wife. Presently a great number
of the inhabitants, at first only the men and a few children, but
subsequently the women also, came crowding round, and a lively
conversation ensued. With the few words at my disposal I
managed to make them understand that I wanted at once one
or two men to take me that same evening to the top of
Mt. Pan, where I wished to spend the night. This was soon
arranged and at 4.15 we set out again. After a quarter of an
hour's walk over hilly ground we reached the foot of the mountain, and from there the ascent is straight and steep. The slipperiness of the ground and my saturated clothes, added to the
disgusting worries inflicted by the patjats[4]) and my anxiety not
to be overtaken by the darkness, made the end of this day most

1) Tikar (Mal.) = Matting of native make.

2) Toewa or toewah (Mal.) = Old, the eldest i. e. the chief of the house.

3) Wooden block, used as a mortar, for crushing rice.

4) Patjats (Mal.) = Leeches with which the humid underwood and the low plants swarm.
They attach themselves immediately to any one passing by and positively infest the forests
of Borneo, more particularly the slopes of the mountains. They are however not found
above an elevation of about 1000 metres. There are several species, three of which are more
common than the others. The commonest kind abounds in the moist forests and swamps,
and is of a dark brown colour; their bite is painless and as a rule the intruder escapes
unnoticed, leaving a tiny red wound which, in the case of Europeans at least, bleeds very
freely. The second species is greenish. These creatures are more active crawlers and do not,
like the former, confine their attacks to the legs. Their bite is painful, and the intruder
therefore is generally detected at once and killed before it has had time to draw the blood.
These two species are of about the same size; when extended about the length and breadth
of an ordinary wooden match. The third species which is only found in the Schwaner mountains
on the slopes of the high peaks in the vicinity of Mt. Rajah, is much larger and causes such bad
sores, that even the natives, who do not mind the smaller kinds, have a great dread of this one.

tiresome and I was well nigh exhausted when I reached the top
shortly before sunset. We found ourselves on a long narrow ridge,
rarely more than 8 metres wide and fairly flat. The prevailing
trend is W. 20 N. While my tent was being put up, I walked a
little way along the ridge and soon came to the conclusion that
there was not a single good point for observation. I sent back
one of my Batang Loepar Dyaks to order ten men from Roemah
Aloe to be here in the morning to cut down the necessary trees.

Before sunrise I was on the ridge again, to select the most **April 12.**
suitable place for my observations. I chose a spot where the
ridge of the mountain was about 10 metres above the average
height and very narrow, whereupon I ordered the Dyaks, who
meanwhile had arrived from Roemah Haloe, to make a clearing
on each side of it. They are used to this sort of work when
making ladangs, and they do it with great speed and dexterity.
They commence by cutting all the heavy trees half through on

Fig. 24. VIEW FROM MOUNT PAN ACROSS SARAWAK.

the side facing the slope of the mountain, beginning at the lowest
point from where the clearing has to be made and gradually working
their way up to the top; the thin trees are not touched at all.
At last the trees nearest to the top are chopped right through
in such a fashion as to make them fall down the incline carrying
along with them the lower standing, partly hewn trees; all fall
in the same direction down the slope, the thinner trees snap off

and are swept down with the rest. With a thundering noise the entire mass is hurled down to the bottom, an inextricable chaos of timber and foliage. At 10 a. m. the clearing was completed, and by noon the mist had sufficiently dispersed to reward us for our trouble by a splendid view both in the direction of Sarawak and of the lakes.

Looking towards Sarawak the eye rests upon a much intersected hilly territory in which a succession of ranges striking in an east to west direction is unmistakable. As far as I could see these hills are not very high; one only which I gauged at N. 20 W. is of about the same height as Mt. Pan. Facing towards the

Fig. 25. Mt. Kĕnĕpai, seen from the top of Mt. Pan.

Dutch side the most conspicuous feature is the sparkling level of the lakes, shut in on one side by the Mĕnjoekoeng, on the other by the Lĕmpai, Sĕligi and the mountains adjoining to the westward. To the left of Mt. Patjoor rises the Kĕnĕpai, which

Fig. 26. View from Mount Pan across the lakes southward.
In the Back-ground Mt. Pijaboeng (1130 M.) in the territory of the Oeloe Embahoe and Mt. Pĕningooen (989 M.) between the Upper-Bojan and the Silat, in the middle the Andesite-cones of the Middle-Embahoe (average 400—500 M.), in the fore-ground Poelau Tĕkĕnang and the great lakes.

seen from here looks exactly like a volcano with a gigantic crater. The peak called Mt. Oelak (1080 metres) and the principal cone of the Kĕnĕpai (1156 metres) appear from this point

Pl. XIII.

DANAU LOEWAR, AS SEEN FROM POELAU MÉLAIOE.

of view to be of equal height. In a south-south-easterly direction
the eye roams over the lakes and the Kapoewas plains, limited
at a distance of about 8o kilometres by the grotesquely shaped
hills on the river Embahoe, behind which in the far distance the
long ridges of the Madi mountains stand out against the horizon.
Before noon I partly descended the mountain on the Sarawak
side, this slope being also very steep, about 30°. The outcome
of my observations both here and also during the ascent and
descent on the southern side is given in fig. 27. The mountain

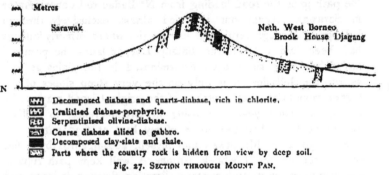

Decomposed diabase and quartz-diabase, rich in chlorite.
Uralitised diabase-porphyrite.
Serpentinised olivine-diabase.
Coarse diabase allied to gabbro.
Decomposed clay-slate and shale.
Parts where the country rock is hidden from view by deep soil.

Fig. 27. SECTION THROUGH MOUNT PAN.

consists of alternating strata of diabase, quartz-diabase, olivine-
diabase, diabase-porphyrite, clay-slate and shale. In various places
the diabase rocks are uralitised or much decomposed and then
rich in serpentine and chlorite. The strike of the beds varies from
N. W.—S. E. to W.—E. and the dip is 70°—80° to the north. Soon
after one o'clock I left the back of the mountain, to start on the
return journey. In our descent we chanced upon the biggest tapang-
tree[1]) I ever saw in Borneo. The plank-shaped excrescences leading
from the trunk to the roots were at least eight metres high and
fully six metres long. The diameter of the giant trunk was two and
a half metres, of course not allowing for the said excrescences.

After our descent we at once struck another route more westerly
than that of the preceding day, which brought us about 4.30

1) Tapang, kind of tree resembling an oak-tree in shape (see Fig. 54). In this tree
the wild bees generally make their nests.

p. m. past Roemah Djajang across some hilly ground into the valley of the Batang Boenoet, leaving the sharp cone-shaped Mt. Sandok close to us on our right (this is presumably the Mt. Patjoer of the topographical map). Wherever the rock was exposed, I could prove that we were in the chert area, the strike of the contorted strata being invariably E.—W. or nearly so. We passed successively the Batang Loepar houses, Roemah Tjaoek, Roemah Moentai, Roemah Mélien and Roemah Patti Moedjap, all situated on or close to the Batang Boenoet. Close to N° Badau the path joins the road leading from N° Badau to Loeboek Antoe in Sarawak. Again our way led almost exclusively through ladangs or through scrub and low jungle, where formerly ladangs had been laid out. In the Batang Loepar lands the primeval forest has long since been exterminated in the plains as well as on the hillsides; it is only on the very steep slopes of the higher mountains, that stretches of the original wood may yet be found, but each year decreasing in circumference. In brilliant moonlight I reached the house of my host at 6.30 p. m.

April 13. At 9.30 a. m. Mr Spaan and myself, accompanied by a few coolies started for Loeboek Antoe. A well kept path runs in a western or west-south-western direction through ladangs and brushwood, slightly undulating up to the water-parting (72 metres high) between the basin of the Batang Loepar and the Kapoewas river. At the beginning we followed principally the strike of the hills which are composed of white or red fractured chert, occasionally intersected with veins of iron-ore. We crossed the boundary between Dutch Borneo and Sarawak at the southern slope of Mt. Tělaga. About $1^1/_4$ kilometres across the frontier, the country rock is serpentine, a little further on it is once more the red Radiolarian chert, the strike being E. 13° N. and the dip 67° S. The road now turned slightly more to the north-west and became more fatiguing because of the many steep and narrow ridges, striking almost due E.—W., which had to be crossed. As we neared Loeboek Antoe the road steadily improved, one might with safety even ride here. Fairly well kept coffee plantations, belonging to

Chinamen made a pleasant variety in the scenery and indicated moreover that we were approaching the end of our journey, which we eventually reached at 12.30 p. m. There is a fort on the high left bank of the Batang Loepar, belonging to the government of Sarawak. It is a small square building containing one cool spacious apartment with earthen floor, alongside of which are the dormitories; on the second floor is another apartment furnished with portholes and where a few blunderbusses are stored. One misses the useful projections of the Dutch koeboes which command the outer walls. This fort is manned by seven native soldiers and a sergeant. The river here runs in a valley, 14 metres deep, but it is only when very high that the water reaches up to the steep bank in all places; as a rule there is a low sloping margin between the bank and the water's edge, on which a few Chinese houses are built surrounded by palm trees. Close to the koeboe one cannot see these houses because of the precipice which hides them from view, although they stand but a few paces distant.

The Batang Loepar makes a sudden bend here, and in the angle the steep bank is so extensively undermined that pieces of soil occasionally break off and fall; hence it is called Loeboek Antoe [1]). The country rock here is a much weathered quartzitic sandstone, it is shown close to the steps leading from the river up to the koeboe, and also a little further on in a small streamlet. In midstream is a broad boulder island, to which I waded out. The flat boulders are of soft sandstone, quartz, quartzite, gabbro, different varieties of diabase and diabase-porphyrite like those of Mt. Pan. We did not find a single piece of the red chert with Radiolaria.

At Loeboek Antoe I slightly added to the finances of the Borneo expedition by a little business transaction, which may serve as an illustration of the singular confusion which reigns here owing to the rapid fluctuations in the price of silver. The current standard coin all over West Borneo is the Mexican

---

1) Loeboek (Mal.) = bend. Antoe or hantoe (Mal.) = ghost. Loeboek Antoe = haunted corner.

dollar, or the Japanese Yen of equal value, both called by the natives *ringgit boeroeng*[1]). This is divided into soekoes (Dutch coins, worth half a guilder), which are subdivided as follows: 1 soekoe = 2 tali (Dutch India coin) = 5 kĕtip (Dutch India coin). The smallest negotiable coin is the duit (the Old Dutch coppers of the seven Provinces)[2]). One kĕtip is reckoned to be worth 12 duits. Small calculations are generally made in a fictitious coin, the wang, its value being reckoned as equal to 10 duits. The comparative value of the soekoe and its subdivisions, and in fact of all Dutch current coins, has remained unchanged, as also that of the soekoe and the Dutch "rijksdaalder", *ringgit baroe* or *ringgit Kompanie*, and of the Dutch guilder, worth respectively five and two soekoes[3]). But the fall in silver did affect the comparative value of the ringgit boeroeng and the soekoe as the former was not protected by the Dutch gold standard coins. Originally a ringgit boeroeng was worth 7, later on 5, and finally 4 soekoes. In the first half of the year 1894 the price of silver was so low that at Singapore and at Pontianak the ringgit was only worth $2^1/_2$ soekoes or a little over. In the neighbourhood of Sintang and Sĕmitau it was still worth 3 soekoes, and in the Upper Kapoewas the value of the *ringgit boeroeng* remained equal to 4 soekoes. This dilatoriness in apprehending the depreciation of silver was particularly advantageous to the Chinese who obtained the cheap dollars in Singapore with which they bought rattan and other forest products in the Upper Kapoewas, and it need hardly be stated that they did all they could to make the Dyaks continue in their belief in the old currency.

A new complication and a new source of profit for the Chinese now arose on the borders of Sarawak. The standard coin in Sarawak is the *ringgit boeroeng* divided into 100 copper cents

---

1) Boeroeng (Mal.) = bird; an eagle being stamped on the Mexican dollar.

2) The "duiten" used in Borneo are almost exclusively those of the provinces of West-Friesland and Utrecht.

3) As a matter of fact the Dutch silver 'Rijksdaalder', because of the double standard, has merely a fictitious value, that is to say, it is reckoned as equal to the fourth part of a gold ten guilder piece, although in reality it is not worth more than the eighth part.

which are coined in Sarawak. These cents were current also in the Dutch Batang Loepar lands and their market value was not depreciated by the fall in the value of the Mexican dollar. In N⁰ Badau and Poelau Madjang therefore, that is on Dutch territory, a ringgit boeroeng was considered to be equivalent to 80 Sarawak cents. And so it came to pass that at Loeboek Antoe I exchanged 4 ringgits for 400 Sarawak cents, and a few hours later at N⁰ Badau I exchanged these again for 5 ringgits. I cannot tell how long this crooked state of things has lasted, but it is a fact that large quantities of Sarawak copper coins have found their way over the Dutch frontiers and that many a Chinaman has chuckled over an easily acquired profit.

Let it be borne in mind that the Dutch rijksdaalder, *ringgit baroe*, and the guilder, *djampal*, did not deteriorate in value, being protected by the Dutch gold standard coin and were worth respectively 5 and 2 soekoes. This fact, the true cause of which neither the Dyaks nor the greater part of the Chinese rightly understood, helped to increase the respect for the Dutch government: the Kompanie¹). Many a time have I watched a Dyak weighing a Dutch 'rijksdaalder' against a somewhat heavier Mexican dollar, trying to find out why the former was worth about twice as much as the latter. I never failed to bring the lesson home that all things proceeding from the Kompanie were *baik betoel*²) and could therefore be trusted better than any foreign importations.

Shortly after 4 p. m. we left the Sarawak territory and after a quick march we reached N⁰ Badau at 6, where I was able to enjoy for some considerable time on the top of Mt. Pérak, the striking spectacle of a heavy thunderstorm breaking out over Mt. Bĕsar.

Of all the passes between the great river-basins of Borneo, the one between the Batang Loepar and the great lakes in the

1) The natives in the interior of Borneo still speak of the Dutch government as the "Kompanie", owing to the rule of the country by the Dutch East Indian Company in the 17ᵗʰ and the 18ᵗʰ centuries.

2) Baik betoel (Mal.) = first rate, of very good quality.

Kapoewas-basin is best known. IDA PFEIFFER was the first to cross the boundary line of mountains, in 1852. Probably she followed the route now familiar to us from Loeboek Antoe, but this cannot be accurately gathered from her description. It is quite evident that her ears were not accustomed to Malay and Dyak sounds; and therefore no great importance should be attached to her rendering of names. She[1]) left the Batang Loepar on January 28th at Bengkalang Sing-Toegang at the foot of mount Sĕkamiel. Bengkalang of course means *pangkalan* (the point where the land route commences, the landing or anchorage place). I have not been able to trace the name of Sing-Toegang nor of Mount Sĕkamiel, perhaps the former is meant for Sĕkoe-wang, in that case she must have reached the Boenoet river near Boenoet by way of Sĕkoewang. After a two days march, of 8 hours each, she came to Bengkallang Boenoet on the Batang Loepar river on the other side of the water-parting. This evidently should be either the pangkalan Batang Loepar or the pangkalan Boe-noet on the Boenoet river. Following that river she reached lake Boenoet, which was full of tree-trunks, not uprooted and scattered about, but firmly fixed in the ground, only they were quite dead and had neither branches nor crowns. Here she probably refers to the inundated forest of the Danau Loepa Loewar, north of the Danau Sĕrijang. A wide watershed or natural channel about a mile long brought her into another lake called Taoman, larger than the first and with a perfectly clear, level surface, from whence she rowed into the Kapoewas. This lake Taoman is of course Danau Sĕrijang, she confused the name of the lake with that of the river Tawang which connects the lake with the Kapoewas.

Fourteen years later came BECCARI[2]) the well-known Italian traveller, who made numerous botanical excursions in Sarawak. He also visited the lake district, starting from Sarawak. It is generally supposed, as for instance in Posewitz' "Borneo", that BECCARI followed the above mentioned route. This however is

1) IDA PFEIFFER, *44*.
2) O. BECCARI, *I*, p. 193. 1868.

incorrect. On May 7th he left Maroep situated much lower down than Loeboek Antoe on the Batang Loepar; he passed the water-parting at an altitude which he estimates at 1200 ft. — probably between Mt. Boewaja and Mt. Njambau — and on the 10th he began to descend the Kantoe, where he lost his sampan by a sudden flood, this causing a delay of two days [1]). On the 13th he descended the river as far as Sĕgĕrat on the Umpanam (by which is meant Empanang) and the following evening he reached lake Lamadgian, which, as the sequel plainly shows, is meant for Madjang, i. e. Danau Sĕrijang. Time, names and circumstances, — particularly the altitude of the mountain-pass and the descent of a dangerous mountain stream which does not exist in the N[a] Badau district — all fit in so exactly that there can be no doubt that BECCARI really did come down the Kantoe and the Empanang and reached Danau Sĕrijang by a roundabout way. BECCARI, however, tells us that he followed the same route that IDA PFEIFFER had travelled by, and that on the way many Dyaks even told him about her; but here he must have misunderstood his informants. CROCKER [2]) visited Danau Sĕrijang in 1880, starting from Loeboek Antoe past N. Badau. He calls the district which he passed through "perfectly flat", which is not correct.

Since the establishment of the military station at N[a] Badau in 1880, and the making of the road between that place and the Batang Loepar, this pass has been so frequently made use of by Europeans, that it is not worth while enumerating the different travellers who during that time have described the road.

The summit of Mt. Badau (138 metres high) in the neigh- April 14. bourhood of N[a] Badau on the other side of the Boenoet river, which, as mentioned on page 92, is formed entirely of slaty jasper with innumerable Radiolaria, had been completely cleared for the journey of inspection of the chief of the topographical service, Colonel BOSBOOM, and so I could from there survey the Batang Loepar

---

[1]) Una piena nel torrente (Kantù) si portò via le nostre barche e ci fece perdere i due giorni seguenti, l. c. p. 204.

[2]) W. M. CROCKER, 6, p. 193—205.

lands to perfection. I spent most of the day there, taking observations and sketching the panorama. From there one can see
the wide expanse of the lake-district in its entirety. It is shut in,
to the north, by the hilly ground of which Mt. Badau forms part,
and which more to the north, and especially to the north-east,
passes into mountain-land, the prevailing trend always being
east to west. Mt. Tadjoem (284 metres) south-south-east of
Mt. Pan was from our vantage ground the most conspicuous
object of the territory. I gauged the mountain at E. 15° N.
Almost facing this hilly ground there rises a range of mountains
trending N.N.W—S.S.E. which forms the eastern boundary of
the lake-district and separates it from the valley of the Lĕbojan.
I will call this the Lantjak range. I observed, from north to south,
Mt. Engkoeni (605 metres) E. 8° N., next, Mt. Sĕmbĕroewang
the highest of all, (752 metres) E. 2° N.; after that, Mt. Mĕlyau
(707 metres) E. 8° S. and finally, connected with it by a low range
of hills, Mt. Sap (320 metres). Mt. Sap, being composed of granite,
has a more gently sloping outline than the other mountains of
this group which are formed of tilted stratified rocks. More
to the south comes the Mĕnjoekoeng-group, standing by itself,
its highest peak being in the direction S. 59° E. Between the
Lantjak and Mĕnjoekoeng mountains there arise in the far
distance on the other side of the Kapoewas plain, the characteristic contours of the terraced or tuff-mountains of the Upper
Kalis and the Upper Mandai. We could clearly distinguish the
Gĕgari-group and the Liang Pata; Mt. Tiloeng remains hidden
behind the Lantjak mountains. To the right of Mĕnjoekoeng
follow Mt. Sĕmoedjan S. 53° E.—S. 35° E. and Mt. Pĕgah S. 47° E.,
encircled near their summits by clear, white, perpendicular walls of
well exposed rock, which bear a striking resemblance to the terraces
near to the top of Mts. Lĕmpai and Sĕligi. The two flanks (not the
middle, highest part) of the Mĕnjoekoeng have a similar terrace
formed by a brilliantly white precipice (see fig. 29). Still more to
the right, quite in the foreground between the lakes, rises Poelau
Tĕkĕnang (140 Metres) S. 36° E. and diagonally behind it the twin

BORNIT SÁLIGI, AS SEEN FROM ROEMAH SÁLIGI.

peaks of Mt. Sĕnarah near Djongkong; in the far distance the mountains of the Upper Embahoe, amongst which Mt. Pyaboeng resembling a volcano, and Mt. Ampan are the most conspicuous features. More to the West comes the Lĕmpai-group, striking E.S.E.—W.N.W., then the Mt. Bĕsar-group striking N.N.W.— S.S.E. as far as the boundary of Sarawak, farther on curving round in a west-north-westerly direction. All these mountains have a great family likeness, Mt. Lĕmpai (fig. 28) is the simplest specimen. In all of them the upper part is formed by one or more terraces, cut off sharply or perpendicularly to the north-east and north, and gently sloping down to the south and south-west. The outlines shown in fig. 28 hold good in general for all the mountains in the Lĕmpai- and the Bĕsar-groups. Later on we shall see that these terraces are composed of beds of sandstone resting unconformably on highly folded stratified rocks.

· I had arranged with the controller of Sĕmitau, Mr VAN VELT- April 15. HUYZEN for the steam-launch Poenan to fetch me on the 17th of April and to take me to Lantjak. As I wished first to visit Mt. Lĕmpai I left Nᵗ Badau on the 15th at 7.30 a. m. together with Mr SPAAN who also had to go to Lantjak on business. After a most enjoyable walk we reached the pangkalan Pĕsaja at 9.30 and a quarter of an hour later the pangkalan Djaroep, about midway between pangkalan Pĕsaja and pangkalan Pinang. Our carriers male and female joined us in small straggling groups, and gradually our two sampans were being loaded with our baggage; about 11 a. m. we took leave of this motley picturesque crowd. The water was low, but fortunately the pintas Pĕnemoi was still navigable so that we reached Poelau Madjang at 5 p. m.

Next morning at 7.30 we pulled across Danau Sĕrijang to April 16. the marshland at the foot of Mt. Lĕmpai, which is always flooded at high water. After a good half hour's walk through mud and over tree-trunks, Roemah Gĕramma at the foot of Mt. Lĕmpai, was reached at 8.45 a. m. The house was old and dirty, and, like most of the houses in this part of the Batang Loepar lands, it is built on posts which stand not more than about 2 metres above the ground. The interior of the house is so low, that a European cannot stand upright in it. To get to Mt. Lĕmpai we had to go through the house and soon we were climbing up the mountain by the way of the ladangs on the north-east side.

Mt. Lĕmpai is composed of strongly folded beds of serpentinized diabase-tuff and diabase-tuff-breccia, the rock being well exposed in several natural cuttings of the little streams on the northern slope. The strike of the strata is here W.N.W., the dip S. 3° W. at an angle ± 75°. Thick layers of friable yellowish-white sandstone lie unconformably upon the tuff-breccia, the sandstone of the lower strata being in several places coarse grained and allied to conglomerate. The strike of these beds is nearly due east and west, the dip 9° to the south. These sandstone beds cause the sloping incline on the southern side; and the steep terraces on the northern side of the mountain, where the beds are broken off, are formed by their outcrops. (See fig. 28). On the northern side the line of demarcation between the sandstone and the older formation is at a height of about 110 metres. The topmost ridge of the broken sandstone terraces, viz. the crest of the mountain, is not equally high in all parts, and one can distinguish a western and an eastern peak. The latter gives a grand view over the lakes, the former overlooks Mts. Sĕligi, Singkadjang, and Bĕsar

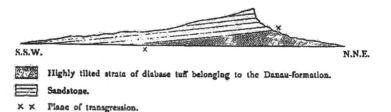

S.S.W.   N.N.E.

Highly tilted strata of diabase tuff belonging to the Danau-formation.

Sandstone.

× ×   Plane of transgression.

Fig. 28. Section through Mt. Lĕmpal

which, as can be clearly seen from here have all the same build as Mt. Lĕmpai. One can see distinctly that the sandstone beds overlie the older formation, and the almost vertical escarpment on the northern side changes into a gentler slope directly the lower limit of the sandstone terraces is reached. On Mt. Singkadjang moreover, one can see the sandstone beds alternating with softer beds, probably composed of finer sandstone and shale, the latter always forming a gentler slope in the outline of the mountain. These gentler slopes are not covered with forest and utilised as ladangs, but the sandstone walls either expose the glittering white, naked rock, or else they are thickly wooded, a fact which I had already noticed through my field-glass from Mt. Badau. On Mt. Lĕmpai itself the entire southern slope is cultivated[1]), but on the steep northern incline an occasional clump of the original forest has been left standing.

The heat in the rice-fields on Mt. Lĕmpai and again on our return journey in the open sampan on the lake, was almost unbearable. On this day, one of the hottest I have ever experienced in Borneo, and the only one on which the sky continued almost entirely cloudless from 10 a. m., I registered the following:

7 a.m.	temp.	25° Cel.	Wind east, slight.			
8.30 „	„	27°	„	„	„	„
10 „	„	30°.5 „	No wind.			
11.30 „	„	32° „	„			
1 p.m.	„	35° „	„			
3 „	„	36° „	„			
6 „	„	30°.2 „	„			

In the shallow lake the water becomes very heated on hot sunny days, so that on this particular evening the temperature of the water was 28° C. A good strong breeze however cools it fairly quickly again. At midnight the temperature was still 27°, and not till 3 a. m. a strong west wind with continuous rain brought the much desired coolness.

In the afternoon the "Karimata", the Resident's steam-boat, and April 17. the Poenan, respectively anchored off Poelau Madjang. On board the first vessel were Colonel Bosboom, chief of the topographical service and Captain van Enthoven chief of the military topographical brigade in Borneo. On the other was the controller van Velthuyzen who told me that he intended to steam out again directly. After

---

1) I took care to ascertain that on the sloping southern incline of Mt. Lĕmpai quite down to the foot, the rock was exclusively sandstone.

paying a visit to the "Karimata", I went on board the Poenan, and shortly after we steamed in a southern direction across Danau Sĕrijang. One of the many winding water-ways, of which only a few are navigable for small steam-boats, brought us to Danau Sĕntaroem and from there picking our way in between some small scattered islands to Danau Tĕkĕnang. Shortly after sunset a terrific thunderstorm with a heavy squall burst upon us, and we thought ourselves very fortunate in riding at anchor on the lee side of the lofty island Tĕkĕnang; here the gusts of wind could not reach us in their full force.

April 18. Immediately after sunrise I pulled round the north side of the island Tĕkĕnang into a channel which enabled me to approach the high part, Mt. Tĕkĕnang, on its western side. Hewing our way through the dense wood we at length reached the top, which was no easy matter as the ground was very slippery and the trees consisted mainly of the wild sago palm called *ransa*, the edges of the leaves of which are covered with sharp prickles [1]). Fortunately the strong lateral roots thrown out from the branches are not so protected, so that one can pull oneself up by them without any risk. I never knew any place so swarming with mosquitoes and kĕriangs [2]), the latter absolutely drowning the human voice with their shrill cries. The slope of the mountain is steeper on the north and north-east side, where the outcrops of broken off sandstone beds form an almost vertical escarp‧ment. The sandstone is crumbly, closely resembling that of Mt. Lĕmpai. It is clearly false-bedded, which probably indicates its deposition either in running water or under the influence of the tide. The strike of the sandstone is about E.—W., the dip 10° to the south. The summit of Mt. Tĕkĕnang is not flat but

1) This wild sago palm, *ransa* or *rangsa*, *Eugeissonia*, which grows on the steep rocky mountain slopes, does not seem to be at all common in the western division of Borneo. Teysmann (57, p. 316) found it on the Pĕnai mountains, and I came across it on Poelau Sĕpandan and also in the Upper Tĕbaoeng and on Mt. Lĕkoedjan on the borders of the western and eastern division of Dutch Borneo.

2) Kĕriangs, certain kinds of Cicadas, which make a very fierce noise, much louder than that of crickets.

slopes gently to the west. The eastern edge is 140 metres above the level of the sea, the western 14 metres lower. We proceeded from Poelau Tĕkĕnang by Danau Sĕntaroem and Danau Soembai, thence through a veritable labyrinth of water-ways to Danau Loewar. It is most difficult to find one's way amongst these winding water-ways edged by miserable looking scrubwood everywhere exactly alike. We had secured one of the few guides of the district, here called "loes", a corruption of the Dutch word "loods" (pilot) on board ship, who always knew, whatever the height of the water might be, by which water-ways communication between the lakes was possible for our small steamer. The Danau Loewar is the largest of the West Borneo lakes, the surface of the open water being as much as 87 square kilometres. These great lakes were discovered by the Dutch in 1823 and in VETH's classical work [1]) we get, chiefly from the reports of van LIJNDEN and GROLL [2]), a correct description not only of the position and the names of the principal lakes and islands, but also of their character and significance. In the very centre of Danau Loewar there is a small rocky island called Poelau Mĕlaioe, 14 metres high, after which the whole lake is often called Danau Mĕlaioe. It is entirely composed of the same kind of folded rocks, serpentinized diabase-tuff and tuff-breccia [3]), which also form the base of Mt. Lĕmpai and which, as previously noted, crop out on its northern slope, below the sandstone. The diabase of Poelau Mĕlaioe is very rich in glass and the rock may be called in some places vitrophyrite-tuff.

It is not easy to define the strike and dip of this contorted and altered rock, but on an average we may accept a N.W. strike, the invariably high dip however, varying in direction. On the shore I found some pebbles and small boulders of very fine-grained sandstone exactly like the sandstone on the east cliff of Poelau Tĕkĕnang. That these pebbles are well rounded is

---

1) VETH. 64 I, p. 27.
2) D. W. C. van LIJNDEN and J. GROLL. 35, p. 537.
3) It is from this locality that I have given to this type of rock the name *Poelau-Mĕlaioe rock*, vide p. 88.

easily accounted for by the heavy waves which in a strong gale disturb the lake; it then becomes dangerous and often even impossible to cross it in a rowing boat. The conjecture may be correct that these sandstone-boulders are the last remains of a sandstone-formation which formerly, as on Mt. Lĕmpai, capped the older formation, although the possibility should not be excluded that pieces of sandstone may for some reason or other have been deposited there by the natives, thus e. g., finely grained sandstone is often used both by Dyaks and Malays to sharpen their parangs (swords) on. The island is covered with low brushwood amongst which we found some delicious pine-apples. The view is very extensive and may be called the finest in the whole of the lake-district.

Plate XIII shows a part of this view looking north. To the left in the foreground is Mt. Sĕpandan on the island of that name, surrounded by submerged wood and open water; to the right the high mountains of Lantjak, Mt. Engkoeni (605 metres) and to the right of that the colossal twin-peaked Mt. Sĕmbĕroewang (752 Metres). To the left and more in the background near the horizon rise the boundary mountains of Sarawak, the Tintin Kĕdang, the western continuation of which we found at Mt. Pan.

From Poelau Mĕlaioe we steered for Poelau Sĕpandan the highest part of which on the north side of Mt. Sĕpandan, rises precipitously out of the water. The portion which is usually inundated is entirely without vegetation and the rocks are covered with a black coating which at the time we were there lay like a fringe, 2 metres wide, all round the island. The first explorers supposed this sharply defined fringe to be a tidal terrace and derived from it the erroneous conclusion, that the area of the great lakes had been covered recently by the sea.

The entire island is formed of beds of uralite-diabase alternating with layers of uralitised diabase-tuff and tuff-breccia. The position of the strata is difficult to determine; the strike is about E. 4° N., the dip 85° S. The rock is cleaved into thick beds, the cleavage-planes traversing obliquely the planes of bedding, the strike of both systems of planes being the same. On and near the summit

(114 metres) is high forest, and along the slopes shrubs with a goodly number of pine-apples. The highest eastern top of the island is composed of a coarse diabase resembling a gabbro, intersected with numerous quartz-veins (I, 789, 790).

Upon my return to the boat the pilot was sent ashore to fetch April 19. coolies in the kampong Lantjak. A deliciously warm, really tropical night was followed by a still warmer day with a slight south-wind. The pilot did not return till 2 p. m. with the news that the Lantjak house was deserted, all the inhabitants having moved to the north-western outlets of the Toenggak Bĕlawan. He had secured a few Dyaks from there, with whom I went ashore. The shore is here separated from the open water by a wide edging of swampy brushwood. At 4 p. m. I reached an empty and partly collapsed house 1½ kilometres to the east of the real Lantjak house, where I took up my abode. As I intended to climb Mt. Sĕmbĕroewang the next day I had at once a path cut through the dense underwood of the old ladangs which surround the lower part of the slope of the mountain.

The night had been clear, and heavy dew had fallen, so April 20. that we were soon wet through when at 7 next morning we set out for the foot of Sĕmbĕroewang, following the narrow path through the dense brushwood 5 or 6 feet high. We went first in a westerly direction as far as the Lantjak house, then we turned to the right, following for a considerable distance the valley of the S. Djaoeng, which arises in a deep chasm on the south side of Mt. Sĕmbĕroewang. Half a kilometre below the house, the brook Djaoeng joins the Lantjak (see sketch opposite page 110)[1]. In the bed of the Djaoeng I discovered the same rocks which I found afterwards on the top of Mt. Sĕmbĕroewang, and in addition, large pieces of uralitised diabase. Leaving the valley, we now ascended with much difficulty the most western spur of Mt. Sĕmbĕroewang, facing south. At an altitude of 325

---

1) On the topographical map (weg- en rivierkaart XVIII, g. 1 : 50 000) this brook has no name, but another right hand tributary which runs into the Lantjak a little further upstream, is there called Djaoeng.

metres we reached the sharp crest of this spur; the solid rock, vitrophyrite, is well exposed in distinct beds at an altitude of 340 metres. The strike of these beds is E.—W., with a considerable dip to the north. Our way now led through high jungle, where we shot a young deer, and where the Dyak coolies managed to kill a Sawah snake more than 10 metres long; its flesh is a delicacy to them. At 10.30 I reached the point, 500 metres high, where the edge of the spur joins the mountain. I gave my instructions about arranging the night-quarters and ascended the highest peak (752 metres) having started my Chinaman and a few Dyaks in advance to try and secure a good view for me by chopping down some trees. It appears that the central part of the Sĕmbĕroewang mountain is composed of diabase (I, 813, 820) varying in texture from compact to coarse-grained and arranged in thick beds, the strike of which is E. 20° S., the dip 8°.3 S. To the south-west it is superseded by banded chert followed by diabase-tuff, after which comes, farther south, vitrophyrite and vitrophyrite-tuff; then follows probably the uralitic diabase, great boulders of which lie in the bed of the Djaoeng. This succession can be seen distinctly at the places indicated on the sketch and it is confirmed by an examination of the lower part of the steep precipice (50 metres high) which cuts off the valley of the Djaoeng from above. It requires some experience in mountain climbing to make one's way up and down this precipitous height, where orchids with fine big yellow flowers grow in profusion. Near and on the summit there were several fresh bear tracks, from which animals the mountain derives its name [1]. The top of the mountain commands a fine view of the surrounding country, but not until after 4 p. m. did it become sufficiently clear to be appreciated. Particularly striking both from the top and from the bivouac is the view over the mountain land of the Upper Lĕbojan and the Upper Embaloeh. The latter appears to be a typical mountain range, consisting of a succession of sharp-crested ridges, striking E.—W. These ridges gradually increase in height to the north and north-east. The mountain range

---

1) Sĕmbĕroewang from Saran Bĕroewang = bear-hole. Derivation given to me by the natives.

Molengraaff. Borneo.

Diabase.   Chert.   Diabase-tuff and   Vitrophyrite and   Alluvial sandy and
                                tuff-breccia.   vitrophyrite-tuff.   marshy soil

rises quite suddenly from the Kapoewas plain. From my bivouac I could discern with the help of my telescope the Koeboe at Djaweh on the river Lĕbojan, and in the foreground, deep down in the valley of the Empasoek, surrounded by bright green ladangs, lie the houses of the Batang Loepars who formerly occupied the houses at Lantjak. About 30 metres below the bivouac one gets a beautiful view of Mt. Mĕnjoekoeng (see fig. 29) [1]). This illustration shows clearly that the sides of the mountain support nearly horizontal terraces which most probably

PĔNGALANG 480 M.　　MĔNJOEKOENG 630 M.　　PANGĔBAS 520 M.

Fig. 29. MT. MĔNJOEKOENG, SEEN FROM THE SOUTHERN SLOPE OF
MT. SEMBĔROEWANG.

are of sandstone, like that found on Mts. Lĕmpai, Sĕligi, Tĕkĕnang, etc. Mt. Sĕmoedjan and Mt. Pĕgah possess similar banks of sandstone, broken off all round and exhibiting far-shining, white, vertical cliffs. I was told that near Mt. Pĕgah, coal seams have been found in the sandstone. At Mt. Mĕnjoekoeng the terraced sides join the central many peaked mountain-group by means of low ridges; they therefore appear to be standing alone and have separate names, Mt. Pangĕbas and Mt. Pangĕlang. The central group, the real Mĕnjoekoeng, is of granite. I did not personally ascend this mountain but a Batang Loepar of the house at its foot was ordered by the aspirant-controller SPAAN to knock off pieces of rock at different places on the northern slope, and near the top of the highest hill in the group, the one which in our illustration is marked Mĕnjoekoeng. I must add that the information given by Dyaks has proved with a very few exceptions to be perfectly reliable. The fragments which he brought me

---

1) Seen from the North, as in this illustration, the terraces appear to be horizontal, but from other points of view where I could examine the mountain from the East or from the West, a slight gradient to the South is distinctly visible.

were all tonalite containing tourmaline, tourmaline-granite and tourmaline-fels, and this, together with the fact that the boulders which Mr SPAAN had collected for me in the brook near to the house consist for the greater part of the same kinds of rock, justifies me in the conclusion that the central group of Mt. Mēnjoekoeng is formed of tourmaline-granite.

April 21.   In the course of the forenoon I returned to the deserted house near Lantjak, where I found a letter from Mr SPAAN, informing me that he was making a journey of inspection to the houses by the river Empasoek, and would like to row with me next day across Danau Loewar to Pamoeter. In the afternoon I went half way up the Mt. Toenggak Bēlawan, where on the northern side we had to cut our way through an almost impenetrable jungle. This part of the mountain is formed of diabase.

April 22.   Early in the morning I went with Mr SPAAN to the pangkalan Lantjak and then rowed round Poelau Sēpandan to Danau Loewar, from where we steered south. Leaving Poelau Mēlaioe to our left, we reached pangkalan Pamoeter at 2 in the afternoon. A good hour's walk partly over batangs took us to the Dyak house of the same name, numbering 15 pintoes. It had struck me more than once, and especially when on Danau Loewar, that the hills in the neighbourhood of Pamoeter had much softer and more rounded outlines than the mountains near Lantjak. I found on my excursions in diverse directions, always starting from Roemah Pamoeter, the explanation of this fact, in that the solid rock here, is, without exception, granite. Mt. Pitoeng or Pamoeter consists of porphyritic amphibole-biotite-granite, which is particularly well exposed in the vicinity of a small waterfall in the neighbourhood of Pamoeter house and close to the bathing place belonging to it. Mt. Sap is also composed of granite rock, huge blocks of which lie scattered about in the forest all along the slopes on the north side.

The population of Pamoeter was most friendly and in the evening all the inhabitants squatted round us in the large public apartment which was lighted up with smoky, flickering, dammar-torches [1]). We learned that there was much want in this settlement. The rice harvest had failed several times, and as the inhabitants were too poor to pay the exorbitantly high prices which the Chinese and more especially the Malays demanded of the Dyaks for rice in times of scarcity, they had, for more than a year past, been able to procure rice only now and again, and had been compelled to live principally on the fruits and roots of the forest. There is much distress in the Batang Loepar lands, and altogether prosperity among the Dyaks is at a low ebb. This is accounted for by the incredibly slovenly way in which the ladangs are laid out. The wood is chopped down anyhow, and burned, as far as it will, during the dry months of August and September,

Fig. 30 A LADANG NEAR PAMOETER.

and then without any further preparation of the soil, the rice is planted out amid this chaos of sound and charred trunks and

1) Dammar — Resin of the dammar-tree.

branches. After this they consider the ladang finished [1]). If the earth
be loose, the crops succeed fairly well as a rule in the fertile forest
soil, if, on the other hand, the ground is hard and clayey, as is
the case in the weathered granite soil near Pamoeter, the harvest
is generally poor, whereupon that plot of ground is deserted, a fresh
piece of virgin forest is cleared, and the same course pursued over
again. Now as the Batang Loepar lands are, relatively speaking,
somewhat densely populated, and the Batang Loepars, to avoid
conflicts with other tribes, have been prohibited by the Government
from dwelling or laying out ladangs, outside the territory allotted
to them, the result is that the primeval forest, except on the very
steepest mountain slopes is here well-nigh exterminated. This
year they seemed particularly hard pressed, and so the inhabitants
of Lantjak decamped to the valley of the Empasoek, within the
territory of the Embaloeh-Dyaks, while part of the inhabitants
of the eastern shore of Danau Loewar had moved to Mt. Měnjoekoeng, which did not belong to the Batang Loepar lands
but was as yet uninhabited. A careful investigation into this
matter, was the object of Mr SPAAN's journey, and it proved
that it was indeed pressing need which had induced the Batang
Loepars to overstep the borders of their territory. As I was
informed later on, the Government decided to leave them in
possession of the good ground where they now had established
themselves. One of the most radical means to promote the well-
being of the Dyaks would be to teach them the rudiments of
rational agriculture, and then to make them relinquish this system
of reckless tillage. However any one acquainted with the con-
servatism of the Dyaks and their strong adherence to the
customs of their forefathers, will know at once, that such a
radical change would require a stronger and more constant
European influence than can possibly be exercised by the few
Dutch officials. No help can be expected from the Chinese and
the Malays; the former devote themselves exclusively to trade,

1) Fig. 30 gives a view of such a finished ladang.

and as for the Malays, their influence at all times is for the bad.

Mr Spaan left next morning for Mt. Ménjoekoeng, while I ^{April 23.} ascended the southern summit of Mt. Sap. Persistent fog and drizzling rain prevented my studying the view from this mountain, a thing which I had much looked forward to. Without further delay I returned to the pangkalan Pamoeter and made up my mind to return at once to Poelau Madjang, as we had nearly come to the end of our provisions. Arrived on Danau Loewar a westerly wind arose, and the rain came down in torrents. Soon after nightfall we passed Poelau Batik where I found a few pieces of biotite-granite close to the shore, very much like the granite found on Mt. Sap. Whether these blocks had been brought here or whether granite forms the foundation of this island, I cannot say for certain, as darkness prevented a more careful investigation. Personally I incline to the former alternative as the soil is not composed of granite-sand but is a hard yellowish-white clay, worn down perpendicularly by the action of the water, as is the case with all the low lying islands in this lake-district. We reached Danau Sérijang just before 10 p. m. in heavy rain and with a strong west-wind blowing; it was with the utmost exertion and not without danger of capsizing that we ran our little boat ashore at Poelau Madjang at 10.45 p. m.

The next morning was fine and bright and at 9.30 I set ^{April 24.} out for Génting Doerijan intending to visit Mt. Séligi and Mt. Bösar. We rowed straight across Danau Sérijang, steering for the place where the rivulet Kélian empties into the lake. Owing to the low state of the water we had to disembark at a place nearly an hour's walk in an easterly direction from the pangkalan (landing-stage) used at high water. We had now to walk through the swampy forest and over the same ground, which at high water can be traversed by sampans. Before long we had to wade through a rapid stream of fresh clear water called the Kélian, which rises on Mt. Patjoor. The Sépan house was deserted and in ruins; we met the head of it close by and he told us that his people were all stricken down with malaria and had retired

to the ladang huts on the southern slope of Mt. Sĕligi; he himself, although suffering from the fever, was one of the few still
able to go about. He declared that none of his people would
be capable to take me up to Mt. Sĕligi. I gave him all the
quinine I could spare with the necessary instructions for a rational
use of it, and continued my toilsome march. Toilsome indeed
it is to a booted European, for almost the whole of the so-
called military road from Gĕnting Doerijan to Danau Sĕrijang
is made of tree-trunks of all sizes and dimensions and all degrees
of slipperiness, chained together and resting here and there on
shaky supports high above the swamp. To a European not
possessing any great equilibristic talent, a road of this kind is
a veritable trial; the natives however call it a "djalan ennak" [1])
and go over it with their loads as merrily as if it were a
well-paved highroad. After a while I had the mis-fortune to
fall from one of these tree-trunks into a hole in the ground,
thereby spraining my right foot, and it was with much difficulty and pain that I dragged myself to the empty Roemah [2])
Sĕdempa. This house is built at the north-east foot of Mt. Sĕligi
and from the outer verandah one gets a close view of the steep
side of the mountain. One can easily discern in the uppermost
part of this steep and scantily clothed slope, three or four
thick beds of sandstone, their outcrops forming as many perpendicular terraces (see Pl. XIV). Below these terraces the slope
of the mountain becomes more gradual, and shows plainly the
deep fissures made by the watercourses. As I had to give up
all idea of climbing the mountain myself, I sent my Chinaman
with a few coolies to collect samples of the solid rock from the
base of the mountain and also from the higher terraces, and I
superintended their labour through my field-glass. The sandstone
from the terraces of Sĕligi proved to be a fine-grained variety,
the clay, cementing the quartzgrains, being full of small pyrite
crystals. When weathered, the sandstone turns to a brownish yellow

---

1) Djalan ennak = a delicious road.
2) Roemah (Mal., Dy.) = house.

owing to the decomposition of the pyrites and becomes incoherent. I was unable to ascertain the nature of the rock underlying the sandstone on the lower part of the slope of Sĕligi, but the numerous boulders of amphibole-porphyry in the bed of a little stream that rises on the mountain and runs past Roemah Sĕdempa led me to suppose that the same rock prevails there. The whole territory at the foot of the steep northern incline of Mt. Sĕligi is entirely covered with massive blocks of sandstone which have broken off from the terraces and fallen down.

From here to Gĕnting Doerijan the track is a batang-road, April 25. leading partly over marshes and partly over forest-land. Owing to my sprained ankle it took me $3^1/_2$ hours to travel the 6 kilometres of this road. Gĕnting Doerijan consists of a few houses besides the koeboe, which is built on the same plan as the other koeboes already described. This settlement is situated on the Tĕlijan river on the road between Mt. Patjoor and Mt. Bĕroewi, which forms the communication between the Batang Loepar lands and the valley of the Ĕnsana and the Ĕmpanang, formerly inhabited by the Kantoek Dyaks, but now deserted. This koeboe was built to prevent the Batang Loepars from making inroads among the Kantoek and Kĕtoengau Dyaks. The place is well chosen for it lies on a saddle between the Lĕmpai, Singkadjang, Sĕligi and Bĕroewi mountains on the one side, and the range of the Patjoor, Pangoer Doelang, and Bĕsar mountains on the other. The latter range is called collectively Mt. Bĕsar.

My badly swollen foot compelled me to take a day's rest. The April 26. boulders and pebbles in the bed of the Tĕlijan near the koeboe chiefly consist of sandstone; but besides there occurs a dark-violet chert which with the naked eye cannot be distinguished from the chert of Mt. Sĕmbĕroewang. In this chert however I found traces of Radiolaria, which I looked for in vain in the rock of the Sĕmbĕroewang mountain. Occasional fragments of porphyry may also be seen, closely allied to hornblende-andesite. Thus the pebbles consist of a mixture of constituents derived from the Danau-formation and from the newer sandstone beds resting unconformably on it.

April 27. At 7 o'clock the next morning I left Génting Doerijan in order to ascend Mt. Béroewi, which took up a great deal of time, as it was hard work to clear a passage with our choppers through the dense underwood. It soon became evident that Mt. Béroewi is of the same type of build as Mt. Lémpai. The upper part consists of sandstone beds gently sloping to the South, but broken off precipitously towards the North, where they form steep terraces. The mountain has two peaks separated by a little stream which running from south to north falls rapidly down the incline. In the upper part of its course within the sandstone area, the bed of this brook forms a gigantic natural staircase. At 9 a. m. I reached the highest eastern peak (215 metres) and I had to wait till noon before the clouds lifted sufficiently from Mt. Bésar, to enable me to take a photograph from which the outlines of fig. 31 are derived. From this point of view it is quite evident

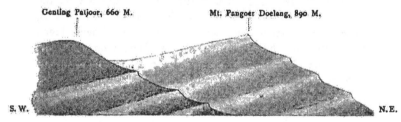

Genting Patjoor, 660 M.   Mt. Pangoer Doelang, 890 M.

S. W.   N. E.

Valley of the S. Bésar.

Fig. 31. THE BÉSAR HIGHLAND, SEEN FROM THE EASTERN TOP OF MT. BÉROEWI (215 M.).

that Mt. Patjoor, Pangoer Doelang, etc. do not form a continuous chain trending N.W.—S.E., but rather that they constitute a range of separate mountains with an average strike E. 15° S. They all slope gradually on the southern but have a steep incline on the northern side. They are all crowned with terraces gently sloping southward, while on the north and east sides the steep white sandstone cliffs can be seen at a considerable distance. On the northern incline these sandstone terraces form gigantic

BOEKIT BĔSAR, AS SEEN FROM GĔNTING DOERIJAN.

staircases the steps of which have been entirely cleared of trees and used for ladangs, but close up to the steep precipices separating the steps a few high trees have been left standing. This however I had already noticed from Poelau Madjang (see panorama, page 21 of the atlas). From the west side of Mt. Béroewi one gets a good view of Mt. Kénépai, across the valley of the Empénang. Viewed from here this mountain looks more than ever like a volcano (Fig. 32). We returned to Génting Doerijan about 1 p. m. in torrents of rain which continued all day.

Fig. 32. MT. KÉNÉPAI, SEEN FROM THE EASTERN TOP OF MT. BÉROEWI.

It happens occasionally in Borneo that after heavy night-rains, April 28. the atmosphere is particularly transparent. In the foreground every leaf then sparkles in the morning sun as if studded with thousands of pearls, while in the background, and as far as the eye reaches, every object can be discerned, distinctly yet softly outlined, which gives to the tropical landscape an indescribable charm. It was on a morning of this kind that I took leave of the neighbourhood of Genting Doerijan and Plate XV represents the last impression which I carried away with me as a remembrance of the place. What followed after that, the march back over the batang-road, with my lame foot, partially *hors de combat*, is best passed over in silence. There were times when I heartily envied our brown brethren, who being so much better equipped for a life in the jungle than we are, marched over their 'djalan ennak' with evident satisfaction. However I reached Poelau Madjang in the afternoon, safely and not too much exhausted.

There I happened once again to meet the aspirant-controller SPAAN who had meanwhile returned from his expedition to Mt. Ménjoekoeng. He told me that after he had left Pamoeter he passed through the Pintas Kawi to the Lébojan, and that having rowed up stream for some distance he spent the night close to the river Tĕlatap. The next day he reached the pangkalan satengah [1]) at 2 p. m. and had then another good hour's walk to the Dyak house at the foot of the mountain, and from there the top of Mt. Mĕnjoekoeng can be reached in three hours. He brought me from there the tourmaline-granite and sandstone, already mentioned on page 114. That same day I left Poelau Madjang for Sĕmitau where I arrived next day at 11 a. m. after spending a most delightful and, for Borneo, exceptionally clear night on Danau Sĕrijang and the Tamang river.

The lake-district or the Batang Loepar lands, is the most westerly and lowest part of the great lowlands of the Upper Kapoewas, stretching away from 110° 50' E. to 113° E. from Greenwich, i. e. over a distance of 134 kilometres with an average breadth of 54 kilometres, sometimes on both sides, sometimes exclusively on the north side of the Kapoewas river. The average height of this area is between 34 and 50 metres above sea-level. The lake-district is clearly separated from the rest of the Upper Kapoewas lowlands, by a continuous range of mountain-groups i. e. the mountains of Lantjak and the Mĕnjoekoeng group stretching in a north to south direction from the mountains on the boundary of Sarawak to close up to the Kapoewas, and this gives it a kind of geographical independence.

The oldest formation known in this area is a system of strongly folded strata, striking almost due east and west, consisting of quartzite, chert, shale, diabase, diabase-tuff and tuff-breccia, diabase-porphyrite, and serpentine. Nowhere in the lake-district did I come across any true crystalline schist. This formation seems to be the northerly continuation of the rocks in the Sĕmitau hills, which are folded with the same strike. The amphibolitic and chloritic schists in those hills represent most likely the oldest core in this succession of folded rocks. The whole group of folded strata, occurring in

---

1) There are there three pangkalans: the pangkalan ocloe for high water, the pangkalan satengah for an average state of the river, and the pangkalan Uir for low water.

the lake-district, I have placed together under the name of the *Danau-formation*. The beds of chert, hornstone and different kinds of jasper containing Radiolaria lie probably fairly high in this system, above the chief mass of diabase-tuff and breccia, although in their turn they are overlain by diabase and diabase-tuff. In a few places I came to the conclusion that chert would be found alternating with diabase tuff and breccia on more than one horizon, but in consequence of the intensity of the distortion and plication and the scantiness of rock exposures I could not give this as my decided opinion, nor could I for the same reason declare with certainty that the Radiolaria rocks occupy a high position in the system of folded layers, and I can only say that as far as my observations go, I am inclined to believe that it is so. In the whole of this system traces of organisms can only be detected in the said chert and jasper with Radiolaria, which are so well represented in the neighbourhood of N̄ Badau. These Radiolaria, according to Dr. JENNINGS HINDE, belong to a pre-Cretaceous age and, probably to the Jurassic period, so that the Danau-formation must be at least Jurassic.

There are in this system a few granite mountains; Mt. Ménjoe-koeng, and a little farther away to the West, strictly speaking outside the limits of the lake-district, Mt. Kĕnĕpai, which, I take it, must be regarded as intrusive bosses. Note for instance the occurrence of tourmaline in the granite, and the presence of pure contact-metamorphic rocks at the foot of Mt. Kĕnĕpai, such as andalusite-hornfels (comp. pag. 40). These intrusive bosses now greatly protrude above the deeply eroded layers of the Danau-formation, in which at one time they were completely embedded, they are therefore with regard to the latter in exactly the same relative position as the granite bosses in Cornwall to the Devonian Killas, or the granite of the Table mountain near Cape Town with regard to the Malmesbury-greywacke and clayslates which in the Cape-Flats have been carried away by erosion down to the sea-level. The more recent beds of sandstone lie unconformably upon the older formations. In some places, beds of clay and coal-seams are intercalated in this sandstone, which up to the present time are being worked for instance near Mt. Sĕgĕrat. This sandstone is probably of neogenic age. Frequent false bedding leads us to suppose that it has, in part at least, been deposited by running water, which is quite compatible with the presence of coal.

Faulting and erosion have greatly altered the appearance of the landscape, where this sandstone occurs, since its deposition. One glance at the map is enough to convince us that the combined action of faulting and erosion must have divided the sandstone plateau, which no doubt had already been upheaved and sloped southwards, into a series of separate areas all sloping to the south. The now entirely isolated sandstone terraces of Mts. Bĕsar, Sĕligi, Lĕmpai, Tĕkĕnang, Pĕgah, Sĕmoedjan, Pangĕbas, etc. probably formed, at one time, a continuous sandstone plateau, extending much farther to the north than it does at present. A very prominent fault appears to run along the north side of Sĕligi, Lĕmpai, etc. Most likely the precipices of Mts. Sĕligi, Lĕmpai, Tĕkĕnang, Pĕgah, etc. are merely the remains of the fractured northern margin of the southern portion of the sandstone plateau, while the northern portion has been let down to a much lower level [1]). These W.N.W.—E.S.E. faults, in conjunction with other faults running in different directions, are possibly the primary cause of the low level of the lake-district [2]). And lastly erosion has taken such fast hold of the entire district that it is only here and there that parts of the Danau-formation as well as of the sandstone plateau have escaped destruction. With the exception of these few rem-

1) We must not forget to point out that while accepting this theory by which the occurrence of the isolated area of sandstone on the south slope of Mt. Bĕsar is readily explained, the question yet remains to be answered where the northern continuation of the sandstone plateau, which, on account of its lower position, might have better escaped erosion should be looked for. One might reasonably look for the reappearance of the sandstone on l'oekau Mĕlaioe, which was hinted at on page 109. For the present it seems to me reasonable to accept the existence of one or perhaps more faults trending W.N.W.—E.S.E. The present incomplete knowledge of details concerning the geological structure of this territory, does not however yet allow of any satisfactory explanation of the causes of the present configuration.

2) It will be shown later on that the peculiarly low level of the whole of the Kapoewas plain, is probably caused by faults.

nants of hills the lake-district lies entirely below the flood level of the Kapoewas river but above its level at low water.

This fact explains why the hydrographic condition of the lake-district is ruled by the Kapoewas river. With the rising of the Kapoewas its waters run through the bed of the Tawang and by a number of natural channels into the lakes, which they entirely fill. Of course the waters coming down from the neighbouring mountains into the lake-district, and banked up by the Kapoewas, largely contribute to the rapid filling of the lakes. Practically therefore the lakes when thus swelled are nothing but the widely extended surface of the Kapoewas itself, „flood-lakes" in the real sense of the word. When the water falls in the Kapoewas, it also begins to fall in the lakes, but much more slowly than in the Kapoewas itself, for it can only run back into that river by a limited number of comparatively narrow channels. If the water in the Kapoewas falls rapidly the current is sometimes very strong in the Tawang and the other connecting waterways. And if the water in the Kapoewas sinks below the average level of the bottom of the lakes and continues at this low stage for any length of time (a few months of drought are sufficient), then the lakes run quite dry and the water only remains standing in a few pools and in the deep beds of the rivers rising on the surrounding mountains, for instance, the Lĕbojan, the Soempa, the Sĕrijang, and others which empty themselves into the Kapoewas chiefly through the Tawang. It is only in years of drought that the lakes run more or less completely dry.

These lakes are of great and beneficial importance as regulators of the state of the water in the middle course of the Kapoewas, for which we take the section between Sĕmitau and Tajan. For after heavy rains the water, coming down in sudden and impetuous floods from the Upper-Kapoewas, must spread over the lake-district, and the force of the current is thus broken, while on the other hand, in the case of a sudden, sharp fall of the water, the lakes, forming a natural storage, feed the Kapoewas for some considerable time, and prevent the water in the middle course from running away too quickly. Below the lakes where the Kapoewas

breaks through the Sĕmitau mountains, the river-bed is comparatively narrow. The difference therefore between high and low
water is here very considerable [1]), and without the modifying
influence of the great lakes, the Kapoewas valley between Sĕmitau
and Silat would undoubtedly be subject to the most terrific floods,
making those districts well-nigh uninhabitable.

The soil in that portion of the lake-district which is usually
submerged is a fine grey or yellowish grey sandy clay, which
dries up quickly when the lakes run dry, and the cracks produced thereby are so numerous and so deep, that the ground is
almost as impassable as the rugged and honey-combed surfaces
of raised coral-beaches, such as may be found on many of the
West-Indian islands. And so it is obvious that when dry the great
lakes of Borneo offer a most serious obstacle to communication.
The soil of the area not usually submerged is sterile, for what else
could be expected from the weathering and decomposition of the
underlying rocks, which are for the most part of quartzite, serpentine, serpentinized and silicified diabase-tuff and sandstone?
The granite bosses alone make a favorable exception; they furnish
a richer soil and make the neighbourhood of Mt. Mĕnjoekoeng
into the most fertile spot of the whole of the lake-district. In the
generally submerged part of the district, the conditions of vegetation are necessarily somewhat peculiar, and form an absolute
obstacle to the growth of several plants, and we cannot therefore
be surprised to find that the inundated forest of the lakes numbers only a very limited variety of trees and shrubs. And yet
this poorness of vegetation, which strikes even a layman, must be
attributed, to some extent at all events, to the poverty of the
soil of the periodically submerged district and the surrounding
hills, a poverty which is apparent also in the geologically allied,
hilly country round Sĕmitau.

---

1) At Sĕmitau ± 17 metres, at Sintang according to observations made from 1877 up to
1894 about 14½ metres.

CHAPTER VI.

MOUNT SĔTOENGOEL, SINTANG, AND MOUNT KĔLAM.

On the 2ⁿᵈ of May 1894, I descended the Kapoewas with two May 2.
small rowing boats from Sĕmitau to Sintang. My original intention had been to visit the Sĕbalang mountains to the South of
the Sĕbĕroewang river upon which I had often gazed from my headquarters at Sĕmitau. But when I was told that an excursion to

Fig. 33. DYAKS OF MT. KĔLAM.

Mt. Sĕbalang would take at least four days, I gave it up and
made Mt. Sĕtoengoel my point of destination for that day at all
events. This mountain is the western continuation of the Sĕbalang
range, it lies much nearer to the Kapoewas and can be reached
from that river within a very short time. From Sĕmitau to the
mouth of the Sĕbĕroewang the Kapoewas takes its course across
undulating ground, low ranges of hills named the Sĕmitau hills,
trending E. 8° S.—W. 8° N. or thereabout.

Just below the mouth of the Sĕbĕroewang-bĕsar¹), the course

---

1) Bĕsar (Mal.) = great, big; sĕni (Mal., Dy.) = small, little, narrow.

of the Kapoewas is obstructed by one of these low ranges and is forced to make a bend to the W.N.W., thus following the trend of the hills. Four kilometres further on is the Dyak house Pagong, inhabited by fifteen Kantoek families; here I procured a guide to visit Mt. Sĕtoengoel. Fully 8 kilometres downstream from Kwala[1]) Sĕbĕroewang the Kapoewas curves to the south-west and is almost immediately checked again by a ridge of sandstone which slightly inclines to the north-north-east and forms a steep rocky ledge, the Batoe Kĕling. This sandstone ridge forces the waters back again in an east-south-easterly direction straight towards Mount Sĕtoengoel, which, as plainly visible from here, belongs to the same range as the Batoe Kĕling. The sandstone extends westward along the Sĕntabai river, and there, according to CHAPER[2]) layers of lignite are found bedded in it. At a short distance from Mt. Sĕtoengoel the Kapoewas, which, as we saw above, has curved back to the E. S. E. again approaches within a distance of $2^{1}/_{2}$ kilometres the mouth of the Sĕbĕroewang, in its meandering course almost forming a closed curve. At very high water, this long (14 kilometres) detour of the river from the mouth of the Sĕbĕroewang to this point can be avoided by using a short water-way through the hills, known as the pintas Marang. Near Mt. Sĕtoengoel the Kapoewas resumes its course in a south-south-westerly direction, and its bed has again the character of a transverse valley. We rowed past the Sĕtoengoel river, and took the little tributary Ankarin or Sĕ-toengoel-sĕni a little beyond, and landed at 3 p. m. about 50 metres upstream. Although the top of Mt. Sĕtoengoel is scarcely a couple of kilometres distant from this landing-place, the difficulties in reaching it proved greater than I had anticipated. Unusually high brushwood (on an average three metres high) much hindered our progress, as we were but indifferently supplied with choppers. By 4.30 we had reached the base of the actual mountain and came to a point where a tributary

1) Kwala (Mal.) = mouth.
2) M. CHAPER. 5, p. 880.

of the Sĕtoengoel flows down in a series of waterfalls over
immense flags of sandstone. We arrived at the top about 5.30
and found that the mountain is mainly composed of layers of
sandstone with a few intercalated shale beds, which, striking E. 10°S.,
dip to the south at an angle of 23°. To the north they form
steep precipices, and this gives Mt. Sĕtoengoel its unilateral structural appearance; all the hills of the Sĕbalang range which follow
Mt. Sĕtoengoel in an easterly direction, have this same characteristic feature. Viewed from Sĕmitau, facing south, one looks
upon these precipitous slopes, and their jagged extremities, the
outcrops of the sandstone-beds, appear as brilliantly white patches between the dark foliage. The sandstone is greyish in colour;
some of the beds contain numerous fragments of shale between
the quartz-grains, which makes the rock pass into greywacke.
Occasionally also, carbonaceous particles are found in them.
EVERWIJN [1]) who also visited Sĕtoengoel, tells us that thin beds
of coal occur in the sandstone of the adjoining Mt. Lilin.

While I was engaged in taking my bearings on the top of the
mountain, the piercing shrieks of the kĕriangs suddenly warned me
that the sun was setting and that I was belated. The journey back
in the dark over the newly cut or trampled down brushwood
was tiring in the extreme, and it was not till 7.30 p. m. that we
regained the sampans and continued our journey. Between this
point and Sintang the Kapoewas makes another detour in order
to avoid the Pĕnai mountains. With many windings the river
compasses these mountains on the west side. The Pĕnai mountains were visited in 1855 by CROOCKEWIT [2]), in 1857 by EVERWIJN [3]),
and in 1874 by TEYSMAN [4]). The two former found that the
mountain is composed of sandstone, and on the map of EVERWIJN
both the Pĕnai and the Sĕtoengoel sandstone are considered to
be of Eocene age.

1) R. EVERWIJN. *11*, p. 25.
2) J. H. CROOCKEWIT. 7, p. 279.
3) R. EVERWIJN. *11*, p. 26.
4) J. E. TEIJSMAN. *57*, p. 315.

May 3. During that night we proceeded as far as the Poeri, a small streamlet which empties into the Kapoewas about 3 kilometres above the Kĕtoengau river. Digging for gold has been carried on near the bank of this small river, where a few Chinese were at work. The alluvial soil is excavated to a depth of 9 metres just above the average water-level in the Poeri, and as is the custom in the alluvial diggings of Borneo, it is washed in a parit (gutter) by a strong current of water from the reservoir which is formed by a dammed branch of the river itself. Below is shown the vertical succession of the beds in these pits. The gravel is composed of different kinds of rocks, which, however, with the exception of quartz and a few cherty rocks are decomposed into a clay, which is sometimes almost pure kaolin. The quartz has become brittle and crumbling, a feature which I noticed in all similar diggings in West Borneo.

SURFACE.

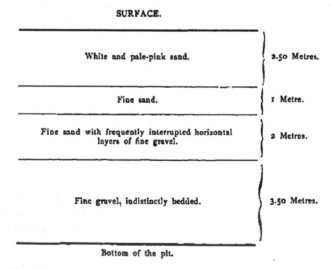

White and pale-pink sand.	2.50 Metres.
Fine sand.	1 Metre.
Fine sand with frequently interrupted horizontal layers of fine gravel.	2 Metres.
Fine gravel, indistinctly bedded.	3.50 Metres.

Bottom of the pit.

The diggers informed me that gold is only found in the two lowest layers of gravel. According to them, these workings yield on an average about 3 boengkol of gold per month, representing

a value of 210 florins or £17 odd., but one cannot attach much
credence to the statements of the Chinese concerning the pro-
duction of their mines.

I continued my journey, and about 2 p. m., when in midstream,
we were caught in a heavy thunder-storm. A terrific squall
preceded it, our frail vessel nearly capsized, and it was with the
utmost difficulty and only by steering close to the bank that
we battled successfully against the fury of the elements. Rowing
on steadily amid torrents of rain we at last gained Sintang at
8 p. m., where I arrived wet and shivering, but was soon restored by
the hospitable welcome bestowed upon me by the assistant-resident.

A slight indisposition detained me here for a few days, after May 7.
which I set out for Mount Kĕlam, accompanied by two officers
and the military doctor from the garrison at Sintang, and it is
owing to this circumstance, that this excursion is recorded with
golden letters in my Borneo reminiscences. The delights of scientific research were enhanced by the pleasures of a prolonged
pic-nic in lovely scenery and in congenial company.

At all times Mt. Kĕlam, by its bold shape and isolated position, has attracted in a high degree the attention of European
travellers in West Borneo. CROOCKEWIT visited and ascended it
in 1855, but he did not reach the top. The first who succeeded
in doing this was the botanist, Dr. GÜRTLER. In February 1894,
Dr. A. HALLIER, botanist of the Borneo expedition, who had
chosen Mt. Kĕlam as one of his botanical stations, climbed to
the top more than once.

From Sintang Mt. Kĕlam may be reached either by way of
the Mĕlawi river, or of the Kapoewas. From the Mĕlawi one
has first to row up the Djĕtah for some distance and then up
the Kĕlam, and at the point where the latter streamlet ceases
to be navigable, a path leads in 1¹/₂ hours to Roemah Lajang
at the foot of Mt. Kĕlam. As a rule, by taking this route from
Sintang, Roemah Lajang cannot be reached in one day; but it
is feasible by rowing up the Kapoewas, and its tributary the
Djĕmĕlah, for some distance. The landroute, however, is in this

case considerably longer, nearly 17 kilometres. It was this route I chose.

So we left Sintang at 7.45 a. m., first ascending the Kapoewas for a distance of about 4 kilometres and then up its left branch the Djĕmĕlah. The banks of this little stream soon become marshy, or even quite flooded. The stream itself moreover frequently bifurcates and occasionally widens out, finally it passes into a laby-rinth of open sheets of water bounded by marshy brushwood and connected by narrow winding channels. The circumference of this tract of water, Danau Djĕmĕlah, depends, like that of all danaus in the Kapoewas district, on the height of the water in this river. Under average conditions the lake ends and the land-route begins about 2¹/₈ kilometres to the west of a small Dyak house, a temporary station occupied by one of the principal chiefs of the Kĕlam Dyaks and therefore called Roemah Toemenggoeng [1]). All around it are pretty extensive rice-fields. The path, which leads from the landing-place, past this house, to Roemah Lajang, at the foot of Mt. Kĕlam, winds across a slightly undulating plain, the soil of which consists of white quartz-sand. The whole of this district has been frequently used for ladangs, but it is now covered with underwood and tall ferns; their shining foliage rea-ches up to a man's head and, strongly reflecting the rays of the sun, renders the tramp over this glittering white sand in the full glare of mid-day, sun no trifling matter for a European. Nearer to Mt. Kĕlam the hills become somewhat higher (30 to 35 metres) and at the bottom of the deep gorges and valleys which separate them there are swamps bridged over by a succession of batangs, to make communication possible.

Close to Roemah Toemenggoeng we got a beautiful view of the mountain which from that point has the appearance of a hugh dome (fig. 34). For a long distance beyond it was hidden from our eyes by woods or hills, until, about 2 kilometres from Roemah Lajang, emerging from low brushwood we suddenly

1) We reached this pangkalan at 9.45 a. m.

found ourselves on a low hill, face to face with the gigantic rock-mass, only one broad valley intervening (Pl. XVI). The impression is overwhelming. From here the mountain looks like a flattened

Fig. 34. MOUNT KĚLAM, SEEN FROM THE WEST.

cupola, rising 900 metres above the valley in front of us, the side facing us being about 3¹/₂ kilometres broad. The lower part of the mountain is clothed with dense forest, above which follows a naked belt or girdle averaging 350 metres in height, the slope of which is so steep that no soil fit for vegetation can lie on it. The bare rock deeply grooved by numberless perpendicular furrows, is exposed over the whole of its circumference. The general colour of this rock-wall varies from grey to reddish brown, while here and there white patches indicate the places where blocks have recently broken off and fallen down. From the slightly arched summit clothed with marshy wood, water oozes on all sides and hurrying down the precipice through the furrows, engirdles the body of the mountain with numberless vertical threads of water, here and there reflecting the sunlight with dazzling effect.

Twice over I revisited this spot during my stay in Borneo (in May and in December), with the special object of photographing this view under various light effects. Each different time of day, each different foreground, each different distribution of light, made me contemplate the giant rock with renewed admiration, and each time my imagination clothed it in some fresh likeness. Sometimes its deeply furrowed and scaly crust reminded me of the back of a giant elephant. Another time when a dainty cloud fringe amalgamated as it were with the body of the rock, I fancied myself on the seashore, and the whole looked to me like a huge billow, ready to break over and bury me in its briny foam. But it appeared to me most imposing and most worthy of its name "Dark mountain" [1]), when a thunder-cloud enveloped its crest and the ragged clouds swept wildly along its bare surface. One's imagination might almost fancy the rumblings of the thunder which seem to proceed from the very bowels of the mountain, to be the groanings of the spirits (hantoes) which are supposed to inhabit it. No wonder the mountain is held sacred in the district. What numberless legends would be attached to this solitary rock as lofty as the greatest heights of the Thüringer Wald, if perchance its appointed place had been in the Rhine district, so rich in folklore?

Comparing Pl. XVI showing Mt. Kélam from the south, with fig. 34 showing it from the west, it is evident that its ground-plan is almost elliptic, the long axis, pointing W. 20° N.—E. 20° S.,

---

1) Kélam (Dyak) = gĕlap (Mal.), dark.

being about 3 times the length of the short one. The average slope of this mountain is remarkably steep, about 45°, it is only at one of the extremities of the great axis of the ellipse, to the westward therefore, that it is somewhat less precipitous, and there only is the ascent possible. The mountain rises nearly on all sides abruptly from a level or slightly undulating plain, only to the W. 20° N. does it slope more gradually into the plain where it continues as a strip of slightly raised ground, which however nowhere exceeds 75 metres in height.

On the 7th of May we reached Roemah Lajang about 2.30 p. m. When still at a considerable distance a melodious rhythmical sound fell upon our ears. It emanated from the house, but when we entered, the noise was decidedly too loud to be pleasant. It was caused by the pounding of the rice. In front of the greater number of pintoes, in the passage leading along the large public room stood two young girls opposite one another on a rice-block. With one hand they brought the heavy ironwood pounder, with great force down into the hole in the middle of the block, while with a dexterous movement of the feet they pushed back into the hollow the grains of rice which had jumped out. When the pounders are being carved, two small four-winged pieces of wood, one just above, the other just below the middle of the pounder, where it is thinner so as to be taken hold of easily, are cut loose in the interior of the staff. These loose bits of wood rise and fall with every up and down movement of the pounder and cause the melodious click-clack. The wooden sticks for planting rice are furnished with a single similar contrivance. I have found these musical utensils nowhere but amongst the Dyaks of Mt. Kélam.

At 4 p. m. we left Roemah Lajang, proceeding along the foot of the mountain to its west side. The ground was strewn with large pieces of rock which had broken off and fallen down. At the place where we commenced the ascent the mountain rises suddenly from the plain with an incline of 35°. The path leads through high forest ground, it is stony and not very slippery, so that the ascent is comparatively easy although rather fatiguing because of its exceeding steepness. We took up our night-quarters at an altitude of about 250 metres, where a little stream forms a waterfall down a perpendicular rock-wall. Our faithful Dyak carriers did not reach our bivouac till some time after sunset.

May 8.   At 8.45 a. m. we resumed the ascent. The track was still very steep and led in an easterly direction, first through forest, then partly over rocky slopes clothed with ferns, partly over naked rocks which in places were so steep as to make the ascent only possible by means of rattan ladders. Shortly after eleven we reached the highest point attained by CROOCKEWIT in 1855. Here his progress was averted by the rock-wall, which, as we saw above, encircles the mountain. In this place the wall is only 50 metres high. We scaled this obstacle by means of a rattan ladder which the natives in search of products of the wood had fixed there. Pl. XVII shows this ladder and the mode of our ascent. It is 47 metres long and is only fixed in the soil at the top, middle, and base, while the rest of it hangs loose against the rock. The rungs are irregular and now and again inconveniently wide apart, often as much as 3 feet. Where the slope is not quite so steep the ladder is stretched so tight against the rock that one can hardly get hand or foothold; where the rock is perpendicular or projects, the ladder swings round while one is mounting in a manner fit to make the steadiest head feel giddy.

When once this ladder is mounted all real difficulties are overcome, for from here a mountain track leads through a boggy forest, in which however no old wood is to be found, which extends over the whole arched summit of the mountain. The wood comes to a sudden stop and is cut short, as it were, on all sides by the above mentioned steep wall of naked rock which encircles the mountain like a girdle. Therefore when emerging from the wood one gets a free and open view in all directions. On the south side it embraces the valley of the Mèlawi from Nª Pinoh to Sintang and its surroundings, including in the background the boundary range between West and South Borneo near the head waters of the Pinoh, and a little to the west the

Pl. XVII.

ASCENT OF THE MOUNTAIN KELAM.

boundary mountains with the Matan district. The most conspicuous features are Mt. Saran (1758 metres) and Mt. Koedjau
(1322 metres). Both these mountains, terraced to the south and
north, have very steep inclines to the east and west.

On the opposite or north side of the mountain, lies the
extensive plain through which the Kapoewas winds its sinuous
course. The most noticeable features in this direction are the
precipitous and isolated Loeït (436 metres) and Rëntap (658
metres) both in the foreground. They might almost be called
miniature editions of Mt. Këlam. Behind them comes the Pënai
range, a low, narrow, winding, mountain ridge, rising abruptly from
the plain and reminding one of the caterpillar shaped constructions
on old-fashioned maps with which mountain ranges used to be
indicated; then follows the Sétoengoel, Sébalang and the other
ranges in the neighbourhood of the Silat and the Sébéroewang;
to the left the mountains of the basin of the Këtoengau; more in
the background Mt. Kënëpai and the mountains of the lake-district.

Plate XVIII represents the border of the wood and the upper
part of the rock-belt on the south-west side of Mt. Këlam. The vega-
tation is of a very peculiar nature; it is principally made up of
Coniferæ, Myrtaceæ and Ericaceæ which fill the air with their
balmy fragrance. In the furrows of the rock one finds here and
there, fine specimens of the Nepenthes and quantities of orchids[1]).

These Coniferæ with their short, gnarled trunks, outposts, as
it were, of the forest, grow but very slowly in the crevices of
the rock and look much older than the slim tall stems growing
in the rich soil on the top. This fact agrees with the story of
an old Dyak, who told me that twice in his lifetime the forest
on the top had been destroyed by fire. The upper part of the
mountain was then burnt entirely bare, except at the edges, where
the Coniferæ rooted in the crevices in the upper part of the naked
perpendicular rock, had escaped destruction.

According to CROOCKEWIT one of those fires broke out in 1845

1) About Mt. Këlam and its vegetation see: II. HALLIER. *16*, p. 430, and *19*, p. 18.

and GAFFRON in 1847 saw the top of Mt. Kĕlam entirely devoid of vegetation. But this does not strictly agree with SCHWANER's statement[1]) that in 1848 he saw its brow well wooded.

Both Pl. XVII and Pl. XVIII give a very good impression of the external appearance of the Mt. Kĕlam rocks. No doubt the surface is very deeply grooved by gutters but the rest of it is perfectly smooth. The rock is spotted over and covered with fantastic figures traced by the growth of numerous lichens. There is no sign of stratification anywhere, one might say that the entire mountain is one solid mass of rock. The mass as a whole peels off like a bulb, so that flat or arched flakes parallel to the outer surface of the rock detach themselves in various places; this desquamation is due most likely to the sudden changes in the temperature of the rock shortly after sunrise and after sunset[2]).

Mt. Kĕlam is a solid mass of micro-granitic quartz-porphyry, of a pure white colour, and with a very thin greyish-brown weathered crust. This quartz-porphyry is of a very peculiar texture, microscopically more like sugar-grained quartzite or Itacolumite than suggesting an eruptive nature. Phenocrysts are altogether wanting, but with the help of a very strong magnifying glass, one can identify here and there small bunches of flakes of muscovite. A microscopic examination proves that the rock is entirely composed of quartz and feldspar, chiefly plagioclase, besides the bunches of muscovite-flakes already mentioned. Muscovite is the first product of crystallisation, then follows quartz and the greater part of the feldspar. The last to crystallize has been a second generation of feldspar which consequently forms a kind of cement which surrounds and holds together the crystals of the earlier quartz-feldspar generation. The individuals of this second generation of feldspar are also very small and not perceptible to the naked eye. The structure of the entire rock bears a strong

---

1) C. A. L. M. SCHWANER. 55, II, p. 190.

2) In hot dry districts, desquamation is an important factor in the general denudation process. See for instance: J. WALTHER. 66, p. 362.

likeness to that of the ground-mass of a micro-granitic quartz-porphyry.

In my estimation Mt. Kĕlam is part of a boss or of a big dyke of quartz-porphyry, which has evidently so effectually resisted the denuding forces that it now stands in majestic isolation, while the rock through which it has forced its way will be found on a much lower level, most likely covered by more recent deposits.

Judging by their shape and general appearance we may consider that Mts. Loeït and Rĕntap to the north-east of Mt. Kĕlam also consist of quartz-porphyry rocks and are closely connected with Mt. Kĕlam; and this applies with equal of probability to several other more or less isolated mountains of somewhat conical shape, situated in the plain which is shut in by the Pĕnai mountains, and the rivers Kapoewas, Mĕlawi and Kajan [1]).

About 5 o'clock in the evening we saw a thunder-cloud come towards us from our western neighbour Mt. Koedjau, which shortly afterwards enveloped Mt. Kĕlam. Lightning flashed all round, and the rain fell in torrents till 7 p. m., after which it gradually subsided.

We bivouacked on the top, and at sunrise next morning, the landscape below us lay enveloped in a white mass of clouds. At 9.30 the clouds had lifted to halfway up the mountain and by 11 a. m. only the lower fringe of the cloud-bank which by that time was broken in many places, just touched the summit of the mountain. May 9.

When, by one o'clock, I had concluded my barometric observations of the previous 24 consecutive hours, we descended the moun-

---

1) CROOCKEWIT (7, p. 292) describing Mt. Kĕlam, calls the rock a modern sandstone with a large proportion of clay-cement. The rock of Mt. Pĕnai, he also calls a sandstone. This accounts for the fact that EVERWIJN (11, p. 116), who visited the Pĕnai mountains and Sĕtoengoel, but not Kĕlam, painted Kĕlam, Pĕnai and Sĕtoengoel all of the same colour on his geological map, as being all of Upper-Eocene age. This is decidedly incorrect, as the formation of Mt. Kĕlam has nothing in common with the rocks of Mt. Sĕtoengoel and the Pĕnai mountains. Besides which the porphyry of Mt. Kĕlam is undoubtedly much older than the sandstone of Pĕnai and Sĕtoengoel.

tain at a fairly good speed and were back at Roemah Lajang by 3.30 p. m., where in the evening we witnessed the grand spectacle of a thunder-storm breaking over and around the top of the mountain. Mt. Kělam deserves a reputation for affording a capital and interesting bit of mountaineering, yet it presents no real dangers to any one at all skilled in the art. Although Baedeker no doubt would class our mountain under the head of "nur für Schwindelfreie", it is nevertheless true that CROOCKEWIT [1]) in his description of Kělam, which is trustworthy in every other respect, has greatly exaggerated the dangers of the ascent.

May 10. After a night's rest at Roemah Lajang we returned on May 10ᵗʰ by the same route; the Kapoewas was reached before sunset, and the lovely scenery of the water, gloriously illuminated by the setting sun, made the latter part of our excursion up this majestic river as far as Sintang one unbroken delight. Page 21 of the to Atlas shows the panorama of the Kapoewas looking downstream from the left bank and the European quarter at Sintang. To the left, in the foreground, is the fort or benting, situated on the strip of land at the confluence of the Kapoewas and Mělawi, so that the back of it faces the Mělawi. To the right of the benting one perceives on the left shore just below the junction of the two rivers the Chinese kampong with its numerous bandongs (roofed trading vessels) lying at anchor, while at the right end of the panorama the Malay kampong built on the right shore of the Kapoewas is visible. In the corner of the picture one can just catch sight of the mosque, and a little higher up is the residence of the panembahan, a large building with a high stair-way leading down to the water. Along the shore in front of the Malay kampong lie several rafts carrying small floating bathing-sheds.

According to EVERWIJN and CROOCKEWIT beds of sandstone and shale are found on the shores of the Kapoewas and Mělawi in the neighbourhood of Sintang wherever the ground is slightly

---

1) J. H. CROOCKEWIT. 7, p. 288.

raised and not marshy. These beds are either horizontal or dip gently in different directions, occasionally thin coal-seams and beds of shale with shells being intercalated between them. Of these fossils only a few proved to be determinable, and judging from these, all the beds are supposed to be of Eocene age [1]).

Coal as well as fossil shells are also found for example at Tĕlok, on the left shore of the Kapoewas, 8 kilometres above Sintang; again just below Bantok on the right shore of the Mĕlawi, about 15 kilometres above Sintang; further in a few other spots in the Mĕlawi or its tributaries, between Nˣ Pinoh and Sintang, on the lower Pinoh and on the Tampoenak, which latter empties into the Kapoewas 26 kilometres below Sintang.

Many a time I visited the banks of the Kapoewas and the Mĕlawi in the vicinity of Sintang in order to explore the beds with fossils, but the water was always too high and they were entirely hidden from view [2]).

I was more fortunate however when from the 21ᵗ to 24ᵗʰ of August I made a little excursion to the Pinoh river. I was then able to corroborate the fact that on the banks of the Mĕlawi between Sintang and Nˣ Pinoh, no other rock is to be found but the above mentioned shale and soft sandstone, which are visible here and there at the edge of the water. Near Batoe Lintang, about 4 kilometres below Nⁿ Kajan, a ridge of sandstone runs in an eastern direction quite across the river, forming a great obstacle to navigation. Small steam-boats, of not more than 4 feet draught, can navigate the Mĕlawi as far up as Nˣ Pinoh, but at Batoe Lintang, even at an average height of the water, there is only a narrow channel left for navigation along the right shore, while at low water all intercourse by steam-boat is stopped.

The Dutch settlement at Nˣ Pinoh is on the left shore of the Mĕlawi at the confluence of the two rivers. Nˣ Pinoh is the resi-

---

1) P. van Dijk. *8*, p. 147.

2) Van Schelle when he visited the place, in 1879, was no more favoured than I was in discovering these beds with fossils, for the water was then also too high. See C. J. van Schelle. *51*, p. 36.

dence of a controller, and there is also a fort with a partly European, partly native garrison under the command of a lieutenant. It is a charmingly situated though very quiet place. All the business is limited to the Chinese quarter, situated a little higher up on the left shore, numerous vessels laden with forest-produce generally lie in front of it (Plate XIX). The left shore lies low and is often flooded at high water, the right bank, on which the Malay kampong is built, is, on the contrary, high and steep and consists of horizontal beds of soft sandstone.

I rowed up the Pinoh river for some distance; near its mouth the banks were low and marshy, and the water though rapidly flowing was perfectly smooth. About ³/₄ kilometre upstream from

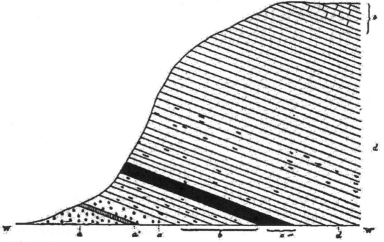

Fig. 35. Section on the left bank of the Pinoh 4 kilometres above its mouth.

a. Hard clay-bed with numerous shells, with an intercalated bed (a′) without fossils.
b. Clay-stone with hard concretions.
c. Flaky coal.
d. Clay-stone with hard concretions, like b.
e. Sandstone.
W. W′. Waterlevel in the Pinoh.

the mouth a ridge of rocks crosses the river in an almost due E.—W. direction. At low water they project above the surface and form an obstacle, insurmountable to steamers and other boats of any size. They consist of grey shale which a little further down on the right shore is overlain by soft sandstone.

Fully 3 kilometres higher up, the left shore is steep and rocky and displays the section represented in fig. 35. Dr. KRAUSE[1]) considers the shells in these beds to be of old-Tertiary age; they represent a fauna from brackish water. The coal cropping out here is of inferior quality. The strike of the beds is W. 30° S., the dip 30° to N.N.W. Sandstone also crops out on the opposite bank, and 1$^1/_2$ kilometres higher upstream clay-stone and sandstone are again exposed on the right bank, dipping about 25° to the north. Beyond this again, the contortions of the river form a loop, so that one can actually save half an hour's rowing by walking 40 metres overland. Fourteen kilometres above N° Pinoh where the river has a much less sinuous course and runs direct from the south, the hilly district commences, and with it an area occupied by granite, the rocks of which give rise to numerous dangerous rapids. The first group of rocks which meets the eye on the left, coming from N° Pinoh, is called Batoe Raoeng and consists of biotite-granite (II, 687). I ascended the top of one of the adjoining hills (75 metres high) and noticed that the prevailing trend of the hills and mountain-ranges between West and South Borneo is from east to west. I had no time to penetrate further into these territories, which, although travelled over by GAFFRON in 1857, are practically unexplored from a geological point of view.

There are no special difficulties or dangers attached to an expedition from N° Pinoh either to Soekadana or to Katawaringin which would be quite easy to manage and would lead to interesting scientific results. From N° Pinoh I took a sampan back to Sintang and was twelve hours on the way.

---

1) P. G. KRAUSE. 27.

CHAPTER VII.

The Embaloeh River [1]).

May 15. On the 15th of May I accompanied the Resident, Mr Tromp, on board his steam-yacht Karimata from Sintang to the Upper Kapoewas, to make a flying geological survey of that river and of the Embaloeh, one of its largest tributaries.

May 16. We left Sintang at 1.30 p. m. and arrived next day about 6 a. m. at Sĕmitau, where we stopped only for a few hours to take on board all the necessaries for the proposed expedition.

May 17. We passed Boenoet by night and at 4 p. m. on the 17th we cast anchor 6 kilometres below Poetoes Sibau. Massive tree-trunks here obstruct the river and make that settlement at low water quite inaccessible by steam-boat. It was so in our case.

May 18. In the morning I rowed up to Poetoes Sibau, where I met Messrs Hallier and Nieuwenhuis who both appeared well-satisfied with the results of their botanical and ethnographical investigations in the neighbourhood of N⁴ Raoen. We were comfortably lodged in the koeboe on the left shore of the Kapoewas. Poetoes Sibau is the highest settlement of any importance on the Kapoewas. It is inhabited by Malays, Dyaks and Chinese. The Malays live on the left shore where the koeboe stands. The Dyak houses are on the right shore, and the Chinese are supposed to live there also, but they remain almost entirely on their large bandongs. Most of the houses on the right shore are fortified and protected by palisades finished off with sharp spikes of bamboo. The inhabitants took this precaution because of the Batang Loepars of Sarawak who used to come here head-hunting, and they were

1) For this Chapter consult Map V of Atlas.

Pl. XVIII.

ASCENDING THE BT. KĚLAM, NEARLY REACHING THE SUMMIT.

still in constant dread of them. This also explains the preference
of the Chinese for their floating dwellings. Of late years however
the populace has become much less timid and now that since
1895 a controller has been appointed, this apprehension as to
their personal safety will soon vanish altogether.

On the 19ᵗʰ of May I left Poetoes Sibau at 6 a. m., and tra- May 19.
velled 7 kilometres down-stream in the steam-yacht Poenan which
was to take me and my bidar to N' Embaloeh. I was accompanied
by HALLIER who was compelled by frequently recurring attacks
of malarian fever to discontinue his botanical investigations and
to return to Buitenzorg. Between Poetoes Sibau and N' Embaloeh
the shores of the Kapoewas are flat and monotonous. Wherever
a bend in the river opens a free prospect to the south, the beautiful mountains of the Upper Mandai come into view. Close to
Poetoes Sibau, the Mt. Tiloeng attracts most attention, but a
little further down-stream the Gĕgari is much more conspicuous.
I passed N' Embaloeh at 1 p. m. and reached a sandbank 15¹/₂
kilometres higher up, about sunset, where I camped that night.

The Lower Embaloeh has a very insignificant fall (compare May 20.
section on page 20 of atlas) and runs through low-lying, marshy
territory, being a part of the great Upper Kapoewas plain. I
found the water low and consequently edged with high, steep
banks of mud and sand. Here and there, through narrow fissures
in the banks, water from neighbouring danaus flows down into
the stream; several of these danaus line the Lower Embaloeh.
The principal ones are D. Sĕngkoedjoe, D. Bangkiling, D. Koeroek,
D. Émpoewan, D. Bélatong, D. Boenjau Kĕra and D. Podjit.

From the Lower Embaloeh at high-water communication with
other rivers is possible. For instance, the Tĕmĕroe river which
empties into the Laoeh river, a side branch of the Palin, can
at high-water, partially discharge itself into the Embaloeh river,
by way of the so-called Pĕlais river. Lalans (a kind of hornet)
and mosquitos abound in these parts and bivouacking on the
sandbanks is an extremely unpleasant necessity to travellers.

The sandbanks in the river-bends are on the land side skirted May 21.

with high rushes. The concave shores facing the bends are deeply eaten out by the stream, so that where the curves are sharp the river assumes very wide proportions. In the course of the fore-noon we passed a couple of small Malay settlements, Marassau and N^a Loewangan, consisting of a few houses only, and further on, a couple of Dyak kampongs, Oelak Paoeh and Rantau Pat, inhabited by Malo-Dyaks. Malo is an abbreviation of Embaloeh, and the river itself is also often called Malo, which name is made use of by many former travellers such as MÜLLER, VAN LIJNDEN and EVERWIJN [1]). Just in the same way as Embaloeh is shortened into Malo, Embahoe is shortened into Maoe. In the vernacular, neither Embaloeh nor Embahoe are ever used.

The Malo-Dyaks are closely related to the Batang-Loepars. Both the Kĕtoengau and the Malo-Dyaks are probably originally descended from the same tribe as the Batang-Loepars. They are like the Batang-Loepars, of a much heavier build, — especially in the lower extremities —, than the Dyaks of other tribes, the Poenans alone excepted. They wear their hair cut very short and flat on the brow. Their houses are not long and as the verandah is very deep, this appendage becomes almost square. The verandah always faces the river. Inside the house are two rows of family apartments separated by a central passage which runs the entire length of the building.

VAN LIJNDEN [2]), who in 1847 went a little way up the Emba-loeh, describes the houses of the Malo-Dyaks as follows:

"The houses of the Dyak-Malo are nearly 100 fathoms long and much deeper than those at Sangouw. They are raised 20 feet or more above the ground. Through the centre of the house runs a narrow passage, on both sides of which are the lawangs or private apartments, both larger and cleaner than those at Sangouw. At certain intervals an open space is left between the

1) VAN LIJNDEN uses both names, Malo and Ambaloh, and writes on his map Ambollo. 31, p. 550. O. VAN KESSEL uses the name Ambalauw. 25, p. 185.
2) D. W. C. VAN LIJNDEN and J. GROLL. 31, p. 578.

apartments, forming a kind of hall furnished with a fireplace and dried skulls. This is the Chief's favourite seat where he holds his council and receives strangers. The house is entirely closed, the only access being by the staircase outside, which leads through a hole in the floor, into the Chief's hall. This staircase consists of a tree-trunk in which steps are cut out, and the upper-most part of which is carved in the shape of a dwarf or a human head. At night the aperture is covered with a heavy board upon which a crocodile (cayman) is carved. Similar representations of crocodiles are found on the walls and doors in these parts as also in the Sangouw and Sintang districts. The building is sur-rounded by a high palisade of ironwood, and many other pre-cautions are taken against the Dyaks of Batang Loepar and the Malays of Boenoet."

This description may still be considered fairly correct for the present conditions, but I saw no such very long houses, as he mentions, on the Embaloeh, and the palisading has almost entirely disappeared.

The kampongs generally contain several houses built in two long rows, running parallel with the river. They are connected with the river-side and also with each other by a roadway of massive batangs. The kampongs of the Malo-Dyaks have a flourishing appearance, and are very often surrounded by groups of areng-palms (see fig. 36).

At a distance of 38 kilometres from the mouth of the Embaloeh river, pebbles make their first appearance in the sandbanks and as we proceeded upstream they increased in number, until finally they altogether took the place of the sand. The pebbles in this part of the river are of white quartz, quartzite, sandstone and sandy slate, all of them generally being intersected by nu-merous quartz-veins. We bivouacked almost opposite the Dyak kampong N⁴ Soengei.

At 9.30 I reached Bénoewah Oedjoeng, the principal Dyak May 22. kampong on the Embaloeh. Just below this settlement is the koeboe of the Dutch Government. It was undergoing repairs, so

I went into the Malay kampong on the right bank and lodged with the commandant of the koeboe, who holds the rank of corporal. Here I learned that for travelling in the Embaloeh above Bĕnoewah Oedjoeng, one is obliged to use a special kind of very long and narrow canoes hollowed out of tree-trunks. These canoes are called boengs, they are of tough, soft, and pliable wood, well adapted to resist the strain of the twists and bumps in hauling up or running down the rapids. Boats of hard and brittle wood are quite useless in the torrential portions of the streams.

At first there seemed to be many difficulties in the way of an expedition to the Upper Embaloeh. No boengs, so I was told, could be had, and even supposing they were procured, there would be no possibility of manning them as a crew of 18 would be required. With a little patience, however, I succeeded in sifting the real from the imaginary difficulties, and I was fortunate enough to secure a couple of boengs. They were, however, not in good condition so I had them caulked overnight. It gave me more trouble to engage a guide and the necessary crew, chiefly because I wished to start immediately. It has ever been my experience in Borneo that the Dyaks make a point of having some time to themselves before the day of departure, and moreover would rather engage themselves for a long excursion than for a trip of 8 or 10 days. Finally I succeeded in engaging two Bĕkatans, one of which was to act as guide.

May 23.   We started the next day at 1 p. m., our crew being made up of 5 Malays from Poetoes Sibau, who had also rowed me as far as here, 2 Malays from Bĕnoewah Oedjoeng, 2 Bĕkatans from the Upper Embaloeh and 2 pradjoerits from Djaweh, therefore, including my boy Aboe, my Chinaman and myself, 14 in all. It was only "à force d'argent" that I persuaded these people to enter my service at such short notice. For the trip which was to take 7 days, I paid the two Bĕkatans 5$^{1}/_{2}$ ringgit each, to each Malay half a ringgit per day, the same as I had paid to the 5 Malays from Bĕnoewah-Oedjoeng.

Bĕnoewah-Oedjoeng is the emporium for the forest produce

gathered in the extensive virgin forests of the Upper Embaloeh. There were heaps of rattan (Mal. rotan) lying about, which is bought by the Chinese and exported to Singapore by way of Pontianak. The profits which the Chinese make in this trade are enormous, and as the little steam-boats on the Kapoewas are also in Chinese hands, all the profit is their own. Whereas the Dyaks who collect the rattan in the forests receive, after months of arduous labour, but a very insignificant sum for their rattan from the Chinese merchants, the Chinese themselves make fabulous prices.

Three months of constant hard work will enable a Dyak to collect enough rattan to sell for 50 or 75 ringgit, and I was told that many Chinese in West Borneo, and especially in Singapore, make some hundred thousand guilders a year by their trade in Borneo rattan.

It would be a good policy on the part of the government to monopolize the transport by steamer on the Kapoewas river, and to encourage enterprising Dutch merchants to devote more atten-tion to the rattan trade. The existing unscrupulous speculation upon the ignorance of the Dyaks with regard to the real value of their forest produce, would then soon be at an end.

We must not however lose sight of the fact that the Chinese merchant thinks nothing of staying for months together at this out-of-the-way station, in order personally to superintend the purchase of the rattan. He has no intellectual aspirations and isolation is no hardship to him, but a European would feel it a very great privation. Yet on the other hand if success attended the undertaking, how soon a small European settlement would change the whole aspect of the place!

The rattan of the Embaloeh forests is gathered principally by Batang Loepars from the Batang Rĕdjang in Sarawak; they tie the bundles of rattan together into a kind of raft on which they descend the river as far as Bĕnoewah-Oedjoeng. A few hundred of these Batang Loepars happened to be on the spot when I visited it. They are a splendidly built and sympathetic looking set of men. I took a photograph of three of them, which is reproduced on plate XX.

Their toewa [1]) (chief), on the right, told me that his tribe on the Batang Rédjang in Sarawak had been having a small head-hunting raid, and that he and his men were anxious to get back quickly as they were afraid that during their absence the Rajah of Sarawak might raze their houses to the ground for punishment. To the right, just above the Malay settlement, are a few Dyak graves hidden amongst the rushes by the river-side. These are little sheds of bamboo, ornamented with beautiful carving and covered with small kadjang mats under which the coffins are placed (plate XXI). The hats of the deceased lie on the top and their clothes hang in front of the shed, while inside other articles of daily use are treasured. The carvings remind one of Chinese designs. The little flags to the right and left of the grave are white, edged with pale-red or vice versa. The standard flag placed on the centre of the grave is blue at the top, red and white below. From here one gets a good view of the Dyak kampong on the opposite shore, as reproduced in fig. 36. More inland is another row of houses, which are not visible in the illustration. These are inhabited exclusively by Malo-Dyaks, while in the kampongs further up-stream some of the nomadic Bĕkatans have settled.

The water was low and therefore favourable for travelling up-stream, but in consequence of this, just past Bĕnoewah Oedjoeng, it became so shallow over the karangans that the boengs had to be poled and towed. There are a few boulder-islands covered with trees and shrubs, in this part of the stream, the most important of which is Poelau Pandjang, one kilometre above Bĕnoewah Oedjoeng. It is 650 metres long by 200 wide. The width of the river at Bĕnoewah Oedjoeng is 120 metres but measured across Poelau Pandjang it is 500 metres. Whereas, when the water is high, small steamers can get up to Bĕnoewah Oedjoeng without difficulty, the shallowness caused by the widening of the stream at Poelau Pandjang is an insurmountable obstacle to the approach

---

1) Toewa = eldest.

of all vessels of any considerable draught. The river-banks are composed of gravel, overlain by sand rich in vegetable remains. We

Fig. 36. DYAK KAMPONG, BĔNOEWAH OEDJOENG.

passed in succession the Dyak kampongs Karam, Boekang and
Bĕlimbis. Just above the latter is another boulder-island ³/₄ kilometre long, which, at extreme low water, is joined to the left bank.
   At Pĕndjawan we had reached the highest point of the boundary-strip between the Kapoewas plain and the mountains. In
this area the river deposits nearly all the boulders which it carries
down from its upper course. The fall of the Embaloeh which
below Bĕnoewah Oedjoeng is about 1 : 10200, increases in this
boundary-strip (the area of the boulder-islands) to 1 : 800. At
Pĕndjawan the real upper course commences, the river-bed gets
narrower, the boulder-islands are less numerous, and the fall ceases
to be regular. The total amount of the fall for the first 19¹/₂
kilometres, viz., from Pĕndjawan to Loang Goeng, is 21 metres,
giving an average rate of 1 : 905. Above Pĕndjawan the river
flows in an almost due southerly direction from the mountains
down into a long rantau, and looking up-stream the chains of
the Upper Embaloeh mountains stretching east and west, and
rising one behind the other, furnish the setting to the most exquisite bit of distant scenery imaginable.
   One more night had of necessity to be spent in the low marshy
brushwood on the left shore of the river just below Roemah Lamau.
That night I endured with more than ordinary equanimity the

torments of the myriads of mosquitos which pester the borders of all rivers in the plains of Borneo. But I did not mind, for the morrow would see me high up amongst the mountains where I should have left these plagues far behind me. I dreamt of the mountains with their delicious coolness, their waterfalls and clear, pure streams, their majestic virgin forests and well exposed sections of rocks. They wafted me a promise of exciting and interesting scientific work, in fact of all the good things which to this day make the remembrance of my travels in Borneo dear to me.

May 24.   At sunrise our little fleet was ready to start and at 7 a.m. we passed Roemah Lamau the highest Dyak settlement on the Embaloeh. Above this point there is not a single house to be seen in the whole of the Embaloeh-basin, and not an inch of ground has been cultivated. At about 8 a.m. we reached Békatan-poort, the place where the river emerges from the Upper Kapoewas mountain range, into the plain. Abruptly the mountains rise up before us like a solid wall. While at Lamau the plain is not more than 60 metres above the level of the sea, a few hundred metres more to the north the river runs in a deep gorge between mountains averaging about 400 metres in height. In fact, the Embaloeh, from its source up to this point, flows through a mountain range, formed by a succession of chains running parallel and trending E.—W. Every where in this range the strong tilt of the beds, their folding and faulting, point to enormous pressure, which, acting in a S.—N. direction, has thrown up the strata and so generated the mountain range itself. The strike of the strata is on an average, from E. 25° N.—W. 25° S. to E.—W., the dip, as a rule very steep to the north, but often also the strata are vertical. In many places they are intersected by quartz-veins running in different directions.

The very first rock-groups we come upon at the banks of the river above Lamau furnish samples of the kinds of rocks which are the chief constituents of this mountain-range. They consist of phyllitic clay-slate with well marked silky lustre, beds of quartzitic sandstone being intercalated occasionally between the layers of

CHINESE KAMPONG NANGA PINOH.

slate. Besides these rocks, greywacke-slate, dark-blue drawing slate and quartzite also occur in this mountain-range, the clay-slate with silky lustre, however, being the main and most typical constituent, for which reason I have given to this entire complex of unfossiliferous strata the name of the "old slate-formation". Map V on page 7 of the atlas gives all the details of the geological structure of the Upper Kapoewas mountain-range as far as they could be gathered from observations made during my journey along the Embaloeh. The localities marked thereon are absolutely correct, and it is advisable to consult this map for the right understanding of the text.

The water was quite low when I arrived, an essential to geological surveying along a river. Of course the vegetation, which hides all details of geological structure, does not reach below the average water-level, so that between that level and the present surface of the water (in my case a strip of about two metres in height), the fresh, naked rock becomes exposed and presents a natural rock-section in which all the peculiarities of the stratification can be distinctly observed. Almost without exception the mountains stand close up to the water's edge, which makes the valley of the river narrow, deep, sharply defined, and withal very picturesque. The current in the Embaloeh is very strong but the difficulties of navigation are not of a serious nature. At low water, in the rapids, the boengs have to be towed over the boulders, but when the water rises, these places are merely indicated by a particularly strong current. At times of flood the current in the Embaloeh is so strong that it is practically impossible to go up the river.

The highly-tilted position of the strata, together with the well defined cleavage of the clay-slate, greywacke-slate and flagstone, have given rise to the development of different river-types, dependent upon the direction of the current with regard to both the strike and dip of the strata. In the gorge of the Embaloeh as well as in that of its tributary, the Tĕkelan, the following types may be distinguished:

Type A. The symmetrical river-type. This type is obtained when the direction of the current crosses the direction of the strike of the beds at a right angle, in other words, when the river cuts right across the beds. Rocks project in this instance from both sides of the river, being those portions of the strata which have better been able to resist the erosive power of the water. Very often, the connection between these projecting rock groups and the shore, is partially or entirely destroyed. Take a case in which the beds are not vertical but dip, for instance, to the south, the river also flowing southward, in looking up-stream the effect is produced as illustrated on Plate XXII. Both sides of the river are skirted by ridges of projecting rocks or rocky islets which turn their smooth surfaces towards us. Were we to view this part of the river from the opposite side, viz., down-stream, the effect would be totally different, for we should then not see the smooth surfaces of the rocks but only their outcrops ¹).

All these projecting rocks or rocky islets, as illustrated on Plate XXII, turn their smooth sides to the south and their steep, angular, irregular flanks to the north. But to the right and to the left the effect remains the same, whether one looks up or down-stream, and the arrangement of the rocks is therefore symmetrical as regards the course of the river. We have therefore called this type A, *the symmetrical river-type.* In this type the deepest and often the only navigable track is in midstream; the sides are always unsafe.

Type B. The asymmetrical or one-sided river-type. In this instance the direction of the current corresponds with the strike of the strata. Plate XXIII illustrates this type with the strata dipping to the south. Whether one looks up-stream or down-stream the one bank always appears paved as it were with big flagstones, while the opposite invariably turns the jagged, menacing outcrops of the strata towards us. There is therefore a marked difference between the right and the left bank, and in consequence I have

---

1) Of course this difference does not exist when the strata are vertical.

named this type the *asymmetrical or one-sided river-type*. In this type, as in the other, the most serviceable track lies usually in midstream. Close to the shore on which the outcrops of the strata stand, the water is generally dangerous, but the opposite side is, as a rule, quite safe. On the whole the water is quieter in this type than in the symmetrical.

Type C. This type is a peculiar modification of type B. When the strata are vertical, there is no longer any difference between the two banks, but on both sides and parallel with the stream, gigantic rock-slabs, often more than 10 metres high, rise perpendicularly out of the water. The surface of these rocks is split into all manner of grotesque shapes, causing them to appear like a row of monuments on either side of the river. I have therefore called this the *monumental river-type*. The map shows where these types may be seen most purely developed. A little beyond Loang Goeng, situated 79 metres above the sea-level, being the highest point in the Embaloeh valley of which the astronomical position is exactly known, these types are particularly distinct.

POWELL [1]) has shown in principle, that it is possible to distinguish different types of valleys, based on the mutual relation between the direction of the stream and the position of the strata through which the river has cut its bed. This problem is elaborated somewhat more in detail in the two first columns of the table on the next page, in which the names adopted from POWELL are indicated by the letter (P). Each of these types however gives a particular stamp to the landscape exhibited by the valley of the river. Thus it may happen that the landscape presents the same aspect whether viewed from up-stream or down-stream, or from the right and left, being in this instance symmetrical in two directions (di-symmetrical); also symmetry may be shown only from up-stream to down-stream or from right to left, or, finally, there may be no symmetry whatever perceptible in the landscape. Moreover from a certain type of valley a well characterized type

1) J. W. POWELL. Exploration of the Colorado River of the West. Smithsonian Institution. Washington 1875, p. 160.

Mutual Relation between the position of the strata and the direction of the stream.	Type of Valley.	Character and type of the corresponding Landscape.
A. Strata horizontal. Changes in the direction of the stream do not affect the type of the valley.	Neutral Valley.	Table Mountain- and Cañon-type.
B. Strata not horizontal. Changes in the direction of the stream affect the type of the valley.		
I. Direction of the stream at right angles to the strike of the strata.		
a. Strata in open folds, dip alternating.	Transverse Valley (P). Diaclinal Valley (P).	Both sides of the river symmetrical with regard to each other. The up-stream part of the valley periodically symmetrical and non-symmetrical with regard to the down-stream part. *Periodical di-symmetrical type.*
b. Strata, dipping in one direction.		
α. The stream runs in the direction of the dip.	Cataclinal Valley (P).	*Symmetrical-type A* (Pl. XXII) or Longitudinal mono-symmetrical type.
β. The stream runs against the dip of the beds.	Anaclinal Valley (P).	Valley periodically narrowed (hard layers) and widened (soft layers), *Coulisse-type.*
c. Strata vertical.	Orthodiaclinal Valley.	The up-stream part of the valley, symmetrical to the down-stream part.
II. Direction of the stream the same as the strike of the strata.	Longitudinal Valley (P).	
a. Strata in open folds, dip alternating.		
α. The stream follows the anticlinal axis of the fold.	Anticlinal Valley (P).	Di-symmetrical type. Water turbulent. Bold landscape.
β. The stream follows the synclinal axis of the fold.	Synclinal Valley (P).	Di-symmetrical type. Water smooth. Gentle landscape.
γ. The stream runs in the direction of the strike between the two axes of the fold.		
b. Strata dipping in one direction.	Monoclinal Valley (P).	*Asymmetrical Type B* (Pl. XXIII) or transverse-monosymmetrical Type.
c. Strata vertical.	Ortho-isoclinal Valley.	Di-symmetrical Type. *Monumental Type C.* (p. 155).
III. Direction of the stream oblique with regard to the strike. This type is rare and can always be resolved into frequently alternating parts belonging to Groups I and II.	Diagonal Valley.	No symmetry; different types follow each other in rapid succession. *Irregular Type.*

of landscape may result. The scheme on the opposite page is obtained by taking into account all these relations and consequences.

About 4 kilometres further on I experienced the first real difficulties travelling up the Embaloeh. In the middle of the river is a rocky island, Batoe Mili, where the current is very strong, and where a sharp bend in the river has created a whirlpool. With the utmost difficulty we succeeded, after many failures, in towing our boengs past this dangerous spot. We bivouacked not far from there on the left shore, where I had a splendid view of Batoe Mili and the seething waters around. That day I had covered a distance of 20 kilometres, and was now in the very heart of the mountainous district.

The morning vapours still hung like a cool, damp veil over May 25. the gorge when at 7 a. m. we resumed our journey. All along the bank, the intensely folded beds are exposed to view. As a rule the folding has been very sharp and perfect, so that in the successive isoclinal overfolds, the dip remains almost the same for a very considerable distance. Often however, and especially where thick beds of greywacke-slate or quartzose sandstone occur, they are cracked and broken, and synclinals and anticlinals follow each other in close succession. One of the best sections is found near N⁰ Engkĕloengán about 1¹/₂ kilometres above Batoe Mili. The vertical section at this spot is reproduced on Plates XXIV and XXV from photographs, and a ground-plan is given on Map V. At the place where the Engkĕloengan empties with a small cascade into the Embaloeh, this river curves round; the direction of the stream changing from W. 15° N. to N. 30° E. Just at the bend, the river-bed becomes considerably narrower, because of a flat rocky peninsula joined to the left shore, which at high water however is entirely submerged. The best point from which to survey the section exposed in the low vertical cliffs on the right bank near N⁰ Engkĕloengan and in the peninsula just mentioned, is from a large sandbank on the left shore, from which point also the photographs have been taken. On the peninsula there are exposed greywacke-slate and clay-slate folded in acute flexures, which

show a tolerably perfect isoclinal structure (see left side of Pl. XXIV).
The trend varies, and beds striking E.—W. have been folded over
those striking E. 20° N.—W. 20° S. (See also the ground-plan).

On the right shore (see Plate XXV) the strata of greywacke-
slate, quartzose sandstone and phyllitic clay-slate are folded in
open flexures with alternating dip. At the anticlinal arches the
beds are either fractured or bent, but in the latter case they are
also cracked at the bends and there traversed by numerous minute
fissures, which have been filled up with quartz. In the third
syncline and the fourth anticline fold-faults have been produced.

The careful reproduction of this section may serve as a type
of the prevailing geological architecture in the Upper Kapoewas
range. The inconstancy of the direction of the strike of the strata
(as shown in the ground-plan on map V) is so common in this
mountain-range, that one can only speak of an approximate strike.
Although the predominating trend of the mountain-range in the
Embaloeh district is undoubtedly about east and west, yet one look
at the map, where, in many instances, I have carefully marked the
exact strike, convinces us at once of the numerous deviations from
this prevailing direction. And, as a matter of fact, the flat bare
rocky islets, which are exposed at low water, show quite plainly
that the intersection of each single layer with the horizontal plane
— i. e. the strike — is an undulating line varying as to its direction
more than once even within the short distance of say, 5 metres.

At noon we passed the mouth of the Géndali river, an important affluent on the west side of the Embaloeh, being navigable
with boengs for a distance of 10 kilometres. The aspect of the
valley and of the mountains which we passed remained the same
all through that day. Phyllitic clay-slate with silky lustre, greywacke-slate, quartzose and arenaceous slate, dark-blue clay-slate
(typical drawing-slate) in alternating beds follow each other in
rapid succession, all the strata being very much folded, highly
inclined, and intersected with quartz-veins. The only noticeable
difference as one proceeds northward is that the steep, highly
tilted strata more frequently dip towards the south than towards

the north, while in the southern part of the mountain range the reverse occurs. We bivouacked near to a large karangan opposite to N° Pĕnoehoet. In the middle of the night we were obliged to shift our bivouac some hundred metres further up, on account of being attacked by myriads of ants of a large and aggressive species.

.The landscape continued as picturesque as ever, and the vege- May 26. tation equally rich as on the preceding days. The thick primeval forest does not quite come up to the bank of the river, but is separated from it by a ledge of rock covered with dense brush-wood, principally rhododendrons.

The rocky islands in the river, those at least which rise above the average water-level, display the same vegetation. The temperature in this mountain-land is delicious and strikes one as being much more uniform than in the Kapoewas plain. Although the nights are not any cooler than at Sĕmitau or at Poetoes Sibau (in the valley of the Embaloeh I registered no night temperatures below 23° Celsius) the day temperature is considerably lower than in the plain, and I registered a maximum of 27° Celsius. The only drawback to this delightful climate is its extreme dampness. On fine days it is generally 10 a. m. before the sun has dispersed the early morning mists in the valleys below.

A few kilometres above our bivouac we reached that part of the Upper Embaloeh where the river runs almost due east and west, parallel, therefore, with the strike of the strata and the trend of the mountain ranges. The dangers of navigation are here less serious than lower down where the bed of the river forms a transverse valley cutting right through the mountain ridges.

We reached N° Tĕkĕlan at 9 a. m. The Tĕkĕlan is a small but very rapid stream 15 metres wide; it rises on the boundary mountains of Sarawak and flows towards the Embaloeh in a south-south-westerly direction. We found the water very low, and at the bends, which are very numerous in this small river, lie high karangans, which on account of the fierce current, it is often very difficult and hazardous work to steer clear of. The geological structure of this area is identical with that on the Embaloeh, the mountain scenery therefore is of a similar nature. The only difference is that here the deep, narrow gorge through which the river has cut its way, makes the neighbouring mountains appear much higher than they really are, and the consequence is that the river-types previously described (pp. 153—155) are here far more typical than in the broader valley of the Embaloeh.

For this reason the photographs taken in the Tĕkĕlan gorge have been selected to illustrate these different types. The higher one goes, the fiercer and the more unmanageable become the rapids in the stream. We met, several times, little groups of Batang Loepar Dyaks from the Batang Rĕdjang in Saràwak, floating down the river on rafts made of the rattan which they had collected in the forests. It is most fascinating to watch these superbly-built people performing their feats of prowess and muscular strength as they run down the riams. Towards evening the scenery became grander, but navigation more difficult, and after passing a most wonderfully beautiful kind of cañon (Pl. XXVI) lined with projecting rocks, covered with tree-ferns, we landed at sunset at the foot of the renowned Goeroeng [1]) Narik, the *bête noire* of this particular excursion. We had that day covered a distance of 16 kilometres.

May 27.   The rocks in the vicinity of the Goeroeng Narik are composed of greywacke-slate which splits easily into thin laminæ, alternating with clay-slate, quartzose clay-slate and phyllitic clay-slate. The strata are highly contorted, the predominating strike is N.E.—S.W., the dip 25° to the south-east. They are moreover intersected by a series of parallel quartz-veins dipping 45° to the south-east (see fig. 37) and by hard bands of rock in which the stratified structure is to a great extent lost, and it is highly compressed, crumpled, and infiltrated with carbonate of line. These hard bands are more than half composed of calcite, while the greywacke-slate in other places shows hardly a trace of that mineral. The hard bands dip fully 40° to the south-east. Neither the quartz-veins nor the hard bands weather so readily as the main mass of the rock

---

1) Goeroeng (Dy.) Rapids, or small waterfall between or over solid rocks.

Pl. XX.

GROUP OF BATANG LOEPAR DYAKS.

and therefore produce ridges on its surface. This phenomenon is probably to be explained as follows [1].

The strata previously tilted and folded by mountain-making forces,

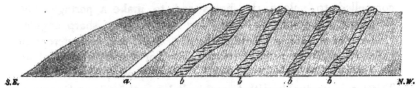

    a.   *b*   *b*   *b*   *b*

Fig. 37. SECTION AT THE FOOT OF GOEROENG NARIK.

The rock is isoclinally folded greywacke-slate, with a dip of 25° to the South-east.
  a. Veins of quartz, dip 40° to the south-east.
  b. Hard bands, dipping S. E. In the hard bands the slate is crumpled and infiltrated with calcite.

acting in a direction from south to north, have been subjected to further earth-pressure from the same direction, resulting in faults and flexures with a south-easterly incline. The fault-planes have subsequently been filled with quartz and calcite, producing veins; in the flexures the connection within the solid rock was not wholly destroyed, but the rock became compressed and crumpled. The finer cracks resulting from this process, were infiltrated with calcite, and thus the flexures were converted into the hard calcareous bands in the rock.

The Goeroeng Narik (Pl. XXVII) consists of a series of small waterfalls of which the third from the bottom is 2 metres high and cannot be navigated. The boengs therefore had to be entirely unloaded and dragged across the rocks for a distance of about 25 metres. This can only be managed on the left shore, and even there the difficulties are very great, as the huge blocks of rock are very slippery and have sharp jagged edges.

With the help of some rough wooden scaffolding we managed, after three hours of hard work, to get the boats safely past the

---

1) On Plate XXVI, giving the view down-stream in the cañon of the Tĕkĕlan from the foot of Goeroeng Narik, the projecting quartz-veins and hard bands are shown in the loose blocks on the left bank.

waterfalls and launched into smoother water. But navigation continued difficult, and twice more on that same day, once at Goeroeng Gĕroegoe and again at Goeroeng Sangkoeh, we were compelled to unload the boats and to make a portage. The Goeroeng Sangkoeh is caused by two rows of sharp angular points of rock rising out of the water, which are the outcrops of beds of arenaceous quartzite, the sides of the open arch of an anticline. At 2 p. m. we reached the highest point where the Tĕkĕlan is navigable with boengs, 2³/₄ kilometres above Goeroeng Narik. We encamped for the night not far from a great boulder-bank on the left shore, just below the mouth of the Pait.

In this bank I found several pieces of coarse greywacke-breccia containing Nummulites. This breccia is composed of quartz-fragments and slightly rounded pieces of clay-slate with silky lustre, greywacke slate, arenaceous quartzite, in short, of pieces of all the different kinds of rock of which the mountains of the Upper Embaloeh are composed. These fragments are joined together by a kind of cement, which in its turn, might be said to consist of very fine greywacke. The cement contains a large proportion of lime and here and there tests of Nummulites are imbedded in it. SCHLUMBERGER has very kindly examined these Nummulites and Plate LII gives reproductions of photographs of these specimens which he has put at my disposal. SCHLUMBERGER concludes that the Nummulites are of Eocene age, but their state of preservation does not warrant his making a specific classification. Now and then examples of *Orbitoides* are also found in this rock. It is a known fact, that in other parts of the Dutch East Indies, a special stage in the Eocene age is distinguished as the breccia-stage, and that lately this same stage has been found to exist in the Sambas district in West-Borneo. Possibly the greywacke breccia with Nummulites of the Tĕkĕlan belongs to this same breccia-stage, although the component parts of the greywacke breccia are widely different from those which form the breccia of West Borneo. I was not fortunate enough to find any greywacke breccia in its original position, but I noted its distribution carefully and came to the conclusion that the boulders of this breccia are carried down by the Tĕkĕlan, so that therefore most likely in the river-basin of that stream there might be a possibility of finding it *in situ*.

I had been told at Sintang by one of the surveyors of the topographical-brigade that Mt. Tjondong, one of the highest peaks of the Upper Embaloeh range, is accessible from Nᵃ Pait and that the summit had been entirely cleared on account of the survey. I had fixed upon this mountain, which is close to the boundary of Sarawak, as the limit of my excursion into the Embaloeh mountains and on May 28ᵗʰ our small caravan commenced the ascent. The path which had to be cut through the dense forest, ran slightly sloping in a northerly direction, across the back of a small spur of the Tjondong mountain between the valleys of the Pait and the Tĕkĕlan. We soon left the Malays from Poetoes Sibau far behind us, and we had constantly to stop and wait for them. This delay prevented us from reaching the top that day, and we had to pitch our tents at an altitude of about 1000 metres, the highest point where water was still within easy reach. Nearly through the whole of the day, our way had led through glorious forests of lofty trees, but when we had reached an altitude of 900 metres, the marshy jungle character began to predominate, the trees were lower and thinner, and covered with thick pads of moss. Thorny rattans twined themselves in between the low brushwood and much impeded our progress. The predominating rock-type continued to be clay-slate with a silky lustre; the average strike about E.—W., the dip moderate (about 35°) to the S. S. W. The night was cool, the minimum temperature being 18¹/₂° Celsius.

At 8 a. m. we reached the summit of Mt. Tjondong 1242 metres high. It is composed of clay-slate. On the occasion of the topographical survey this peak was one of the stations from which observations were made and sketches taken; hence it had been almost entirely cleared of wood. Upon our arrival the top was enveloped in mist, but towards 10 a. m. the extensive panorama began to reveal itself.

It now became clear to me that Mt. Tjondong forms part of a very mountainous district, consisting of a series of rugged mountain ridges separated by deep valleys, trending on an average E.—W. to E. S. E.—W. N. W. The general trend of the entire range may said to be about east and west. Generally speaking the height of the ridges in this mountain range increases towards the east; in a westerly direction I noticed several peaks apparently equal in height to Mt. Tjondong but none surpassing it. Amongst the specially high and noticeable peaks, I observed E. 7° N., Mt. Moebau (1293 metres) and Mt. Lawit (1767 metres). Both in an easterly and a westerly direction, as far as the eye reaches, the mountains which I have called the Upper Kapoewas mountain chain extend, and they preserve their uniformity of character throughout. Mt. Tjondong is not far from the northern boundary of these mountains, towards the north there are a few more straggling chains, and then comes an immense tract of flat or gently undulating ground, the lowlands of Sarawak, behind which in a N. N. W. direction I fancied I could see the Chinese sea. In a N. N. E. direction, however, this lowland is bounded far in the distance by another high mountain range. In some places we could clearly trace the windings of the Batang Rĕdjang through the plain. Towards the south the mountains broaden, and whereas on the one side, Mt. Tjondong is separated from the low land of Sarawak by a few mountain chains rapidly decreasing in height, there extends on the other side between the said mountain and the Kapoewas plain, a long succession of parallel mountain ridges which only very gradually decrease in height. Behind all this and very far in the distance stretches the Kapoewas plain with its few isolated mountain groups such as the Mĕnjoekoeng and the mountains of Lantjak. After my discovery of the greywacke breccia with Nummulites, I had rather expected to find somewhere in the district an area of the table-mountain-type, built up by strata, lying unconformably upon the highly-tilted beds of the old-slate-formation, but I was disappointed in this. Everywhere the mountains of the Upper Embaloeh present the same succession of sharp crested ridges, the characteristic feature of folded and eroded slate-mountains. It is only in the north towards Sarawak, in the river-basin of the Batang Rĕdjang, that there are a few mountains, amongst them Mt. Bakoen, the shape of which, reminds one of the table-mountain-type. Between Mt. Bakoen and Mt. Tjondong lie only three low ridges stretching E.—W.

When on Mt. Tjondong my Bĕkatan guide 'Lon', who, as one of the nomadic Dyaks, was familiar with the whole of the boundary line between the Upper Kapoewas territory and Sarawak, informed me of the different passes used by the Dyaks between the two districts. These are:

I. In the immediate neighbourhood of Mt. Tjondong: from N' Pait one day's march through river-beds and forest-land, then a two day's run down the Kĕtibas to fort Kapit where the Batang Rĕdjang is reached.

II. From pangkalan Para in the Oeloe Embaloeh, one day's journey by land, and one by water, to the koeboe Soeoeng on the Batang Rĕdjang.

III. From pangkalan Lali, in the Oeloe Kapoewas far above N' Boengan, two day's journey by land and four by water down a small branch of the Batang Rĕdjang into that river itself.

At 12.45 p. m. we left the top of Mt. Tjondong, and, following the same track by which we had ascended, reached our bivouac at N' Pait about 5.30 p. m. after a forced march enlivened by a thunder-storm and torrents of rain.

We descended the stream rapidly, and on reaching the water- May 30. falls Sangkoeh and Narik all our baggage had once more to be carried overland, while the Bĕkatan 'Lon', ran down the falls with the empty boengs in the most skilful manner. I reached N' Tĕkalan at 3.45 p. m. and in the evening pitched my tent on the right shore of the Embaloeh, on a boulder-bank which stood one metre above the water. That night a terrific thunder-storm of exceptionally long duration broke out at a short distance from our bivouac over the mountains upstream. At 1 a. m. I was roused by a cry of distress: "Ajar bĕsar, toewan"[1]). The water had already entered my tent. I quickly collected all my instruments, and managed to gain the higher bank, and during that night we were forced gradually to retire deeper into the forest up the slope, before the steadily rising waters.

May 31. In this one night the river rose $4\frac{1}{2}$ metres, and at break of day we saw the clear mountain stream of yesterday, transformed into a dirty yellowish mass of water, rushing past with furious speed and carrying quantities of wood along with it.

Fortunately for us, daylight revealed to us the fact that the only harm done was one night's rest spoiled. Apart from that, the high state of the river was most welcome, it would give us wings. Swift as an arrow we were presently gliding down the stream, carefully avoiding the thick tree-trunks which the stream was carrying along. It was as if a magician's hand were unfolding on either side of us an endless curtain painted all over with wooded scenery; so smoothly, noiselessly and imperceptibly did we glide over the water that I could scarcely realize at times, that it was actually ourselves who were speeding along with the stream at this terrific rate. Bénoewah Oedjoeng was reached at 12.30, our frail vessel had carried us over 54 kilometres in 5 hours time, that is, an average of more than 10 kilometres per hour. Of all the tributaries of the Kapoewas, the Embaloeh is known for the rapidity with which its waters rise at certain times, and also for the fury of its bandjirs[2]). This phenomenon is explained by the course of the Embaloeh. From its source this stream flows for a distance of no less than 105 kilometres almost in a due westerly direction through a longitudinal valley in the Upper Kapoewas chain of mountains, and remains moreover entirely confined to the high-lying district where the rainfall is very considerable. Besides this, the direction of the upper course corresponds with the direction generally taken by the severe thun-

---

1) Ajar bĕsar, toewan: "A flood, Sir".
2) Bandjir (Mal.) = sudden flood.

der-storms in these parts of Borneo; they travel most frequently from west to east. The consequence is, that now and again the total volume of water accumulated during a severe thunder-storm, collects in the upper course of the Embaloeh. Through a comparatively short transverse valley the river afterwards runs straight through the mountains to the Upper Kapoewas plain, and the volume of water which has to be conveyed from the upper course through this channel is, under these peculiar circum-stances, sometimes so enormous, that it causes this amazingly rapid rise and the current to be almost overwhelming in its force.

At Bĕnoewah Oedjoeng my Chinaman received some news which necessitated his return to Sintang. The Chinaman Tim Tjau who brought the message, took his place, but in my expedition to the Upper Kapoewas and the Boengan, I soon found out that he was quite useless, and I was compelled to handle the heavy sledge-hammer myself.

After a night's rest at N° Soengai I returned to the Kapoewas, which we gained about 3 p. m., and at once proceeding up that river we stayed the night at a place six kilometres upstream from N° Embaloeh.

Favoured by continuous fine weather and low water, two days June 1 & 2. of steady rowing brought me once more to Poetoes Sibau, where I arrived on June 2nd at 5.30 p. m. and again lodged in the koeboe. Besides NIEUWENHUIS I also found BÜTTIKOFER here, who had returned from his station on Mt. Liang Koeboeng in the neighbourhood of N° Raoen with a splendid collection of zoolo-gical specimens.

. —

CHAPTER VIII.

The Upper Kapoewas, the Boengan, the Boelit, and the boundary mountains between West and East Borneo[1]).

It had been arranged by the Resident, Mr Tromp, that the controller of the Upper Kapoewas district, Mr van Velthuyzen, should take part in this expedition to the Upper Kapoewas and the Boengan, in order that he might open relationships with the

Fig. 38. The Kapoewas at Poetoes Sibau.

inhabitants of the Upper Mahakkam, who, so far, had been but very little in touch with the Dutch Government. The controller was also charged with the organization and direction of this excursion.

1) For this chapter consult Maps VII[a], VII[b], VII[r], pages 8—10 of Atlas.

DAJAK TOMB NEAR BENOEWA OEDJOENG.

The mountain-land of the Upper Kapoewas is very thinly populated and exclusively by nomadic Poenan tribes frequenting the
forests. The nomadic Dyaks in the Upper Kapoewas district are
generally called Boekats, while more towards the east, in the
boundary mountains of East Borneo, their collective name is Poenans. The highest fixed settlement on the Kapoewas is Lolong near
Nᵃ Era, 54 kilometres above Poetoes Sibau, which is inhabited by
Poenans who have given up their wandering life. A little lower
down is the house Pahi, inhabited by Pĕnihin-dyaks, from the
Oeloe Mahakkam. When in 1885 the Batang Loepars, from the
source-area of the Batang Rĕdjang in Sarawak, invaded the
Mahakkam territory, killed several Pĕnihins and destroyed their
houses, a small number of these Pĕnihins escaped to West Borneo
and settled on the Kapoewas above Poetoes Sibau, where they
now inhabit the kampong Pahi. Further downstream are some
more kampongs, Loensa, Nᵃ Sioet, and others occupied by Taman-
dyaks, a small but very interesting tribe, universally known by
their beautiful garments made entirely of beads strung together.
The large kampongs of the Kajan Dyaks are situated on the Méndalem; these Dyaks form the most powerful tribe in the Upper
Kapoewas district. Of all the Dyak tribes, the Kajans are the
most useful and, in fact, the only fit guides for expeditions in this
district, as they are related to the Kajan Dyaks of the Upper
Mahakkam and they keep on friendly terms with the Pĕnihin Dyaks
of the Upper Kapoewas and with those, who live near the headwaters of the Mahakkam. Moreover there is no feud between
them and the nomadic Boekats of the Upper Kapoewas, and as a
consequence of their frequent journeys into the Upper Mahakkam,
and of their many relations there, they are very well acquainted
with the best land- and water-routes. Perhaps the Pĕnihins of the
Upper Kapoewas would also be suitable guides for expeditions
to the Upper Mahakkam, but there are only a few of them, and
besides this, they are not yet so thoroughly devoted to the Dutch
Government as the Kajan-dyaks, and cannot therefore be relied
upon to the same extent.

Months beforehand, the Resident, Mr TROMP, had informed the Kajans of the projected expedition, and they had declared them-selves ready to accompany the scientific expedition to the head-waters of the Mahakkam. Their wages were to be one guilder a day; the chiefs were to have one ringgit a day. They had prepared twenty boengs for the expedition and these we bought of them for the average price of 15 ringgits each.

When on the 8ᵗʰ of June the controller arrived at Poetoes Sibau, the Kajan-dyaks, faithful to their promise, soon made their appearance with the boats. They were all armed with parangs and many of them had "sumpitans" (blow-pipes).

Our departure was unfortunately delayed by the high state of the water in the Kapoewas. The exceptionally long spell of fine dry weather which had stood me in such good stead when ex-ploring the Embaloeh river, was broken up on the 2ⁿᵈ of June by a severe thunder-storm. For days together the rain came down in torrents and almost incessantly; a west or north-west wind blew all the time. It continued to rain till the 12ᵗʰ and during that time the mid-day temperature was below the average and on most days did not exceed 27.5° Celsius.

During these rainy days the barometer stood about 1¹/₃ to 2 millimetres higher than the ordinary registration without how-ever any disturbance or change in its daily variation. From the 5ᵗʰ of June the water in the Kapoewas rose rapidly, and on the 12ᵗʰ it overflowed the bank upon which Poetoes Sibau is built, but the marshland on the other side had been deep under water some days before. On the 13ᵗʰ the water reached its highest level, about 6 metres higher than the level of the 3ʳᵈ. We may conclude that on that day the whole of the Upper Kapoewas plain was one vast, continuous, sheet of water, albeit shallow in most places and hidden from the eye by dense forest.

On the 15ᵗʰ of June, the Kapoewas, although still abnormally high, had returned within its borders. So we left Poetoes Sibau at 11.15 a. m., the controller VAN VELTHUYZEN, Dr. NIEUWENHUIS and myself, accompanied by WAN ACHMAD the then representative of the Dutch Government at Poetoes Sibau [1]), 85 Kajan-dyaks, with their principal chiefs, KAM IGAU, SĔNIANG, KAM LASSA and LOEDANG, 19 pradjoerits, 5 Malay coolies and 8 Batang Loepars, the latter specially in attendance on the controller, our body-servants and baggage of every description, the principal cargo being, of course, rice, all packed in 24 boats, some aroks, and some boengs.

Five of the boats laden with rice had already been sent on in advance to N᾽ Boengan, and at intervals during our journey similar advanced posts had to be formed, in order to secure the regular victualling of our caravan in · those parts where no supplies of food could be obtained.

No rock *in situ* appears on the Kapoewas during the first day's journey beyond Poetoes Sibau and the adjacent territory continues low and marshy, while the river is bordered by a higher bank. Above Poetoes Sibau the proportion of boulders in the sand-banks at the bends of the river rapidly increases and just past N᾽ Méndalem[2]) we reached the first great boulder-island Karangan Baoeng, the area of which, at an average state of the water, is about $^1/_4$ square kilometre. Further up, the number of boulder-islands and the size of the pebbles thereon greatly increased, a sure sign that we were approaching the mountains which line the Upper Kapoewas plain. The boulders of Karangan Baoeng are of many kinds of rock; most cónspicuous amongst them are several varieties of quartz, chert, hornstone, quartzite, granite, porphyrite, diabase, rhyolite, andesite and basalt. We passed the kampongs Soewai and Mélapi, inhabited by Kajan-dyaks, and encamped on the right shore, $4^1/_2$ kilometres below Ikoe Tambai the settlement of the Taman-dyaks.

The houses on the right shore of the Kapoewas, above Poetoes June 16.

---

1) In 1895 Poetoes Sibau became the station of a Dutch official of the civil service, the controller WESTENENCK.

2) The Méndalem is one of the most important tributaries of the Upper Kapoewas, it has an actual length of 132 kilometres, and its source, like that of the Sibau, is in the boundary mountains of Sarawak.

Sibau are all fortified as a defence against the Batang Loepars of
Sarawak, who until quite lately extended their head-hunting expeditions as far as the Upper Kapoewas. The houses are therefore
enclosed in a kind of palisade of strong stakes, 4 or 5 metres high,
joined together by cross bars of tough rattan, pointed at the ends,
and bent outwards at a
distance of about 75 centimetres above the ground.
Fig. 39 is a sketch of the
not ungraceful shape of
the entrenchment of a
house on the right shore
of the Kapoewas opposite
the kampong Nª Sioet.

Fig. 39. ENTRENCHMENT SURROUNDING A DYÁK HOUSE
ON THE UPPER KAPOEWAS.

a. View in front.   *b.* View from the side.

The first sign of solid rock on the Kapoewas is a coarse-
grained white sandstone, on the right shore near Batoe Boeroeng-
koeng, a little beyond the point from where a path leads from
the Kapoewas to Roemah Pagoeng on the Mĕndalem river, of
which house Kam Lassa is the chief.

Already I caught an occasional glimpse of the mountains which
line the Upper Kapoewas plain; very striking, for instance, is the
view south-eastward just above Poelau[1]) Loensa Ra. Close to the
horizon one can see the table-mountains of the Upper Kĕryau (they
are of the Tiloeng-type and consist most likely of tuff-beds). They
tower above the far extended ranges of hills in the foreground,
belonging to the pre-Cretaceous system (see sketch, page 40).

A little beyond the house opposite Poelau Pahi inhabited by Pĕnihin-dyaks, solid rocks of olivine-diabase appear on the right shore,
while a little higher up, between Poelau Loensa Ra and Poelau
Masoem, the banks on either shore consist of quartz-diabase
containing enstatite (or quartz-norite). We had now come to the
district of boulder-islands, where the river in its passage from the
mountains to the Upper Kapoewas plain, deposits on a com-

1) Poelau (Mal.) = island.

paratively small space the larger number of the boulders which it carries down. In every Karangan the boulders are, on an average,

Fig. 40. VIEW AT POELAU LOENSA ON THE KAPOEWAS IN S. E. DIRECTION.
In the background the presumed volcanic mountains of the Tiloeng-type.
In the foreground hills belonging to the Danau-formation.

of about the same size. They always lie tile-fashion the one overlapping the other, dipping against the direction of the current, and in more or less distinctly marked transverse rows, trending at right angles to the current. Just below Poelau Laap the river is considerably narrowed by a basalt-ridge [1]), which cuts right across it in a S.—N. direction. The outcrop of this basalt, which is rich in phenocrysts of plagioclase and olivine, and moreover contains much spherosiderite, forms on both shores, but especially on the left, a promontory, while its course through the river is marked by a series of rocks, projecting above the water. Just above this ridge of basalt which obstructs the stream, the river has, through side erosion, distinctly widened out into a kind of lake. Near Poelau Lolong there is on the left shore a house inhabited by Poenan Dyaks. This is the highest settlement on the Upper Kapoewas; a short time ago these same Poenans lived a little further upstream near the mouth of the Era river; but that house is now deserted. We encamped on the lower end of Poelau Lolong;

1) This basalt-ridge might be part of a flow of basalt. The whole territory being covered with modern fluviatile deposits no decision on this point could be obtained. In any case I consider it certain that this basalt is connected with the volcanic area on the Mandai, which also agrees with the trend of the basaltic outcrops.

on both banks of the river just above this island there are exposed quartzose greywacke sandstone (partly fine conglomerate), and white quartzose sandstone with an È.—W. strike and varying dip. A little further on cherty quartzite comes in and a kind of hornfels containing actinolite, probably contact-metamorphic rocks, formed under the influence of the granitite, which is found in both sides of the river above Poelau Tĕngkidoe, or of picrite, which forms the cliffs on the right shore opposite this island. The picrite, for the greater part, is altered into serpentine.

Upstream, on the other side of this small granite boss, the banks are somewhat higher and in many places the solid rock is exposed. Here for the first time it becomes possible to indicate approximately, from the section along the river-banks, the succession of the strata, as well as their relations, and in the following sentences I shall endeavour to give, as shortly as possible, all the necessary information on this subject, so as to avoid constant repetitions. The country at this point and higher upstream consists of a system of highly folded strata. The strike may be averaged at about E.—W. and E. N. E., but both strike and dip vary considerably. The strata as a rule are highly inclined and not seldom stand vertical. The oldest group in this complex is composed of clay-slate, chert, hornstone, sandstone, diabase, diabase-tuff, and diabase-tuff-breccia, also gabbro and serpentine, the latter being derived partly from a variety of olivine-norite containing little feldspar, partly from picrite or from hartzburgite. The diabase-tuffs and tuff-breccias, and part also of the diabases, form alternate layers with the sediments mentioned, and are therefore contemporaneous with these. Some of the diabases, and, as far as my observations go, the gabbro and serpentine also, form intrusive masses, sheets, or dykes in the sedimentary formations and are therefore younger than these. They have occasioned rather intensive contact-metamorphic changes in the sedimentary rocks. In contact with the intrusive diabase the clay-slate in more than one place has been converted into a kind of hornfels.

The tuffs and tuff-breccias are often much silicified and then they locally resemble hornstones. The diabases are generally uralitized, and through the influence of pressure have become slaty, which in some parts gives them the appearance of amphibolites. The varieties of diabase-rocks, which are more or less altered by dynamo-metamorphism and silification, which also play a part in the lake-district, I have included in the Poelau Mëlaioe-type. The cherts are sometimes full of biotite and then resemble silicified micaceous clay-slate, at other times they turn to pure jasper and hornstone. The hornstone is sometimes of a milky white colour, often marbled and alternating with ·bright red jasper. The chert, and particularly the jasper and pure hornstone, contain Radiolaria and are often almost entirely composed of the tests of these organisms. In the jasper, the Radiolaria can be detected with an ordinary pocket-lens, and they look like so many round specks, or little grease-spots the size of a pin's head. This greasy lustre is caused by the tests of the Radiolaria being filled with an aggregate of quartz which is coarser than the crypto-crystalline composition of the jasper itself. As mentioned before (page 92) these cherts with Radiolaria are of pre-Cretaceous age. I observed in several places that strata of diabase-tuff and tuff-breccia, containing pieces of chert, lie conformably between these cherts and hornstones.

These pre-Cretaceous strata are overlain by greywacke-sandstones containing biotite; this sandstone contains, besides quartz and plagioclase, numerous fragments of chert. In some places, viz., in the higher strata, the fragments of chert decrease in quantity, and as the proportion of feldspar increases the rock turns to pure arkose (*granite régénéré*). This sandstone again is overlain by quartzose sandstone, in which at one spot near Goeroeng Dëlapan, shells of *Orbitolina concava*, ·Lam. are found, which proves its Cretaceous age. On the maps I have combined the clay-slate, sandstone, chert and hornstone, together with the intercalated beds of diabase, diabase-tuff and tuff-breccia, in the pre-Cretaceous Danau-formation, and I group together the beds of greywacke, sandstone and arkose, with the overlying quartzose sandstone containing *Orbitolina*, as belonging probably to the Cretaceous formation.

The succession of beds as given here, I deduced from the survey along the banks of the Kapoewas and the Boengan, and I only suggest this as being probably correct, for circumstances did not favour me sufficiently to obtain positive proof. To begin with, rock *in situ* is only very rarely exposed along the banks, and each exposure is separated from the next by strips of thickly wooded soil which entirely hides the rock beneath. Then, in the second place, the Kapoewas generally runs in a direction parallel to or only slightly deviating from the strike of the strata, which is unfavorable to the study of their succession. Moreover the entire complex of strata is highly folded, and occasionally fold-faulted, the pressure during the mountain-making process having been so strong as to render both strike and dip inconstant, and to alter the rocks more or less by dynamo-metamorphism. And, lastly, numerous faults, some of which are synchronous with the origin of the folds, and some of more recent date, help to complicate the structural intricacies of this territory.

At 1 p. m. we reached the island Lisoe clothed with lofty trees, which the Kajans look upon as "pantang", and where at their request I did not examine the rocks nor collect any specimens. On the upper side this island terminates in a sheer wall of rock probably of the same serpentinized Hartzburgite which constitutes the solid rock on the right bank.

. Above Poelau Lisoe we first passed a small syncline of chert, and after that, as far as Poelau Balang, the prevailing rocks are diabase, amygdaloidal diabase, diabase-tuff and tuff-breccia, in several places completely uralitized. The type, already known to us from the lake-district, to which I gave the name of Poelau Mělaioe is particularly well represented here. All these rocks have become slaty by pressure, and in consequence have here acquired a great resemblance to amphibolite. In one place lherzolite is found, which by the combined effect of alteration and pressure, has been converted into a kind of talc-schist. From the group of rocks,

Pl. XXII.

Sᵐ Tᴋᴇᴋʟᴀɴ. Sʏᴍᴍᴇᴛʀɪᴄᴀʟ ʀɪᴠᴇʀʙᴀɴᴋꜱ.

consisting of amygdaloidal diabase-breccia, which rise in the middle of the stream, 1 kilometre below Poelau[1]) Balang, there is a splendid view upstream with the picturesque Mt. Noet[2]) in the background.

Fig. 41. THE KAPOEWAS BELOW POELAU BALANG. MOUNT NOET IN THE BACKGROUND.

Fig. 41 gives this view as seen in the evening light shortly before sunset. We stayed that night on a boulder-bank just below Poelau Balang having traversed during the day over 14 kilometres.

Poelau Balang is a beautifully wooded rocky island, composed

1) Poelau (Mal.) = island.
2) This name was given me by the Kajans, but it is not found on the topographical map; perhaps the Mt. Batoe Balang (474 metres) mentioned there is the same mountain.

of greatly altered diabase-tuff. The river-banks above the island
are composed of hornstone with Radiolaria, the well defined
strata being much folded and fractured. Up to this point small
detached rock-groups alternate along the banks with low sandy or
boggy strips of land, where the solid rock is not visible. More
continuous rock-belts begin to appear about $^3/_4$ kilometre above
Poelau Balang where the river is hemmed in by high cliffs of
greywacke sandstone, which always contains biotite, plagioclase,
and fragments of chert. Where the pieces of chert become fewer,
and feldspar and biotite are the more important constituents, the
rock becomes a pure arkose (*granite régénéré*), and can not
then, macroscopically, be easily distinguished from tonalite. The
picture given in Pl. XXVIII of this part of the river, skirted on
both sides by the apparently unstratified blocks of arkose, reminds
one of the well-known and characteristic features of a granite
landscape. At high water strong whirlpools are formed behind the
large masses of rock, and as a consequence of this the blocks of
greywacke sandstone, especially on the side facing downstream, are
deeply eroded and full of pot-holes. Over a distance of 6 kilometres, until close to the island Soewa, the rock on either side
continues to be greywacke sandstone and arkose; the scattered
pieces in the river create rapids in several places, which, however,
are not dangerous.

Above the island Soewa the river bends to the north and becomes considerably narrower. Just at that point the greywacke
sandstone again gives place to the underlying chert beds. And
a little further up, where the Kĕryau joins the Kapoewas, the
river-bed is deeply cut in strata of bright red jasper and hornstone. I determined to go up the Kĕryau for some little distance
more especially to ascertain whether the numerous boulders of
volcanic origin, which are found in the karangans[1]) of the Kapoewas
below N^a Kĕryau, are carried down by that river. We bivouacked
on a large boulder-bank on the right shore of the Kapoewas below

------

1) Karangan = boulder-bank, compare Chapter XII.

N° Kĕryau, 86 metres above sea level. The distance from Poelau Balang, 8 kilometres, had been traversed in 3¹/₂ hours.

It was a delightful morning when, accompanied by Dr. NIEU- June 19. WENHUIS, I set out at 7 a. m. with two small canoes to paddle up the Kĕryau river. The steep banks consist of beautiful folded beds of chert, jasper and hornstone. The river is lined on either side by high hills averaging from 400 to 500 metres, and their steep slopes are clothed with an exceptionally beautiful and thick growth of virgin forest. Here, more frequently than anywhere else in Borneo, I saw gigantic tapang trees, their crowns towering above all the others.

In its habit of growth, the tapang tree reminds one strongly of the oak (comp. fig. 54) but the trunk, seen from a distance, is dazzling white. Above Mata Koewe river the Kĕryau bends to the south and a little further up is the large rock- and boulder-island, Poelau Pĕpanai, opposite the mouth of the rivulet of that name. About 2 kilometres above Poelau Pĕpanai, just below the streamlet Goeroeng Goeroeng, jasper and hornstone again give place to diabase and diabase-tuff (Poelau Mĕlaioe-rock), and immediately beyond all kinds of obstacles arise in the hitherto unobstructed stream, caused partly by rocks in the river, partly by cliffs jutting out from the sides and extending far into the water. Just above N° Djĕmihin a mass of fine-grained quartzitic sandstone stands out in the middle of the river, the beds of which are vertical and strike E.—W. From this rock the photograph reproduced on Plate XXIX is taken, facing upstream. In the middle of the river stands a rock balancing a huge tree-trunk, fully 15 metres long, which the last flood has deposited there. This gives one some idea of the enormous differences in the height of the water in these rivers; as a matter of fact the differences are still greater than this picture leads us to suppose, because at very high water the entire rock seen on the plate is submerged to a depth of several metres. The picture moreover gives a truthful representation of the grandeur of the vegetation in the valley of the Kĕryau.

Just above Poelau Sĕpan there is another bend in the river,

and beyond that a clear stretch upstream of $1^1/_2$ kilometres, then
a kind of table-mountain rises suddenly to view which seems to
shut up the 'rantau'. This mountain is of exactly the same type
as Mt. Tiloeng and the other table-mountains of the Upper
Mandai. As we advanced, other mountains, similar in shape, gradually came in to view, forming a consecutive range of hills trending
almost due E.—W. and all about equal in height. The most prominent are Mt. Todjo (746 metres) and Mt. Ambe (681 metres).
I had already observed these same mountains from the top of
Mt. Lyang-Agang (see page 56), and viewed from that point
their relation to the other volcanic tuff-mountains of the Mandai
district is unmistakable. I have, therefore, not the slightest hesitation in stating that I believe these table-mountains on the
Kéryau river are composed of volcanic tuff and resemble the
table-mountains of the Upper Mandai. The Todjo and the Ambe
are, as a matter of fact, the most north-westerly advanced posts
of the extensive volcanic territory, which comprises a great part
of, if not the entire river-basin of the Upper Kéryau.

At 2 p. m. we reached Poelau Tjombon, a small island built
up partly of solid rock of the Poelau Mélaioe-type, and partly of
boulders. It lies about 12 kilometres above the mouth of the
river and from there we returned to N⁴ Kéryau in a heavy
thunder-storm and torrents of rain, and reached our bivouac at 5 p. m.

A comparative examination of the boulders on Poelau Pěpanai
and Poelau Tjombon in the Kěryau river has led to the following
results. The gravel may be divided into two classes; the one
whose constituents may originally have formed part of the continuation of the system of folded strata through which, in its lower course,
the Kěryau cuts its bed, and the other, whose constituents must
be traced back to volcanic mountains probably like those of the
Upper Mandai, built up of volcanic tuffs. To the first class belong:
chert, hornstone, quartzite, diabase, quartz, proterobase, amygdaloidal diabase, diabase-tuff and tuff-breccia, diabase-porphyry, labra-
dor-porphyry, amygdaloidal melaphyr, hartzburgite, serpentine,
quartz-porphyry, quartz-biotite-porphyry, and quartz-diorite-por-

phyry; to the second class belong: mica-andesite, dacite, mica-dacite, enstatite-andesite, amongst the latter a very beautiful glassy variety, which I had already noticed in all the boulder-banks from Poetoes Sibau to N° Kĕryau, amphibole-andesite and basalt. One cannot however, strictly speaking, classify with either of these two groups, the many loose fragments of quartz-propylite, diorite-porphyry and quartz-diorite-porphyry of andesitic habitus, or a certain polygenous conglomerate containing many small rounded pieces of jasper with Radiolaria, or, finally, crumbly white conglomerates mainly composed of much water-worn quartz-pebbles.

If we compare these materials with those of the boulders from the karangans in the Mandai, the river-basin of which adjoins that of the Kĕryau, we at once detect some differences which enable us to draw certain conclusions. In the first place there is a decided predominance of volcanic material in the Mandai river. In order to explain this phenomenon one might suggest:

1, that the volcanic territory in the river-basin of the Kĕryau is less developed than in that of the Mandai (either vertically, horizontally, or in both directions);

2, that the greatest development of the volcanic territory in the Kĕryau lies in the higher parts of the river-basin, or rather at the head-waters of this river, and that only a small proportion of the boulders from there have been able to reach the Lower Kĕryau without being altogether triturated and ground to powder;

3, that the older formation upon which the volcanic beds rest, rises to a higher level in the river-basin of the Kĕryau than in that of the Mandai, so that whereas the latter river [1]) has, only occasionally, cuts right through the volcanic beds, the Kĕryau in many places makes its bed in the underlying older formation. Possibly these three factors all contribute their share to bring about the different character of the gravel carried down by these two rivers, nevertheless I feel disposed to attribute the largest share

---

1) Of course 'river' stands in this case for the whole system, viz. the principal stream with all its tributaries.

to the last-named influence. It is moreover remarkable that silicified wood, one of the commonest and most conspicuous features amongst the boulders of the Mandai, seems to be altogether absent in those of the Kĕryau. At Poelau Tjombon I only found one piece of wood slightly silicified and entirely infiltrated with iron-ore. This, together with the absence of the many varieties of porous andesite, hypersthene-andesite, basalt, especially the varieties known as "wacke", found in the Mandai, gives us the right to suppose that in the river-basin of the Kĕryau, except perhaps high up near its source, tuffs, analogous for instance to those of Mt. Lyang-Agang, do not play any conspicuous part. I should therefore not be greatly surprised if subsequent discoveries were to reveal that Mt. Todjo and Mt. Ambe, (fully convinced as I am that they are volcanic tuff-mountains), differ nevertheless very decidedly as to their constituents from the tuff-mountains of the Mandai.

Granite and quartz-porphyry which play a comparatively important part in the boulder-banks of the Mandai, are altogether absent in the Kĕryau, from which fact we may conclude that probably they do not occur in the river-basin of the Kĕryau. Again in the Kĕryau we miss both the coal and the modern sandstone which in the Mandai river comprises the coal beds; we must however be careful not to jump to the conclusion that because boulders of these rocks are not found in the Kĕryau, they are therefore altogether wanting in its river-basin, for we must remember that both sandstone and coal are easily triturated and reduced to sand when carried by running water.

Certain types of these rocks are peculiar to the river-basin of the Kĕryau only, and are found in the boulder-banks of the Kapoewas exclusively below N" Kĕryau. Thus, for instance, it is in the Kĕryau territory, where we have to look for the origin of a particularly beautiful variety of Harztburgite [1]), not unfrequently met with amongst the boulders in the Kĕryau, as well as for some varieties of enstatite-andesite and basalt. The last-

[1) Schillerfels (German) or glinsterklip (Dutch).

mentioned rocks are never lacking in any of the boulder-banks of the Kĕryau and the Kapoewas below Nª Kĕryau, but not a single piece has been found higher upstream in the Kapoewas.

The Kĕryau is, next to the Boengan river, the principal affluent of the Upper Kapoewas. It has its source in the boundary mountains between West and South Borneo, and from there and across this mountain chain one can reach the source of the Batang Moeroeng. Its entire length is 78 kilometres, and its width at the point where it empties into the Kapoewas, is 42 metres. It is navigable (although in some parts the passage is fraught with great difficulties) for boengs up to a distance of 37.5 kilometres from the mouth, and from that point a Dyak footpath leads through the mountains to the Batang Moeroeng. A sedentary population is altogether lacking in the river-basin of the Kĕryau, but according to rumour there are a good many nomadic Poenans in the forests of the mountain-land near the head-waters of this river. Of all the different excursions which might be projected through Central Borneo, none certainly could promise more interesting geological results, than a trip from the Upper Kapoewas through the valley of the Kĕryau, and then across the mountains to the Batang Moeroeng, and along this river to Moeara Teweh. This route leads almost entirely through a territory geologically totally unexplored, and the general build of the adjacent districts gives strong support to the suggestion that all the formations which help to build up Central Borneo must be traversed in following this route. It would not be an easily accomplished journey however. In the first place it would be very difficult to procure boatmen and carriers for such an expedition, then the victualling question would have to be planned most carefully, and lastly one would have to be constantly on the alert against the Poenans.

We left our bivouac at 6.30 next morning and paddled and June 20. poled as far as Nª Mĕdjoewai where we arrived at 12.30 p. m. Above Nª Kĕryau the boating up the Kapoewas is rather monotonous, for the banks are generally low, and the higher ground lies at some distance from the river. We passed four islands surrounded by boulder-banks, viz., Hakat, Běléboenoeng, Ingai and Bétoeng. With the exception of a few small rapids, in the vicinity of these islands the route presents no difficulties. The predominating rock from Nª Kéryau to just above Nª Těpai is jasper and hornstone with Radiolaria, followed by amygdaloidal diabase, tuff- and tuff-breccia; close to the island Hakat, clay-slate, arenaceous clay-slate and sandstone make their appearance. The rocks are again strongly folded, and the strike oscillates between E. 20° N. and E. 40° N. On account of the strong plication neither strike nor dip remains constant, for even short distances. But as the general strike is about parallel to the course of the stream, it is not surprising that until Nª Medjoewai is reached, the three previously mentioned rock-types, the chert, the diabase-tuff, and the slate-sandstone-type, constantly alternate on the banks. For details I again refer to the map VIIᵇ. Just above Poelau Běléboenoeng there had evidently been quite recently, a rather serious landslip. Close to the left bank stands a steep hill with its beds of chert dipping towards the river at an angle of 75°. A portion of this hill has slipped along the plane of bedding, thereby exposing a bare, smooth, rock-surface, while on the bank of the river a heap of débris has accumulated. A careful study of my maps will show that in several other places landslips have been marked; as a matter of fact they take a very prominent part in the levelling process of the Upper Kapoewas mountains, more particularly in the chert district. Taking into consideration that in this formation the strata as a rule are highly inclined, that the chert-beds are often separated by very thin intercalated clayey layers, that the rock, moreover, cracked in several directions by the folding process, allows a rather easy percolation of water and lastly that by the constant and regular heavy falls of rain all the cracks and fissures must of necessity be generally filled with water, — it is not surprising that slides along the plane of bedding and hence landslips should be very common in the chert areas.

A little below Poelau Ingai one can catch to the north a very

Sᵀᵉ TÉKÉLAN. ASSYMETRICAL RIVERBANKS.

picturesque glimpse of the high mountain-range in which the Mĕdjoewai has its origin (fig. 42). Below the mouth of the

Fig. 42. THE MOUNTAINS ALONG THE UPPER MĔDJOEWAI.

Mĕdjoewai, on the right bank of the Kapoewas, there is an enormous boulder-bank which at low water fills up three-fourths of the width of the river and is 260 metres long.

When on the Kĕryau I had noticed that only a limited number of the numerous volcanic rocks found on the boulder-islands of the Kapoewas have been carried down exclusively by the first named river. Many other very characteristic types, such as a brecciated rhyolite, rhyolite-pitchstone, etc., I looked for in vain in the Kĕryau although they appear frequently in the Kapoewas. I now intended to visit all the important tributaries of the Kapoewas, to enable me to localise the areas of the occurrence of the most characteristic boulders, in the first place of those of the volcanic rocks. With this purpose in view I paddled up the Mĕdjoewai for some distance. In the lower course this river has cut its bed in highly folded strata of chert and jasper with Radiolaria. The whole of this narrow and deep ravine presents at low water a succession of exposures affording most beautiful and instructive illustrations of folded, twisted, contorted, cracked, sheared and overthrust strata, the results of mountain-making force. Particularly high and large boulder-banks are found at the bends of the river, and the size and quantity of these boulders prove that the Mĕdjoewai in proportion to its width, possesses very considerable transporting power which must be attributed to the steep fall of its bed. A careful examination of the karangans proved that volcanic rocks are altogether absent in the basin of the Mědjoewai river. Amongst the boulders I noticed diorite, diabase, amongst which were very beautiful coarse-grained varieties, enstatite-diabase, uralite-diabase, vitrophyrite, gabbro, amphibole-granite and amphibole-biotite-granite, serpentine, chert, amygdaloidal diabase, diabase-tuff, quartzitic sandstone and clay-slate.

The boulder-bank near N˙ Mědjoewai is almost exclusively composed of rock fragments carried down by the Mědjoewai, thus, for instance, granite blocks are very plentiful therein, whereas they become very scarce as soon as one goes up the Kapoewas, and above N˙ Lapoeng, I did not come across a single piece of it; if granite, therefore, is found at all higher up in the karangans of the Kapoewas, it certainly must rank amongst the very rare occurrences. I did indeed find a tonalite much higher, which had been carried down by the Boelit river, but this was a different kind of rock, being so completely altered and rendered slaty by dynamo-metamorphism that macroscopically it could not be distinguished from tonalite-gneiss or feldspar-amphibolite.

At 3 p. m. we again left N˙ Mědjoewai. The predominating rock continues to be chert until just above N˙ Rangin, where quartzose sandstone and clay-slate occur, in some places alternating with chert. About one kilometre below the Matang river diabase and diabase-tuff gain the upper-hand, and the banks are principally composed of them till close upon N˙ Měnsikai, where we made our camp about 5.30 p. m., partly on the right shore of the Měnsikai, 87 metres above sea level, partly on a rocky island at the mouth of the river. In spite of a delay of over $3^1/_2$ hours we travelled that day 14 kilometres.

June 21. The Kapoewas had risen a good deal, which was decidedly unfavourable to geological observation. It rained hard that night and continued to drizzle till 9 a. m. Just above N˙ Měnsikai several rocks, composed of uralitized diabase and diabase-tuff appear above the water; a little further on a ridge of rocks built up of cracked and folded diabase-porphyrite with cataclase-structure extends across the river. Higher up again and as far as N° Mēnsikai, clay-slate and arenaceous slate predominate, the strata generally being vertical with an average strike of E. 30° N.

Above N° Mēnsikai, besides clay-slate, quartzose sandstone and diabase also occur and in one place amphibole-porphyry was found. At N° Goeng, clay-slate, chert, jasper, and hornstone again prevail. Some little distance from this place we were agreeably surprised by the appearance of a group of Boekats on one of the

Fig. 43. FOLDED STRATA OF CHERT AND JASPER NEAR l'OELAU BOEKAT.

projecting rocks on the left shore. Of all the Dyaks with whom I have come in contact in Borneo I consider these Boekats to be the finest specimens in build and physical strength. The well developed muscularity of their thighs is particularly noticeable in contrast with the slender build of the other Dyaks. There are just a few of the Batang Loepars of the Batang Rēdjang in Sarawak whom I encountered in the Embaloeh valley, who can compare with them in this respect. Their mode of life accounts for this physical feature. As their food consists of fruit and roots from the forests, and the flesh of all kinds of animals which they kill with the poisoned arrows shot from their sumpitans, they are always forced to walk over considerable distances, and to follow their game through the forests. They traverse the most difficult ground with unparallelled rapidity and this fact no doubt accounts for the splendid development of their lower extremities. The Boekats now before us were pretty nearly as white as Europeans, probably to a great extent, the effect of their continuous sojourn in the thickly shaded woods, where it is quite the exception to be exposed to the glare of the sun. The men

are handsomely tattooed; a very complicated design ornaments
their chests, and shoulders, and is continued behind in traceries
right across their shoulder-blades [1]). The skin of the lower jaw is
also similarly adorned, which at a distance looks as if they had been
lathered with coloured soap ready for shaving. I draw attention
to this jaw decoration because, so far as I am aware, the Bĕkatans and the Boekats are the only tribes who indulge in it. There
is no trace of it among the Poenans of the Upper Mandai. One
of the Kajan Chiefs, LOEDANG, who escorted us, was also adorned
in this fashion; but then he was a Bĕkatan by origin, and
through intermarriage had become chief of one of the Kajan-kam-
pongs. These particular Boekats were pleasant and well-disposed,
and they were delighted with some tobacco which we gave them
as a parting gift.

Paddling or poling, according to the state of the stream, we
reached at 3 p. m. the Lapoeng, and pitched our tents about a
kilometre further up near to a large boulder-bank on the right shore
of the Kapoewas. Heavy rain fell during the evening and throughout that night. We had accomplished that day over 17 kilometres.

June 22. During the night the water rose 1.20 metres, and the Kajan
chiefs declared that with the river in this state it would be impossible to pass the renowned Goeroeng Dĕlapan, which begins at
a short distance from there. We therefore remained that day in
our bivouac and I made use of this opportunity to explore
a little way up the Lapoeng, which empties into the Kapoewas
a little lower down. The bed of this river is deeply cut in sandstone and sandy clay, the strata being in vertical position with
the strike N.E.—S.W. To ascend this streamlet is very awkward
indeed, and I had only got one kilometre beyond the mouth
when my progress was stopped by a waterfall $1^1/_2$ metres high.
I left my sampan and proceeded on foot, following the stream

---

1) The women, whom however I only saw at a distance, did not appear to me to be
tattooed. I took photographs of this group of Boekats and also of some beautifully tattooed
men, who, next day, visited us in our bivouac, but the negatives, like so many others,
were totally spoiled by the extreme dampness of the climate.

for some distance, and discovered that a ridge of thick and very hard beds of greywacke sandstone in this system of vertical strata forms a kind of bar and produces the cataract. Above the cataract the water runs smoothly for some distance. In the cataract the strata strike E. 20° N. The greywacke sandstone contains pieces of chert and a good deal of plagioclase; the considerable proportion of feldspar gives it a great resemblance to arkose. I found no volcanic rocks in the Lapoeng river, but in this case I am not prepared to state definitely that they are not found in the basin of the Lapoeng, because the greater number of the boulder-banks were submerged which, otherwise, I might have examined. It rained almost incessantly that day.

Next day the water had slightly fallen, and at 7 a. m. we set June 23. out with a difficult piece of work before us, for the series of falls which we reached in about an hour's time rank amongst the greatest obstacles to communication by water in the Upper Kapoewas district. When the water is high the falls are simply impassable, and even at low water one runs the risk of losing the sampans. They told me that during the topographical survey the whole fleet of one of the surveyors came to grief here, the boats sank or else were dashed to pieces, and the surveyor had to go back and descend the river on a raft of tree-trunks. The river-bed is full of sharp-edged rocks, between which the water forces its way with a thundering noise. Fantastic groups of rock jutting out from the steep banks, occasionally obstruct the passage and create dangerous whirlpools. The name Goeroeng Dělapan [1]) would indicate the existence of eight falls, but in reality it is one immense rapid, nearly 3 kilometres in length, which, to judge from the most dangerous parts, one might just as well divide into 12 or 15 falls. When the falls are reached it is imperative to empty the boats entirely and to drag them along by rattan ropes. The natives, however, and particularly the Malays, have a habit of delaying the unloading as long as possible. They did so in this

---

1) Dělapan (Malay) = eight.

instance; and while I was looking after the careful removal of my instruments our servants actually attempted to cross the lowest fall with the cook's sampan fully laden. We had to reap the reward of their foolhardiness. The sampan filled with water, capsized, and almost all our kitchen utensils, my shoes, most of our lamps, the greater part of our stock of waterproof canvas, and many other almost indispensable articles of our outfit, disappeared in the depths below. Although somewhat late, this catastrophe effectually roused all hands to a sense of their duty, so that all went well afterwards, and by 1 p. m. we were once more settled in our loaded sampans, the falls behind us, and making for N⁴ Boenjan. In cases of this sort it is always advisable to look after a few of the most important things personally and to leave all the rest to the natives. When once roused from their lethargy they are most careful and exact, besides which the roar of the water makes all commands inaudible, and a continual interference on the part of Europeans, irritates the natives, and brings them in to a state of temper called by the Javanese "bingun", something between perplexed and obstinate.

From N⁴ Lapoeng up to the Goeroeng Dëlapan, the banks of the Kapoewas are composed of alternating strata of clay-slate, arenaceous clay-slate and quartzitic sandstone, probably the continuation of the strata in the bed of the Lapoeng river. Near the Goeroeng Dëlapan thick beds of silicified diabase-tuff and hornstone with serpentine are intercalated between the rocks mentioned above, and these more resistant beds are in reality the obstructions which cause the falls. The strike of the strata averages N. 10° E., being almost at a right angle to the direction of the stream. The harder beds have thus formed dams across the stream, while, by lateral erosion of the interstratified softer sandstone, the river has widened itself somewhat in the form of a whirlpool. The most dangerous spots are those marked on the map in Roman figures. The fall at IV bears the name of Riam Péring on the topographical map. The safest passage in these rapids is near the right bank. Above Riam Péring arenaceous clay-slate is found, and a little further up

on the left shore, 1¹/₂ kilometres above Riam Pĕring, greywacke
sandstone occurs, similar to that in the Lapoeng, and in it I
discovered tests of *Orbitolina*. This discovery was of great importance, because it proves with certainty the occurrence of strata
of Cretaceous age in the Upper Kapoewas district.

The remainder of the voyage was uneventful, and at 2 p. m.
we arrived at the confluence of the Kapoewas and the Boengan,
9¹/₂ kilometres above Nᵃ Lapoeng.

At the point where the Kapoewas and the Boengan meet the June. 24
rivers are about equal in width. The coolies sent on in advance
from Poetoes Sibau with a cargo of rice, had made our bivouac on
the narrow strip of land between the two rivers, this being 9 metres
high and 107 metres above sea-level. And a splendid place for
a bivouac it was, as it commanded a grand view of the Kapoewas
downstream with Mt. Tanah Koeban (958 metres) in the background. In the morning I went a little way further up the Kapoewas in a small boeng. As far as Nᵃ Tandjan the rock on
both sides is greywacke, greywacke-sandstone rich in kaolin,
and quartzitic sandstone. Further up, bright red jasper and chert
appear, the strata again being highly folded and the strike varying between E.—W. and E. 20° N.—W. 20° S. This chert continues to be the country rock up, to the highest point reached by
me. Before long the passage became more difficult, and not far
from the goeroeng Moenhoet, which is supposed to be the most
dangerous series of rapids in the Upper Kapoewas, I was compelled to turn back, as the crew of one boeng was not strong
enough to overcome the difficulties which lay before us. Above
Nᵃ Boengan the fall of the Kapoewas rapidly increases, and the
average direction of the stream slants with respect to the trend of
the hills and to the strike of the strata of which they are formed.
The difficulties here become of a more serious nature. A series of
rapids and falls occurs between Nᵃ Boengan (107 metres above sea-
level) and the pangkalan Djĕmoeki (230 metres above sea-level)
where the Kapoewas becomes absolutely unnavigable. The entire
length of the Kapoewas between pangkalan Djĕmoeki and Nᵃ

Boengan is 76 kilometres (Compare Atlas, XX). The most dangerous falls are the Goeroeng Pantak, 2 metres high, the Goeroeng
Matahari, 4 metres high, and the Goeroeng Moenhoet.

From pangkalan Djĕmoeki a Dyak footpath leads, in a two
or three day's journey, across the boundaries of Sarawak to a
rivulet belonging to the river-basin of the Batang Rédjang; this
river itself can be reached in four days. The little that is known
about the head-waters of the Kapoewas above Nᵃ Boengan we
owe to the topographical survey; it is stated to be a high mountainous district, absolutely devoid of more or less permanent human
settlements. The prevailing trend of the mountain chains throughout the district is in an east to west direction. The topographical
map shows quite clearly that there the tributaries of the Kapoewas run from east to west, and consequently form longitudinal
valleys. The Kapoewas itself meanders greatly in its upper course,
and for long distances forms longitudinal valleys running from
east to west, which are connected by short transverse valleys
running north and south.

The development in length of the stream therefore is very
great. The source of the Kapoewas is only 19 kilometres from
pangkalan Djĕmoeki, as the crow flies, but the actual length of
the river between those two points is 49 kilometres. For the
entire course of the Kapoewas river the same proportion is 585
to 1143. On the return journey I paddled for some distance up
the Tandjan river, the bed of which is deeply cut in jasper
and chert rocks. The head-waters of the Tandjan are in
the highest part of the Upper Kapoewas mountain-land. The
average height of the ridges of the mountains surrounding the
valley of the Tandjan is at least 1200 metres, and several
peaks exceed 1500 metres. The highest point in the Upper
Kapoewas-range (with the exception of the mountains of the
Raja-territory, this being moreover the highest mountain-land in
West Borneo) is in the basin of the Tandjan river (1960
metres) on the boundary-line of West and East Borneo, therefore in the watershed of the Kapoewas and the Mahakkam.

Descending from this high mountain district the Tandjan rushes
with a considerable fall and in a comparatively short meandering
course down to the Kapoewas. As might be expected the water
of this mountain stream was deliciously cool, 22° Celsius, while the
temperature of the water of the Kapoewas, just above Nª Tandjan,
was 23.1° Celsius.

In the boulder-banks of the Kapoewas above Nª Boengan, and
also in those of the Tandjan, there are many varieties of chert,
white vein-quartz, sandstone, coarse polygenous sandstone, and
greywacke-breccia containing numerous fragments of clay-slate,
conglomerate, and moreover a small proportion of pebbles of
diabase, diabase-porphyry and amygdaloidal diabase.

A careful examination of these boulder-banks led to the conclusion that volcanic rocks are altogether absent here, and consequently that the enormous quantity of volcanic boulders and pebbles found in the Kapoewas below Nª Boengan, have exclusively
been carried by the Boengan [1]).

The Boengan is the principal tributary of the Upper Kapoewas. June 25.
It owes its origin to the confluence of a much ramified system
of tiny mountain streams arising in the volcanic Müller mountains. In the head-waters of the Boengan itself these mountains
form part of the watershed with the river-basin of the Upper
Mahakkam. Should this same mountain-land of the watershed be
traversed at the head-waters of its principal tributary, the Boelit,
one comes to the head-waters of the Kassoo, one of the tributaries of the Mahakkam, from whence, the natives told me, a
comparatively easy mountain-path leads to the river-basin of the
Batang Moeroeng (Barito).

The actual length of the Boengan is about 87 kilometres.
Its general course is nearly S.E.—N.W.; the current therefore
is either oblique or perpendicular as compared with the general
trend of the country, which condition is favorable for the creation
of rapids and cataracts.

---

1) Below Nª Kéryau the fragments of rocks of volcanic origin carried by the river of
that name of course help to swell the number.

At 7 a. m. our small fleet came up the Boengan, the weather being deliciously cool, 23.5° Celsius. About 400 metres above N° Boengan, a rock jutting out from the left shore extends a good way into the river; it consists of amphibolite containing glaucophane. Taking into consideration that the entire district bears traces of the effects of very intensive earth-pressure, it is quite possible that this amphibolite is a dynamo-metamorphically altered igneous rock. At any rate, I am not inclined to call it a true crystalline schist. Geologically it would be well-nigh impossible to explain the occurrence of a small patch of crystalline schists amidst a complex of Cretaceous and pre-Cretaceous sediments.

Somewhat higher up, jasper and hornstone with Radiolaria again appear, followed, $^1/_2$ kilometre further on, by uralitized quartz-diabase rendered slaty by pressure, which in its turn is succeeded by chert and hornstone (often with Radiolaria), sandstone and greywacke, which often alternate in the river-banks. There is one variety of hornstone found here, milk-white in appearance which is particularly rich in tests of Radiolaria; in this rock the proportion of silica is as much as 97 %. An analysis made by Mr HONDIUS BOLDINGH has given the composition of this hornstone (II, 153) as follows:

SiO_2	97.19
Fe_2O_3	1.12
Al_2O_3	1.45
CaO	0.05
MgO	Traces
K_2O } Na_2O	1.15
H_2O, constitutional,	0.28
H_2O, hygroscopical,	0.06
	101.30%

The whole of this complex of strata is highly folded, the strike varying between N. 20° E. and E. 20° N., and the dip being always very steep. Near Riam-Bi there is on the right shore a

beautiful and well-exposed rock section, showing a dyke of
diabase cutting through beds of jasper.

Until close to Riam-Bi the river-bed, although well strewn
with rocks, is not dangerous to navigation, but from Riam-Bi,
where silicified diabase-tuff and diabase again make their appearance, the condition of things changes and strongly projecting
rocks are the cause of the very perilous narrowing of the river
at Goeroeng Matap. Although the rapids here are not of great
length and the precautions are not as strictly adhered to
as at the renowned Goeroeng Bakang they are nevertheless rather
dangerous, especially going downstream. The scarped rocks
which form the bar allow the water an exit only in one place
wide enough for a boeng to pass through, but it is imperative
when running downstream to give the boeng a dexterous turn
which is not always successful, and many a boat capsizes or is
smashed to pieces here against the rocks.

As we rowed up the Kapoewas and the Boengan I was more
and more impressed by the unrivalled beauty of the forests.
The peculiarity which gives them an altogether different stamp
to the forests lower down the stream, is their wealth of climbing
rattans found here in abundant variety, their waving plumes
towering above the crests of the highest trees. In the low lying,
more populated, and entirely safe districts the best kinds of rattan,
have long since, for the greater part, been exterminated, but here
one gets some idea of the wonderful value of the virgin forests
of Borneo, and one realizes whence the heavily laden bandongs which come down the Kapoewas, derive their precious
freight.

It is lamentable to see that the natives gather the rattan
in such a rough and reckless way, that it has practically the
result of destroying it. The glorious and valuable rattan sëgah,
once a common feature of the forests almost everywhere in
the hilly and mountainous districts of West-Borneo, is now
almost extinct. Only the Oeloe Kapoewas, because of its reputed
unsafety, has not yet been deprived of this treasure. In fact

so common a product is it on the Boengan, that the bale-bale[1])
we slept on at night were made entirely of this precious material.
Let us hope that ere long a stop may be put to the reckless
way, in which the rattan, and still more the gëtah, are destroyed
by the natives, and that a rational system of forest exploration,
on the same plan as that in Java, may be adopted in Borneo,
and supported by the government. Moreover a wise interference
on the part of the government might have a most beneficial
effect upon the prosperity of the Dyaks.

At 4 p. m. we reached Nᵃ Poenoe, 11¹/₂ kilometres above
Nᵃ Boengan. Here at the confluence of the Boengan and the
Poenoe, the river has a considerable width. The low strip of land
between the two rivers ends in a large boulder-bank, which is
only dry at very low water. I poled a little way up the Poenoe;
its banks are composed of fissile clay-slate and quartzite, the
strata being vertical. Although the large boulder-bank near Nᵃ
Poenoe, (the pebbles of which are carried both by the Boengan
and the Poenoe) is rich in volcanic rocks, and contains numerous
fragments of a banded variety of rhyolite and of rhyolite-breccia,
one looks in vain for any appearance of volcanic rocks in the
boulder-banks of the Poenoe river even at a short distance from
its mouth. The character of the boulders is similar to that which
we observed in the Tandjan: chert, sandstone, quartz-conglomerate and breccia pre-dominate; I found also a few fragments of
diabase. The Poenoe is a swift stream with plenty of water; its
bed is known to have a considerable fall and awkward rapids
which reach even close to its mouth.

It rained heavily that night; the natives quartered on the
bank were in the middle of the night forced to seek the boats
or the higher shore. By morning the karangan was almost entirely
submerged.

June 26.　At 7.30 the signal for departure was given. The river-bed

1) Bale-bale (Mal.) = stretcher.

which is very deep just above N⁸ Poenoe, cuts through rocks almost entirely covered with vegetation, which consist at first of chert, later on of altered diabase and silicified diabase-tuff. At 8 a. m. we reached the Goeroeng Bakang; here in one of the later months of the year 1825 a drama was enacted which has gained a morbid renown in the history of the exploration of Borneo. When GEORG MÜLLER, the intrepid explorer, had reached this point on his journey from Koetei along the Mahakkam from its head-waters and past the watershed, he and all his companions, with the exception of three Javanese, were most cruelly murdered by the Pénihin-Dyaks, when they attempted to cross the Goeroeng Bakang [1]).

A better place could scarcely have been fixed upon for the carrying out of their villanous scheme [2]). The wild foaming waters shoot down with tremendous fury, whirling in the deep holes which cut up the river-bed. These holes are caused by the rocks jutting out on all sides and narrowing the stream. The right shore is altogether inaccessible. The steep slippery rocks are covered with thick underwood close up to the precipitous edge. The left shore is rather flatter and deeply grooved, and numerous scarped cliffs protrude from the banks until nearly in midstream.

Across these rocks and for a distance of 300 metres the luggage has to be carried, and this is no trivial matter; the rocks being most awkward in shape and worn quite slippery with the action of the water.

The roaring of the water drowns the orders shouted at the pitch of one's voice; a nameless dread seizes the European, as he crawls or slides over and under the slippery rocks or tries

1) The natives including the Kajans assured me that GEORG MULLER had certainly been murdered here. Strictly speaking, the fact is not positively ascertained, but the unanimous testimony of the natives, together with what was already known about the matter, are sufficient to prove the fact that the murder was perpetrated near the Goeroeng Bakang.
Much has been written about it. See for instance J. HAGEMAN. 15, p. 487, and P. J. VETH. 64, II, Chapter VII. These give a complete record of the event.
2) Plate XXX gives a picture of the cataract (sketched from a photograph) looking upstream. The morning mists hang over the water.

to get a foothold on the masses of drifted wood, hemmed in and piled up between the rocks. Small blame to him should the thought cross his mind that, however well armed, his life would not be worth much, should the guides have any treacherous design against him.

Fortunately we had nothing of that sort to fear, our faithful Kajans were a very safe protection. The water, however, caused us far greater anxiety, for even the empty boengs were tossed about so mercilessly by the angry waves, that every moment I expected to see them go to the bottom. And this actually did happen to my own boat; at one of the most perilous points it was swamped by the waves, the rattan rope snapped through the additional weight, and in a moment the frail vessel was dashed to pieces against the rocks, not a vestige did we ever see of it again. All the other boats fortunately escaped unhurt, and by arranging the luggage somewhat differently we managed to keep one boeng free for my own use, and so this small catastrophe did not cause much delay.

The origin of these perilous cataracts is explained in the diagram,

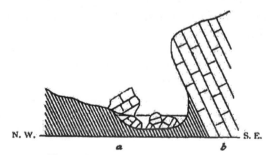

Fig. 44. Section across Goeroeng Bakang.

a. Shale.
b. Silicified diabase and diabase-tuff. Big blocks of this rock narrow the bed of the stream.

fig. 44. The strike of the strata is N. 20° E., the dip very steep towards E. S. E. The high banks on the right shore are of diabase. This diabase is crushed by earth-pressure, the numerous fissures are filled with quartz and serpentine and the whole rock moreover is more or less silicified. It is therefore a variety of the type which I have called the Poelau Mĕlaioe-rock. The left shore consists of fissile, sometimes slightly sandy shale, much decomposed and more or less silicified [1]).

The direction of the current is almost parallel with the trend of the rocks, and the river cuts its bed just along the line of junction between the diabase and the shale. The shale, yielding more easily to the eroding power of the water, is consequently more rapidly removed than the diabase, the latter has been undermined, and now forms on the right shore a high overhanging rock-wall. Large blocks of diabase occasionally fall down from this high bank into the stream below; the river-bed is covered with them, and these blocks form a constantly renewed impediment to the flow of the water.

At 12.30, having left the Goeroeng Bakang behind us, we were once more safely embarked and within half an hour we came to the Langau river. I now became aware that the sudden rush of water which had so startled us at N' Poenoe, must have come from the Langau, for even now this stream was full of yellowish, muddy water, in striking contrast to the dark hued crystal-clear water of the Boengan. Just above the entrance to the Langau a ridge of diabase-porphyry runs right across the Boengan and causes a rapid fall which it is not easy to cross. Our tents were pitched a little higher up on the right shore, $1^1/_2$ kilometres above N' Poenoe.

In the afternoon I paddled up the Langau which near its mouth has cut its bed in diabase and diabase-tuff, which is silicified and in consequence in some places resembles chert.

At a distance of 250 metres all progress by water is stopped by a fall, $3^1/_2$ metres high. A rock-bar runs here right across the river-bed, it consists of dark hornfels with serpentine, lying

---

1) This shale is possibly in part much decomposed and altered diabase-tuff.

at the contact between the diabase and a complex of alternating strata of clay-slate, with numerous hard concretions, and a kind of sandstone, passing into quartzite. The entire system of strata is vertical. The great hardness of the bed of hornfels, the contact-rock, has produced the waterfall. I went a little further on foot in a southern direction, through high, dense brushwood over untrodden, difficult ground until I again struck the Langau. The river was still very swollen, so that only a small portion of the boulder-banks was visible above the surface. Unfortunately only a limited number of boulders could therefore be examined, but not a single one of volcanic origin was discovered. And this exactly agreed with my observations in the Lapoeng; the Langau and the Lapoeng have their sources close to each other, and there can be scarcely any doubt that the head-waters of both lie to the north of the northern boundary line of the volcanic territory.

On the return journey we were attacked in the forest by a swarm of wasps which make their nests in the tree-trunks. Their sting is very painful, but in my case it did not produce fever as the natives predicted it would. One gets used to the stings of wasps in the mountain districts of Borneo. If one happens to linger on a boulder-bank in the daytime, which I was often obliged to do, face and hands are soon covered with insects preying on the sweat. One of the commonest is a small black bee, which however never stings and appears to be unarmed. Then there is a kind of wasp, slightly large in size, which stings badly as soon as it is at all interfered with. This latter insect I noticed only in the very mountainous districts, as for instance on the Boengan never lower than N⁴ Poenoe.

June 27. We broke up at 6.30 a. m. For the first part of the way the rocky shores are of clay-slate with hard concretions, alternating with quartzite, like the rocks on the Langau river above the cataract. About ³/₄ kilometre upstream, hornfels occurs, followed by diabase and diabase-tuff-breccia. Poling up the river is most trying here because of the quantity of rocks in the river-bed. The goeroeng Roeroei

Pl. XXV.

Fold-faults in the Sth Embalden.

caused some considerable delay; it is a small fall over a ridge
of rocks which keeps back the water and forces it to find an
outlet through a few available apertures. We had to unload
all the boats and make a portage over the tree-trunks laid across
the rocks on the right shore. Many a boat is lost in this goeroeng, one of ours amongst the number, which had been sent on in
advance from Nᵃ Boengan to the Boelit, it was laden with rice
and several karongs were lost on that occasion. The rocks in the
goeroeng Roeroei are of silicified diabase-tuff [1]) forming beds
trending E. 10° S., with a very steep dip to the south. Not far
from here the river runs for some distance parallel with the
strike of the strata, and has widened considerably through lateral
erosion in the beds of soft shale and sandstone which in a southerly
direction, succeed the rocks of Poelau-Mélaioe-type. The fall of the
river is very small here as the water is backed up behind the obstacles of the goeroeng Roeroei and in consequence a great many of
the boulders carried by the river are deposited here and have formed
big boulder-banks. In the widest part of the river is a very
large boulder-bank, Poelau Daroe. Some parts of Poelau Daroe
are covered with vegetation, but at high water it is entirely submerged, and the river looks then more like a lake. Above P.
Daroe the river makes a sharp bend and then follows a long
clear stretch known as the rantau Pandjang [2]), with smooth, deep
water and hardly any rocks showing above the surface. This
area appears for the greater part to consist of clay-slate, shale,
sandstone, and quartzite.

Above Nᵃ Ranai, rocks again appear more frequently above
the surface, and the river cuts through folded beds of sandstone, generally containing particles of coal, and in which also
undeterminable casts of shells are found. In one place on
the left shore I discovered thick beds of a kind of coarse-
grained crystalline limestone, in which the microscope distinctly

---

1) This rock, Poelau-Mélaioe-type, contains here much secondary calcite.
2) Rantau Pandjang (Mal.) means a long straight part of a river.

reveals organic remains, which however cannot be determined with any certainty. I presume that this limestone lies unconformably on the slate and quartzite, but conformable to the sandstone[1]) which a little higher up appears again on the opposite shore. The section is too indistinct however to be quite sure of it. The above mentioned sandstone with flakes of muscovite and particles of coal continues to be the predominating rock until 600 metres below the Loengoe river, where rocks of the Poelau-Mêlaioe-type again come into view. We bivouacked near to a beautiful group of these rocks, having traversed that day 10½ kilometres.

June 28.   At 8.45 a. m. we started again. About half a kilometre above N⁎ Loengoe, just below the goeroeng Bênalau or Alau, the Poelau-Mêlaioe-rocks are again superseded by sandstone alternating with clay-slate and quartzite. Upon a small rocky island, 1½ kilometres below N⁎ Boelit, I saw a bed of melaphyre lying conformably on the sandstone. This is probably an intrusive sheet between the beds of sandstone. The general strike of the strata continues to be east and west, with deviations to E. 30° N. and E. 10° S. As usual they are folded and the dip is very high, nowhere less than 60°. We encountered no great difficulties on the way, and at 11 a. m. we arrived at the entrance of the Boelit.

The view looking up this streamlet from the Boengan is one of the most picturesque bits in the Upper Kapoewas district. The Boelit is arched by the crests of colossal trees, all slightly bending towards the river, and this forming a gigantic "berceau". At the entrance is a bed of sandstone rocks in between which the foaming, bubbling waters force their way. There are only two openings between the rocks wide enough to allow sampans to pass through at low water. The Boelit is much narrower than the Boengan, and bears in every respect the character of a small mountain stream.

With a heavy fall of rain the water in the stream rises rapidly, and this, together with the pretty considerable fall in its bed, gives to

---

1) On the maps this sandstone is considered to form part of the Cretaceous formation.

this streamlet an enormous transporting power; in times of drought
the water falls with equal rapidity, and nothing remains but an arrow,
shallow gutter, flanked on both sides by a mass of boulders, which
in the bends have been heaped up to a tremendous height. So
far our one dread had been to find the river very much swollen,
but it was the other extreme which proved to be our enemy,
for at low-water this streamlet is so shallow that loaded sampans
have to be dragged for a considerable distance over the pebbles
at the bottom of the river. This is not only terribly hard work,
but wears out the bottoms of the boats very quickly. We feared
to have to resort to this method, but decided to wait one day
in hopes that the water might rise. A few sampans with rice
were sent on in advance, in order to get our bivouac ready
near the pangkalan Mahakkam [1]).

In the morning I paddled up the Boengan. Up to the highest June 29.
point that I reached (2 kilometres above N^a Boelit) the bed of
the river is cut in sandstone. The boulder-banks of the Boengan
above N^a Boelit are, like those of the Boelit, rich in volcanic rocks;
in both I observed rhyolite-breccia, banded rhyolite, fine grained
white rhyolite-tuff and dacite. In the Boengan, glassy varieties,
such as perlite, pitchstone, and obsidian are found only in small,
much worn pebbles, while in the Boelit, even close to its
mouth, these types are represented by numerous boulders. In
the Boengan, on the other hand, there is a beautiful hornblende-
andesite found in large blocks with the large phenocrysts of black
hornblende arranged in one direction in a reddish-grey groundmass.
Very numerous boulders of this rock, often as large as one
tenth of a square meter, are found in the Boengan whereas
this rock is very scarce in the Boelit. I noticed that the largest
boulders in the banks of the Boengan have very often deep
potholes ground into them *in situ*, which proves that they must
have been lying on the same spot for a long time.

---

1) Pangkalan Mahakkam, the place where the Boelit ceases to be navigable, and where
the footpath leading to the Oeloe-Mahakkam commences.

As the greatest difficulties of our journey were now overcome, and the transport of our greatly diminished provisions did not require so many boats, we dismissed the chief KAM LASSA with his 21 men.

Some Boekats temporarily residing on the Boengan, came that day to visit us in our camp. The women carry their children in a sling on their backs. In the case of very young infants, one may see the helpless little head hanging down over the side, dangling up and down with every step the mother takes, but these little ones very soon learn to improve their own condition, and it is quite an ordinary thing to see infants of but a few months old, clasping their little legs tight round the mother's waist and accommodating themselves entirely to every movement of her body.

June 30.   I went on in a small boeng, manned by three Kajan-dyaks, one pradjoerit, and my Chinaman, to explore the Boelit. The direction of the boat was entrusted to the Kajan Mělino one, of the most powerful and most handsomely built Kajans of the Měndalem, distinguished in his tribe for his knowledge and intrepidity in the ascent of the dangerous mountain streams in the Upper Kapoewas district. A few slight showers had swelled the river a little, but the dragging of the boat through the numerous rapids was nevertheless a tremendous piece of work. Until fully one kilometre above the entrance, the Boelit runs over sandstone, succeeded by chert and jasper with Radiolaria, the beds of which as well as of the sandstone are almost vertical with an average strike E. 10° N. These in their turn give place to rocks of the Poelau-Mělaioe-type, which here consist of diabase alternating with diabase-tuff, partially silicified and resembling hornstone, and again this rock has given rise to rapids. The most awkward of all is the goeroeng Boewang; from the bottom and along the shores a series of rocks project above the surface of the water, through which the mountain stream rushes on in wild fury. Of course all boats have to unload here and a portage has to be made over the rocks.

At 12.30 these falls were fortunately passed. A little beyond goeroeng Boewing, our eyes obtained for the first time in a south-eastern direction a glimpse of the Liang Boeloek, one of the columnshaped limestone mountains, which contribute so much to the picturesque beauty of the Boelit valley.

A little further up, the river turns to the south-south-west, and another of these monoliths (Pl. XXXI) rises just in front of us. The natives could not tell me the name of this mountain, so I found myself justified in giving to it the name of Cornelia-peak. Decomposed amygdaloidal diabase is exposed along the banks in the neighbourhood of this mountain, the many fissures are filled with calcite, and the entire rock moreover is, in a high degree, impregnated with this mineral; soon afterwards the river-bed is cut through limestone. To the south of Mt. Liang Boeboek the banks consist of crushed diabase, entirely infiltrated with calcite, soon again succeeded by limestone. The most picturesque spot in the Boelit valley is unquestionably at this place, where the river runs between the Liang Boeboek (430 metres) and the Liang Mahang (337 metres). Both these limestone-peaks are vertical or overhanging, and unscaleable not only from the side towards the river but from every other direction.

The high rock-walls which we saw straight in front of us quite close to the banks are entirely covered with brushwood, amongst which hanging, climbing, and swinging plants are conspicuous. These mountains are moreover full of caves, with openings on both sides, the entrances being partially closed by stalactites. It gives one the impression as if the whole mountain, from top to bottom, were wrapped in grotesque draperies of varied colours, the rich tints of the foliage vieing for the pre-eminence with the dazzling white stalactites. One can climb up Liang Boeboek for some little distance and I managed to enter one of the caves about 30 metres up the mountain. Plate XXXII[1]) shows how these caves are on the outside curtained off, as it were, by stalactites.

1) This picture is drawn from a photograph, taken from the interior of this cave.

Through the apertures between the stalactites one can see the trunks and the crowns of the trees surrounding the mountain. These caves are the abodes of bats, and their floors are covered with a thick layer of strong-smelling, powdery, guano. The small cavities at the sides shelter a host of nests of swallows and bees.

The strata of limestone of Liang Boeboek are vertical and their strike is N.W.—S.E. conformable to the beds of the Poelau-Mělaioe-type which flank them on either side.

Just above Lyang Mahang the Boelit runs through a narrow gorge between very high rocks, consisting of serpentinized augite-porphyry containing olivine. Although the current is extremely swift here, the passage is not particularly dangerous, provided one has a fair knowledge of the locality. At a short distance from the left bank are two more peaks of limestone, extremely like one another, I therefore gave them the name of "the Twins". At 5 .p. m. I reached a suitable place for our night-quarters, nearly opposite the large limestone-mountain Liang Bara (528 metres), 7 kilometres from Nª Boelit.

July 1.   In the morning I went downstream again, in order more minutely to explore the caves of Liang Boeboek and Lyang Mahang already referred to. There I came across the other members of our party, who were also on their way to the bivouac at Liang Bara. The Kajans told me that all the limestone peaks in the Boelit valley are unscaleable with the exception of the most westerly, the Liang Kaoeng, which can be scaled from the valley of the Loengoe.

July 2.   In the course of the afternoon Dr. NIEUWENHUIS and myself visited the twin-topped Liang Bara. We crossed the river close to our bivouac and had a path cleared which took us straight up the mountain in a south-eastern direction over steep beds of limestone. We succeeded in reaching the saddle between the two peaks, 125 metres above the river, and climbed the mountain from there until the perpendicular rock-walls stopped us. From a geo-logical point of view the Lyang Bara is most instructive, as it reveals more distinctly than any other mountain in the Boelit valley the relation between the limestone and the rocks of the Poelau-Mélaioe-type. Figure 45 gives a diagram matical repre-

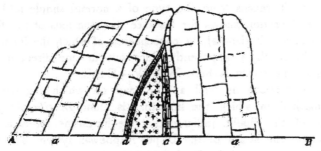

sentation of a vertical and a horizontal section of the mountain. These show that the Liang Bara is a single anticline,

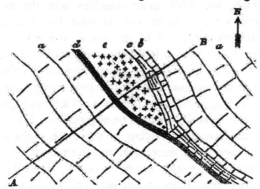

Fig. 45. VERTICAL AND HORIZONTAL SECTION ACROSS THE NORTHERN PART OF MT. LIANG BARA.

a. White limestone, in thick beds.
b. Impure, grey limestone.
c. Laminated, hard, dark clay-stone.
d. Decomposed augite-porphyry, strongly infiltrated with calcite.
e. Augite-porphyry.

the saddle being open and faulted along the axis, which consists of pure white coarse and indistinctly stratified limestone resting on distinctly stratified, clayey, grey limestone, beneath which decomposed augite-porphyry, perhaps also tuff, forming. the inner core of the mountain, makes its appearance. The two sketches show the intensity of the folding of the strata, and the differences from the type of a normal simple fold. The horizontal section shows for instance that the axis of the fold is by no means horizontal, in consequence of which the inner core of augite-porphyry apparently thins out in one direction. The strike of the strata is here W.N.W.—E.S.E.

These sections as well as the map on which the positions are marked in which the limestone is found *in situ* along the banks of the Boelit, indicate that the limestone forms a thick deposit in the strata of the Upper Kapoewas system, and particularly in those of the Poelau-Mĕlaioe-type; and that its beds, like all the others of this system, are highly folded and vertical or nearly so. This quite coincides with the fact that the limestone peaks range in an east to west direction, conformable to the predominating trend of the strata. We may consider that we have here to do with a lens-shaped mass of limestone, because in the first place its extension in an east to west direction is limited to 10— 15 kilometres, and also because only a single range of these rocks is known. Supposing that this were not a lens-shaped mass but a more extensive deposit, it would be only reasonable to expect that, considering the intensive folding of the system, the rock ought to crop out more than once, and therefore that more than one range of limestone peaks ought to exist. But all the natives agree in stating that there is but one, and my own observations corroborate theirs. For, as will be seen presently, I had ample opportunity of surveying a considerable portion of the district from the summits of Liang Tibab, Mts. Nopin and Lĕkoedjan; and I never saw more than this single range of limestone-peaks, so conspicuous by reason of their strikingly characteristic shape. If there had been any other mountains, at all like them in shape, they could not possibly have escaped my notice. We may therefore conclude that the limestone-peaks of the Boelit are parts of a lens-shaped deposit of limestone, which is sharply folded and there fore forms a saddle, the sides of which sometimes meet, as in Liang

Pl. XXVI.

GORGE IN THE TËKËLAN-VALLEY BELOW THE NARIK-RAPIDS.

Bara, in which case the series of limestone rocks is single, and sometimes they are some distance apart because of the connecting arch being lost through denudation. Then the augite-porphyry underlying the limestone becomes exposed as is the case between the Liang Boeboek and the Cornelia-peak, and a double series of limestone-peaks is found. The fact that these two cases occur at a comparatively short distance from each other, although the peaks are all of about equal height, is easily explained as soon as we take into consideration, that in consequence of the strong mountain-pressure the axis of the anticline is not a horizontal but a strongly undulating line. We have seen how in the case of Liang Bara the undulating course of the axis of the anticline could be noticed directly. In an east to west direction we may estimate the diameter of the limestone lens to be about 10—15 kilometres; taking this direction for the length, we can only speak of the breadth as being unknown but most likely also considerable; the average thickness, according to my observations may be estimated at 200 metres. The limestone itself graduates in colour from grey to pure white, and it is, as a rule, perfectly crystalline. Although microscopical examination distinctly reveals traces of organic remains, I have nevertheless not been able to determine these with any certainty. In spite of a most careful search I could nowhere discover any distinct fossils.

In the evening there were a few good showers and the water rose rapidly to a height of 2 metres.

In the morning it had fallen again, but it was still of a dirty July 3. milky colour, which points to the existence of limestone higher up in the river-bed. In other cases where there was no limestone, I always noticed that the water in the mountain streams, when rising, was of a muddy yellow colour, but that, with the fall of the water, this troubled appearance soon diminished or altogether vanished. A strong and prolonged discolouring of the water and a well-marked milk-white colour are caused by fine suspended particles of lime.

We left at 6:30 a. m. Above our station Lyang Bara, diabase-

14

tuff [1]) or other varieties of the Poelau-Mělaioe-type occur in the
river-bed, alternating with limestone as indicated on map VII^c.
One kilometre below N^a Hangai the bed of the river is considerably narrowed by projecting rocks of diabase-tuff-breccia,
causing rapids which are difficult to pass. For the rest the water
runs smoothly and the banks are quite low near its edge.

Where the rocks in the river-bed are of limestone, they
have often been worn by the water into the shape of toad-
stools. This could only be effected provided the level of the
water generally remained at the same height, i. e., not higher
than the eroded portions of the limestone cliffs. And this agrees.
with what the natives said; they declared that the Boelit, as a
rule, is low, but occasionally, with very severe rains, it reaches a
much higher level for a short time. When in a normal condition
the mountain streams of Borneo are fed by the numerous springs on
the slopes of the mountains, where the ground is saturated with
water. This supply never fails throughout the year and results in a
steady average height of the water, which, as we proved in the
Boelit, may lead to a well marked erosion of the rocks up to
a certain level. As a rule, the rainfall has a local character and is
of short duration, it very seldom lasts for many days at a time. The
mountain streams swell suddenly, and they then contain enormous
quantities of water mixed with boulders and gravel, but they return
rapidly to the ordinary level, and even during the rainy season
when the showers are much heavier and more frequent, this
relation remains, generally speaking, fairly well the same. But
in the middle and lower part of the rivers the effects are altogether different. The water-level is there regulated by the rainfall throughout the river-basin, and during the rainy season, i. e.
the time of heaviest downpour, the quantities of water supplied
by all the tributaries, one after another, go on accumulating.

---

[1]) The tuff exposed as solid rock $1^1/_4$ kilometres below N^a Hangai on the shores of the
Boelit, consists of particles of diabase, augite, feldspar, and limestone, cemented by calcite.
As secondary minerals it contains serpentine, chlorite, and calcite.

until the water in the principal river, for instance in the Kapoewas, swells to such a height, and the current becomes so strong that it would appear madness to attempt ascending the mountain streams. And yet this is quite a mistake. The real difficulties lie in the lower part of the upper course of the large rivers where we find already rapids and a considerable fall of the river-bed, and where the water-supply is due to a great number of confluents which consequently keeps the level always high in the rainy season. The goeroeng Dĕlapan, for instance, in the Upper Kapoewas, is generally impassable during the months of October to January. Little streams on the contrary, such as the Boelit, are navigable in any season, provided one keeps a good look-out for the sudden floods (bandjirs) so frequent in these parts, and waits in some safe corner until they have subsided.

Just above the Hangai there are a few ladangs near the banks of the river, the first cultivated ground which we set eyes upon in the Upper Kapoewas district after entering the hilly territory above Nᵃ Era. A Malay, named Adam lived here, and he had quite a harem of Dyak women to keep him company. This man had to take refuge here because he was under a severe sentence for illegal transactions against the Dutch Government. We were told that he gave himself out to be the representative of Dutch administration in these parts, and that he did not scruple to extort fines from the Boengan Dyaks in this assumed capacity, but needless to say, he pocketed the fines himself. Upon our arrival he fled, and we met his wives and children who had been sent downstream by him in a sampan. Shortly after, at 10.30 a. m., we reached the place where the Boelit ceases to be navigable, 4 kilometres above the station Lyang Bara, and we selected this spot for our bivouac. It is called pangkalan Mahakkam, i. e., the landing-place where the path leading into the Mahakkam district commences. This pangkalan is situated 205 metres above sea-level, and about 1037 kilometres distant from Pontianak, measured along the stream.

NIEUWENHUIS and myself ascended the Lyang Tibab (750 metres) July 4.

to the south, starting from our bivouac at 7 a. m. It rained heavily all the time, and we had to climb up a very steep mountain-spur between the Boelit and the Kateh. At 9.30 a. m. we reached the so-called top, which is only a somewhat higher point of the ridge extending in a southern direction. The rain never abated and we could see nothing of the surrounding country, so we did not stay long and were back again in our bivouac at 11 a. m. In my diary I find the following entry for this day:

"A mountain ascent in heavy rain, is decidedly unique in its way; the path, bad at any time, is to day slippery in the extreme. We are in no time wet through to the skin, sweat and water trickle down our bodies, and the leeches (patjats) have a fine time, and consequently occasional bloodstains begin to enliven the dullness of our mud saturated clothes. But we reach the top at last, to find, alas, nothing but heavy grey clouds and dripping trees. After half an hour's rest we turn back in despair, cold and miserable, and then after a series of slips and tumbles, with face and hands scratched by the treacherous rattan branches, we regain our bivouac, trembling in every limb, and chilled to the bone, but worst of all with our instruments half-ruined by the damp. Altogether a miserable failure."

Lyang Tibab consists of diabase[1]) the same as that found in the Boelit near pangkalan Mahakkam. There the river once more cuts through limestone, next to which comes, in the river-bed, quartz-diabase, followed by augite-porphyry with much secondary calcite. These rocks form vertical banks striking N.E.—S.W. (see map VII[c]).

July 5. On this day we again ascended Lyang Tibab, and reached the top about noon. Jupiter Pluvius was still in a contrary mood, and we had to remain inactive till 4 p. m., the rain sometimes coming down in torrents, sometimes more gently, until at last our patience was rewarded by a tolerably clear look-out, and I

--- -- -- ---

1) The specimens collected on Lyang Tibab are so decomposed, that I could not determine all of them with sufficient certainty.

was able to study the surrounding mountain-land. From Lyang Tibab one gets an exquisite view, on the one side of the volcanic Müller mountains close by, and on the other of the Upper Kapoewas mountain-range of which, strictly speaking, Lyang Tibab forms part. In the Müller mountains one can clearly discern two divisions of very different structure, viz., to the west a terrace- and table-land, comprising the tuff mountains of the Kĕryau and the Mandai, and to the east adjoining it, groups of more isolated volcanic mountains. Amongst the latter Mt. Tĕrata (1467 metres) forms the principal figure in the landscape, round which the others group themselves. Plate XXXIII gives a picture of the mountain from a photograph taken just before sunset; while fig. 46 is made up from a series of sketches, taken when-

Fig. 46. Mt. Tĕrata, seen from the top of Lyang Tibab.

ever a break in the clouds revealed some portion of the mountain scenery. Mt. Tĕrata is a rock-mass with pure white slopes, which, for the greater part, are perpendicular and treeless. From our stand-point it resembled a denuded volcano, with a crater the shape of a horse-shoe opening towards the north, and a similarly shaped valley opening towards the south. This valley is flanked by two peculiarly pointed rocks standing sentinels at the entrance. To the left of Mt. Tĕrata and in the back-ground is the boundary chain of mountains between West and East Borneo, and in the far distance some of the mountains of the Oeloe Mahakkam may be distinguished. To the right, and at a short distance from Mt. Tĕrata, rises Mt. Pĕmĕloewan (1340 metres) in shape like a column, which, looked at from a shorter distance, must be a more imposing feature in the landscape than even Mt. Térata itself, on account of the peculiarity of its shape.

S. 20 E.

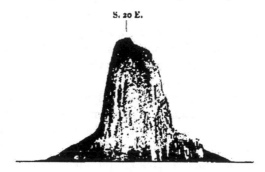

Fig. 47. Mt. Pemêloewan (1340 M.) seen from the upper ridge of Lyang Tihan.

The view to the south-west is also very interesting. In the far distance is the table-land of the Upper Kéryau, and the sharply delineated Mt. Mënakoet (1210 metres); more in the foreground Mt. Sara (1317 metres) with its perpendicular glistening rocky escarpments. The former represents the type of the western section of the Müller mountains, the latter the type of the eastern division. Particularly beautiful in shape is Mt. Hariwoeng (1243 metres) to the west-north-west of us, between the Lapoeng and the Langau. Its isolated position and sheer precipitous cliffs incline one to take it for a volcanic mountain, this however is in all probability not the case, since I have not been able to trace any igneous rocks either in the Lapoeng or in the Langau. Mt. Hariwoeng, moreover, lies a good deal to the north of the volcanic Müller mountains. The series of limestone-peaks along the Boelit river can be seen to its full extent; from Lyang Kaoeng to Lyang Bara they occupy the central part of a longitudinal valley, extending in an east to west direction, through which the Boelit pursues its winding course.

July 6. Our boats had all been pulled ashore and turned bottom upward; everything we could possibly dispense with was stored

away in a "pondok", and so we commenced our journey overland to the Oeloe Mahakkam district. The first day we had a
good deal of difficulty in persuading the Dyak carriers to take
up their burdens, and it was 8 o'clock before all were ready to
start. The Kajans, like all other Dyaks, carry luggage on
their backs in baskets (bérioet), and sometimes, when they have
an extra heavy load, they adjust a sling round the forehead, to
ease the weight. Their baskets are smaller than those of the
Batang Loepars and Oeloe-Ajers. We had to ford the Boelit
just above our bivouac, and from there a forest-path cuts off
a large curve of the stream and then runs parallel with it for
some distance, over the rocks on the left bank. The river-bed
is here full of rocks and sometimes very narrow, so that fierce
rapids follow each other in very close succession. The predominating rock is tonalite, rendered slaty by pressure, and macroscopically resembling tonalite-gneiss and sometimes amphibolite. Sometimes the darker constituents, amphibole and biotite, and the minerals generated by their alteration, are almost entirely absent. In the
neighbourhood of N⁴ Kateh we had to ford the river four times,
which was hard work in the strong current. On the whole, however,
the route was tolerably easy, although slippery through the continuous rain. In some places where we had to climb over steep
rock masses, the Kajans had on the previous day constructed
rattan ladders or stretched rattan ropes across to haul us up;
this was done for the express benefit of the controller, who was
rather stout and not in good trim for climbing. Our bivouac
had been prepared on the left bank of the Boelit half a kilometre above N⁴ Kateh, where I landed at 10.30 a. m. On the
same side just below our bivouac was a huge boulder-bed which
I visited in the afternoon.

Next day at 10 a. m. I continued my journey, first through July 7.
the bed of a small rivulet, the Lobang Tanah, and then up a
very steep but quite passable mountain-track until we reached
an altitude of 200 metres above the camp; and there a number
of fallen trees caused an obstruction which slightly delayed us.

Up to this point the rock *in situ* is decomposed diabase and diabase-porphyry. Keeping on in an easterly direction we now went down-hill and through the bed of another streamlet which is cut in clay-stone. Lower down-stream a cataract is formed over beds of polygenous conglomerate, in which I found pebbles of diabase, limestone, chert and hornstone. These beds are succeeded by alternate layers of polygenous sandstone and conglomerate. All these strata are vertical and trend east and west; sometimes, they dip steep to the south. Following the bed of the Roeran, we again struck the Boelit at an elevation of 261 metres above sea-level, 31 metres higher than Nª Kateh; thus, by crossing the mountain, we had cut off a considerable curve of the river. Our path now led by the right bank of the Boelit, first across large masses of diabase-porphyry, and then over beds of coarse polygenous sandstone and greywacke, in which, occasionally, large rounded pieces of limestone and hornstone with Radiolaria may be seen. The sandstone trends E.—W. and dips steep to the south. Twice more we had to ford the stream and we reached Nª Banjoe at 1 p. m. We bivouacked there at the confluence of the Boelit and Banjoe rivers, on an open space with a grand view across the Boelit. Close to Nª Banjoe lies a good sized, thickly wooded island, called Poelau Banjoe. At low water, as we found it, the portion of the river which skirts the right side of the island, was almost entirely dry, and offered a capital opportunity for the examination of the boulders which the Boelit carries down from its upper course (see the plan on map VIIᵇ).

A comparative examination of the boulder-beds, from Nª Boelit up to this point, showed plainly that volcanic rocks increase as one gets further up-stream. For instance, below the pangkalan Mahakkam, limestone is not uncommon amongst the boulders, but higher up only a single piece was discovered in a boulder-bed near Nª Kateh. Tonalite, schistose by pressure, was not found any higher than in the boulder-bank near the pangkalan Mahakkam; the proportion of diabase also, both as regards variety as well as

A PORTAGE ALONG THE NARIK-RAPIIS.

quantity, gradually diminishes as one advances upstream. Round about Poelau Banjoe I found, besides sandstone and a few pieces of amygdaloidal melaphyr, quartz-diabase, kersantite, quartz-porphyry, and quartz-diorite-porphyry, only boulders of rocks of volcanic origin of the following types: banded rhyolite, very often snow-white by kaolinization, rhyolite-breccia, rhyolite-pitchstone with breccia-structure and often with enclosed fragments of other rocks, perlite, dacite in different varieties, amphibole-andesite and biotite-andesite. However it must be borne in mind that many of the volcanic rocks, such as dacite and biotite-andesite, both with regard to their general appearance as well as to their manner of decomposition, bear such a striking resemblance to the so-called older igneous rocks, that it still remains a matter of doubt, whether they ought not rather be called quartz-diorite-porphyry and kersantite. Varieties of rhyolite and rhyolite-pitchstone predominate however in the boulder-banks of the Upper Boelit and especially at and beyond Poelau Banjoe, in such a marked way, that we may safely conclude that the high mountains, Mt. Térata, Mt. Péméloewan and others, where the tributaries of the Boelit have their origin, are composed of volcanic rocks of the rhyolite-type.

At the lower end (boentoet) of Poelau Banjoe the river-bed is narrowed by rocks of greywacke-sandstone, protruding from the left bank, and on the opposite side thin coal-seams are found in this sandstone. The strike is E. 30° N., the dip 70° to the southeast, but a little farther on a curious discordance appears; (see plan A and section A on map VII³). The strata of the same rock, a polygenous greywacke-sandstone with a few thin coal-seams, stand at the other side of the fault perfectly vertical, the strike being E. 5° N. This greywacke-sandstone contains fragments of chert and limestone, also a certain amount of plagioclase, which makes it allied to an arkose. At the upper end (kapala) of Poelau Banjoe large masses of a very coarse conglomerate stand out in the bed of the Boelit, composed of boulders of silicified clay-slate, quartz, hornstone, diabase, and limestone, cemented together by a sandy rock containing biotite and feldspar very similar in appearance to the greywacke-sandstone just referred to.

July 8. At 9 a. m. the controller arrived with a number of Kajans and the luggage which had been left behind. In the afternoon we were visited by some Boengan-Poenans who resided temporarily in the Oeloe-Boelit. They had been summoned by the controller. These men are tall and gaunt, almost all of them suffer from 'loesoeng', and they are not nearly so well built as the Poenans of the Upper Mandai or the Bockats on the Kapoewas. Their chief, LAKKAU, was suffering from a painful disease of the eyes, and, as represented on fig. 48, Dr. NIEUWENHUIS tried to give him some relief by the application of cocaine. The efficacy of the "obat Belanda" [1]) proved by the instantaneous abating of the pain, made a very deep impression upon this chief and his party. They informed us that it was quite out of the question to ascend Mt. Terata, which I had contemplated doing, because from time immemorial it had been declared "pantang" and had never been set foot on, either by themselves or by any one else, and there was not even a path leading up to it. No more could the Kajans be persuaded to escort me and cut a way through; they said that the Boelit itself beyond N° Banjoe was also declared boeling [2]), because of the death of some relative of LAKKAU's having occurred there.

July 9. Heavy rains had swelled the river considerably. LAKKAU came to see us again, and, softened by our friendly treatment, gave his consent to our going a little way up the Boelit, on condition that we should neither break nor carry away any of the rocks. This we agreed to, and so the geological result of this little

---

1) obat Bŭlanda (Mal.) = Dutch medicine.

2) Mountains, rivers or forests which have been declared "boeling" or "pantang" may not be set foot on. Trespassers bring misfortune upon the tribe which has instituted the prohibition; they run the risk of being killed. Friendly Dyaks may often save their lives by paying a fine. A place may be declared "Boeling" as a sign of mourning and this is therefore only a temporary interdict. "Pantang" may last for ever. Mt. Terata is pantang and has never yet been set foot on by any man.

excursion was of necessity a very poor one. The path is almost all the way a 'djalan ajer", i. e., one has to wade it. In some

Fig. 48. LAKKAU, CHIEF OF THE BOENGAN POENANS.

places the solid rock is sandstone, but more often the bed of the river is strewn with blocks of volcanic origin, and as far as we went, hardly any solid rock is to be seen. On the left shore are a few very primitive 'pondoks' in which the Boengans used to live.

July 10. At 7 a. m. we started, N⁴ Boewang on the Leja river, being our destination for this day. At first our way led through the Banjoe, sometimes over the 'outcrops' of the almost vertical sandstone banks, sometimes through deep pools which have remained stationary in between the rocks. The prevailing strike of the beds of sandstone is E. 30° N., the dip, at first about 80° to the south-east, diminishing as we got further away from N⁴ Banjoe. This sandstone alternates with clay-stone and sometimes with beds of conglomerate. Our path led across one of the northern spurs of Mt. Těrata, and I expected therefore to find volcanic rocks *in ·situ*. I was not disappointed, for when we had got ⁸/₄ kilometre beyond N⁴ Banjoe our progress was stopped by a large heap of rocks in the river-bed consisting of andesite derived from a stream of mica-andesite which we found on the left shore, lying unconformably on the arenaceous clay-stone and sloping gently down towards the valley. At the plane of contact I found no traces of alteration in the clay-stone.

Presently we left the Banjoe, a steep footpath leading up the mountain along the left bank. At an altitude of 450 metres we traversed one of the tributaries of the Banjoe which rushes down in a beautiful cascade across thick beds of arkose-like greywacke-sandstone dipping about 25° to the south. We soon struck the Banjoe again, which stream, narrowing as we proceeded, we forded well nigh up to its source; sandstone continued to be the country rock. One more steep climb through the wood, and at 8.45 a.m., we crossed the watershed between the Boelit and the Leja at an· altitude of 555 metres. We then came upon a brooklet, the Kawau, which empties into the Leja, and we followed its course. From here, volcanic rocks cover the ground, lava streams and tuffs become more continuous over a large distance, and the older formations, sandstone, clay-stone, hornstone etc., are only visible at rare intervals. First we crossed a flow of rhyolite, the portions which had escaped erosion being about 4 metres thick. In its centre this flow consists of rhyolite-pitchstone with a few perlitic cracks; lower down the rock gradually passes into real perlite, and at the bottom of the flow the perlitic cracks are arranged very regularly and perpendicular to the plane of contact in such a way that this lower sheet of the flow (6 centimetres thick) is built up of small columns, each being 6 millimetres in thickness, giving in miniature the well-known columnar structure of basalt [1]). We might, in fact, call this "micro-columnar structure".

This flow of rhyolite rests on rhyolite-tuff with horizontal stratification (fig. 49). There are some small outcrops of the unconformably underlying formation, represented by hornstone with Radiolaria. Then follows a stream of biotite-rhyolite, underneath

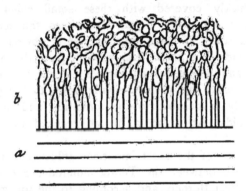

Fig. 49. SECTION ACROSS A FLOW OF RHYOLITE
IN THE BED OF THE KAWAU.

a. Rhyolite-tuff.
b. Rhyolite-pitchstone with perlitic and micro-columnar structure.

which greywacke-sandstone is visible. On the return journey I ascertained in one place the successive order of the volcanic rocks to be as follows: 1, biotite-andesite, 2, obsidian and 3, rhyolite. Then follows again a more or less decomposed variety of banded biotite-rhyolite. But the section on page 19 of the atlas and the map VII[c], will give a better idea than any lengthy description, of the way in which, now and again, the quartz-sandstone or the clay-stone is

---

1) This reminds one of SCROPE's illustration of a "prismatic obsidian passing into globiform", from the isle of Ponza. (G. POULETT SCROPE, 54, p. 105.

exposed, and also of the way in which, in most places, these formations are covered by flows of rhyolite, obsidian, perlite, dacite
and mica-andesite. In between two lava-flows which appear to
have taken a north-westerly course, the fall in the Kawau is as
a rule very slight, and marshes are formed, which however are
easily forded. Thick beds of recent alluvial deposits composed
of clayey sand with small pebbles have formed here, and I found
on the slopes very fine specimens of miniature earth-pyramids.
Every pebble rests on a little column of clayey sand about 1—3
decimetres high, and the steep slopes on either side of the
rivulet, thickly covered with these small columns, resemble
sometimes a miniature organ, sometimes the turreted battle-
ments and walls of some toy-castle. It is obvious that a place
like this, under the thick foliage which perfectly shelters it from
the wind and where consequently the drops of water from the
tree tops always fall perpendicularly down to the ground, must
be eminently favourable to the formation and preservation of
earth-pyramids. At 11.30 a. m. we reached the left bank of
the Leja.

This is one of the most important affluents of the Boengan;
its valley lies between the Térata mountains and the boundary
ranges of West Borneo, and according to the topographical map
the Leja arises on the latter, in the vicinity of Mt. Lyang Tja-
hoeng (1394 metres). It is a true mountain-stream, its bed being
strewn with large boulders and rounded blocks. Near to N"
Boewang its width is 30 metres, while the Boewang itself is only
12 metres wide at the confluence. After fording the river our path
led along the right bank over very rough ground. The rocks of
which the hills here are composed (see sketch B on map VII") are
andesite, dacite-tuff and rhyolite-pitchstone with breccia structure.
About 1 p. m. I arrived in our bivouac at N" Boewang, richly
laden and in high spirits; but tidings awaited us here which
somewhat damped my enthusiasm.

SIGAU, the son of KAM IGAU, who with twelve Kajans had been
sent on by the controller as messenger to the Upper Mahakkam

district, had just returned with apparently alarming news. Rumours
had been spread that it was our intention to invade the Upper
Mahakkam district with some two thousand men, to oppress and
punish the Pénihin Dyaks settled there. The messengers brought
word that the Pēnīhin Dyaks had gathered on the Mahakkam
under the notorious chief BĚLARI, to prevent our passing through.
They were, so ran the report, encouraged and egged on by
about 200 Malays armed with rifles, from the Siang Moeroeng
district. These Malays are hostile to the Dutch Government,
because they are adherents of the would-be claimant sultan of
Bandjermassin, GOESTI MAT-SEMAN, who is not acknowledged by
us, and who therefore had retired with his followers into the highlands of the Barito, in the Siang Moeroeng district. They told us
moreover that the chief of the Kajans on the Mahakkam, KWING
IRAN, had declared the Mahakkam river "boeling", owing to the
death of his grandson, and that he would not allow any one
to pass.

We had now reached an altitude of 430 metres above sea- July 11.
level, and the nights were beginning to get chilly. In the nights
of the 10th and 11th July the minimum thermometer registered 19°
Celsius. I visited several karangans in the Leja, and in the first place
the large boulder-island just below Nᵃ Boewang (see plan B on
map VIIᶜ), to which I gave the name of Poelau Kam-Igau after
the principal chief of the Kajans. With the exception of a few
varieties of sandstone, the boulders here are exclusively of volcanic origin and chiefly of the same types as those found in the
Boelit. Rhyolite, rhyolite-pitchstone and rhyolite-breccia predominate, but besides these I found in the Leja several varieties of
dacite, amongst which a beautiful biotite-dacite with large phenocrysts of quartz and feldspar is particularly striking. The phenocrysts of quartz have the shape of very regular bipyramids. On
the return journey I found pretty much the same kinds of rocks
when I visited some boulder-beds and boulder-islands in the Leja,
above Nᵃ Boewang. Just below Poelau Kam-Igau the bed of the
stream is much narrowed by rocks, which might with equal

correctness be called decomposed amphibole-dacite, or quartz-amphibole-porphyry. If our attention is principally directed to the microscopical character and composition of the rock, the latter qualification is better suited; but if, on the other hand, more importance is attached to the general appearance, the former name should be used in preference.

July 12. At 7 a. m. we left Nᵃ Boewang. Our path took us through the bed of the river Boewang and uphill. The solid rock along the banks is a grey sandstone, occasionally interstratified with clay-stone. The strata as a rule dip gently to the north or north-north-east. About 200 metres above Nᵃ Boewang a beautiful section is exposed on the left bank (section B on map VIIᶜ). There is a fault in the sandstone, the eastern portion being thrown down. The flexure of the strata in the immediate vicinity of the fault is well-marked and strongest in the down-throw portion. Andesite banks, probably the remains of flows of andesite[1]) which have escaped erosion, now and again overlie the sandstone, and they frequently produce rapids and waterfalls. About one kilometre above Nᵃ Boewang an andesite occurs, which encloses large pieces of greywacke-sandstone. This andesite bears the same ancient stamp as the other andesites and dacites in this district, and taken individually might easily be determined as kersantite or porphyry. It encloses however fragments of an arkose-like greywacke-sandstone which in every way corresponds to the same kind of sandstone often referred to as occurring on the banks of the Kapoewas, and in which, near the goeroeng Dĕlapan, tests of *Orbitolina* were found, proving thereby its Cretaceous age. Our eruptive rock therefore must be either of Upper-Cretaceous or else of Tertiary age, which justifies its determination as andesite, this being moreover supported by its general appearance and mode of occurrence. As the sketch clearly shows, the path often leads through the Boewang, which, close to its source, meanders a great deal. About

1) Some of these heaps of andesite blocks are no doubt derived from dykes of andesite which cut through the sandstone, but the place was so ill-suited for careful investigation which might have led to a satisfactory decision, that this question must remain undetermined.

Pl. XXVIII.

The Kapoewas-river above Poelau Balang.

11 a. m. we reached the divide between the Leja and the Abang; at this highest point the solid rock is once more greywacke-sandstone, whilst in the narrow cañon of the Lĕboewang, a tributary of the Abang, a biotite-augite-andesite is again found *in situ.*

My observations between the Banjoe and the Leja, and again here, had convinced me of the fact that on the highest points of the hills separating the valleys, the solid rock is not of a volcanic nature, but that on the slopes and in the valleys the bottom rock is frequently composed of eruptive material; and this proved to me conclusively that the distribution and shape of the valleys at the time when the volcanoes in the Müller-mountains were active, must have been, roughly speaking, the same as they are now; another convincing proof of the correctness of the theory that these volcanic mountains, notwithstanding their far advanced and intense denudation, are, geologically speaking, of comparatively recent age.

At noon we reached the bivouac on the Abang Kĕtjil; it was buried under lofty trees in a narrow ravine through which the foaming brooklet also sped its course.

At 2 p. m. I ascended Mt. Nopin, taking a track which at first followed the left bank of the Abang Kĕtjil, traversed that brooklet and then led in an easterly direction up a very steep slope until we reached a spur running in a N.—S. direction, the highest point of which (989 metres) is more particularly known as Mt. Nopin.

The view from this stand-point is fairly satisfactory all round. At a short distance from us in the north-east and on the water-shed, rises the Lĕkoedjan, with every appearance of a volcano, the crater of which is crumbled down and now forms a gigantic valley, in shape like a horse-shoe, opening towards the W.N.W. On the other side of the watershed, in the Mahakkam district, I located one mountain at a comparatively short distance, E. 15° N. This mountain is about the same height as the Lĕkoedjan, and like the latter resembles a volcano, with an outlet to the north-west, and similar steep, white, slopes, these being characteristic of the volcanic peaks in the eastern division of the Müller mountains.

Mountain chains near the sources of the Tandjan-river.

Mt. Lèkoedjan
E. 40° N.

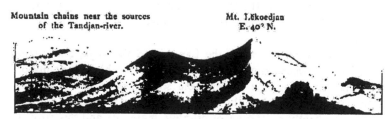

Fig. 50. MOUNT LÈKOEDJAN, SEEN FROM THE TOP OF MOUNT NOPIN.
To the left are the mountain ridges of the Upper-Kapoewas-System.

The natives called the northern portion of this mountain, Mt. Měnětekai, the part facing us, Mt. Pènaneh. Between the Lèkoedjan and this latter mountain one gets an extensive view of the Oeloe Mahakkam, the ground being fairly even, until the view is intercepted by a table-mountain which I located E. 25° N. To the east, i. e., to the right of Mt. Pènaneh, one looks across a far stretching, tolerably low and undulating track of ground, resembling somewhat a table-land. But the view in this direction was unsatisfactory, and even with a good field-glass I could not form a distinct idea of the landscape. One thing however is certain, that in this easterly direction there are no high mountains. The valleys of the Pènaneh, the Kassoo, and the Mahakkam are situated in this district. Due S. E. I noticed in the Mahakkamland, probably in the river-basin of the Kassoo, a few fairly high and steep mountains. Then follow, right to the south, the complex mountain-range of the watershed and the deeply cut valley of the Leja. To the south-west the view is greatly obstructed by the colossal bulk of Mt. Tèrata which from this point of view, is most imposing. The slope turning towards the Leja-valley is extremely steep, almost vertical, and intersected by many high waterfalls plunging into the depths below. The top of Mt. Pěměloewan is entirely hidden behind Mt. Tèrata. More to the west, in the head-waters of the Langau, Mt. Sara

(1317 metres) attracts special attention; from our stand-point it looked like an old battlemented citadel. To the left of the

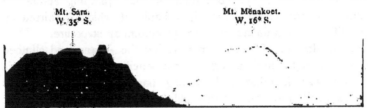

Mt. Sara.
W. 35° S.

Mt. Mĕnakoet.
W. 16° S.

Fig. 51. THE MÜLLER MOUNTAINS, SEEN FROM MT. NOPIN.
In the distance the terraced tuff mountains along the Kĕryau.

Lĕkoedjan and in a northern direction follow the ranges of the Upper Kapoewas mountains at the other side of the deep valleys of the Abang, Bĕtjai and Boengan rivers. These mountain ridges trend E.—W. or E.N.E.—W.S.W. In the direction N. 20° W. one can see distinctly seven separate ridges, which, like the side-scenes of a stage, succeed and partially hide one another. The most distant and highest of them are the Kĕrihoen and the Upper Tandjan mountains. The top of Mt. Nopin consists of much weathered and kaolinized mica-dacite. We found exactly the same rock near our bivouac, only in much fresher condition, for here the weathered crust is constantly removed by the wear and tear of the series of little waterfalls formed by the Abang Kĕtjil. Higher up in the bed of this brooklet, claystone with indurated concretions in almost horizontal layers is exposed under the dacite. The many loose blocks in the stream are composed either of biotite-dacite or of biotite-andesite.

After a pleasantly cool night, (minimum temperature being July 14. 19° Celsius) we resumed our march at 6.30 a. m. intending that day to take leave of the river-basin of the Kapoewas, and to make for the eastern division of Borneo. First of all, our way led across a northerly spur of Mt. Nopin and from there to the Abang Bĕsar, following a series of little falls, all of them together about 13 metres high. The solid rock consists of a most beautiful mica-dacite with large phenocrysts of quartz and felspar, corresponding to the dacite which I had often found in loose pieces in the Leja. We continued our way, jumping across the boulders in the stream, the right bank of which is skirted by high cliffs of mica-dacite with rough columnar structure.

At an altitude of 718 metres we left the Abang and climbed up by the right bank along the mica-dacite rocks while a little brooklet was foaming and rushing past us, which eventually empties itself into the Abang below. The biotite-dacite is here covered with a thick weathered crust, the greater portion of the rock being altered into grey clay containing numerous crystals of pyrite. After a while we followed the course of another brooklet, the Pare, also a side branch of the Abang. Here the solid rock was at first dacite, but at a height of 745 metres it gave way and clay-slate appeared, dipping gently E. 20° N.

After the difficult ascent of a very slippery incline I reached, at about 8.30 a. m., the division between West and East Borneo, situated at an altitude of 830 metres above sea-level. The solid rock in this spot is plagioclase-arkose in horizontal strata. When my native companions arrived here, they were visibly affected, and began to cut from certain trees little bits of wood which they took with them to exorcise the evil spirits, which, on the foreign soil, might bring misfortune upon them. Descending on the other side we came before long upon a rivulet with an easterly course, evidently therefore flowing into the Pĕnaneh. The bottom rock in this brook is a finely granulated biotite-andesite. At 9.15 a. m. we struck the Pĕnaneh itself, as yet an insignificant mountain stream, and soon after we came to the bivouac which had been prepared by a part of our people, sent in advance. Close to our bivouac the stream contains large five- or six-sided columns of biotite-andesite which are from 80 centimetres to 1 metre long, and about 40 centimetres thick. After half an hour's rest I started again in order to ascend Mt. Lĕkoedjan (1190 metres) situated on the water-parting. Our Boekat guide called this mountain Mt. Tĕboeng. From the divide we had to begin with a very steep climb along Mt. Apoen, and soon reached the foot of Mt. Lĕkoedjan, which proved to be composed of coarse greywacke, greywacke-sandstone and arkose in horizontal beds. At an altitude of about 920 metres the sandstone disappears under a covering of volcanic material, at first loose, much weathered, tufaceous matter, gradually passing into rocks of pure white kaolinized dacite. We climbed up by a very narrow spur running about S.W. to N.E., and were constantly checked by perpendicular crags. These are thickly covered with the wild sago palm, "ransah", and it was only by clinging to the aerial roots pending from the branches that we succeeded in pulling ourselves up to a height of 1100 metres.

It did not seem feasible however to reach the highest top (1190 metres). But this was of small moment, for the view from our present vantage ground left nothing to be desired. We stood upon a narrow, thickly wooded, mountain ridge, from 1 to 3 metres wide, bordering on the west side upon the deep horse-shoe shaped valley which I had already noticed when on Mt. Nopin, and where a tributary on the right side of the Abang Bĕsar has its source. On the east side we were separated in a similar manner from the valley of a small affluent of the Pĕnaneh by very sheer precipitous slopes. Mt. Lĕkoedjan is evidently entirely composed of dacite resting on greywacke-sandstone and arkose, which formation, protected as it is against erosion by the volcanic rocks, is here found at a great height, at least 930 metres high. To my mind the gigantic horse-shoe shaped valley of Mt. Lĕkoedjan opening towards the west is not a ruined crater; erosion in reality has already progressed too far, to enable one rightly to distinguish the portion of the relief which belonged originally to the volcano, and that which is the result of denudation. In fact, correctly speaking, volcanic mountains like Mt. Lĕkoedjan, Mt. Tĕrata, etc., can not strictly be designated as volcanoes; at most they might be called volcanic ruins. If I were to draw a parallel, I might say that erosion and denudation have done their work here, as thoroughly or perhaps more so than in the "Habichtswald" near Cassel, where it is found equally impossible to mark the exact spots where the craters of the volcanoes were originally situated.

The panorama from Mt. Lĕkoedjan is a most extensive one. On the one side the view extends far over the Upper Mahakkam territory, which, as seen from here, does not appear to be much broken, the panorama in this direction being, so to speak, cut in two parts by Mt. Pĕnaneh which rises in the foreground at a comparatively short distance from us. This mountain, my guide informed me, is "pantang" for all Dyaks. The entire build and general appearance of Mt. Pĕnaneh is so like that of the other eruptive mountains, that I have no doubt whatever, that, like those, it is composed of igneous rocks. In the far distance behind Mt. Pĕnaneh, in an E.N.E. direction, some isolated mountains with white precipitous slopes, may be observed in the Mahakkam district; I imagine them all to be volcanic and to form part of the eastern continuation of the Müller mountains. My guide told

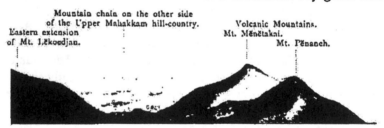

Mountain chain on the other side
of the Upper Mahakkam hill-country.
Eastern extension
of Mt. Lĕkoedjan.
Volcanic Mountains.
Mt. Mĕnĕtakai.
Mt. Pĕnaneh.

Fig. 52. HILL-COUNTRY OF THE UPPER MAHAKKAM, SEEN FROM THE TOP OF MT. NOPIN (989 METRES).

me that the Batang Mahakkam flows in a N.W.—S.E. direction close behind Mt. Pĕnaneh.

To the south is the mountainous district of the watershed between the Kapoewas and the Mahakkam, which stretches out eastward into the head-waters of the Oeloe-Kassoo. Through an opening between these mountains and Mt. Tĕrata, one can look up the valley of the Leja, at the end of which, far away in the distance, one can distinguish a twin-topped conical mountain, the Mt. Atoe-Atoe, (sighted from the Lĕkoedjan at S. 10° W.); according to my guide this mountain is in the Siang Moeroeng district. The south-western section of the panorama is entirely taken up by strangely shaped mountains, more or less detached, the huge proportions of Mt. Tĕrata, standing out in the foreground; the others varying in shape, some resembling truncated cones, others like ruined citadels of gigantic size. The bare, steep, white or red rock-faces of all these mountains sparkle in striking and pleasing contrast with the all-pervading green, and give colouring and variety to the scene. This is the volcanic district of Central Borneo, called the Müller-mountains. Due west follows the deeply cut valley of the Boengan and Boelit, rendered most delightfully picturesque by the rugged limestone peaks, standing in a line, like so many pure white columns, in the very centre of the Boelit vale.

To the north-west and north the scenery is altogether different. As far as the eye reaches, range upon range of hills appear in view, every one of them trending either E.—W. or E.N.E.— W.S.W. They are all thickly wooded without a single break, the highest of them, the Oeloe Poenoe and Oeloe Tandjan are enveloped in clouds, and the whole makes a grand but decidedly gloomy and almost ominous effect. This is the mountain-chain of the Upper Kapoewas, the original home of the Boekats.

In the afternoon, at 3.30, we were back again in our bivouac at the foot of the Pĕnaneh. In the evening there was a heavy thunder-storm, the rain continuing the greater part of the night.

In the morning I went once more across the water-parting to July 15. our bivouac on the Abang Kĕtjil, in order again to ascend Mt. Nopin, where I wanted to take some photographs, the light having been unsatisfactory on the previous occasion. The path was very wet, and the Abang Bĕsar, which we were obliged to ford, had risen considerably, so that our journey was a very fatiguing one. However I had reached the top by 12.30 and took some photographs which have enabled me to give the correct form and outlines of these mountains in the sketch on the opposite page. Returning in the afternoon to the Pĕnaneh, I was met in the valley of the Abang

Bésar by the "controller" who told me that he had broken up
the bivouac at the Pénaneh, and had given orders that all our
men should return that evening with the luggage to the Abang
Kětjil, which accordingly was done. I felt very much annoyed.
In the first place I was surprised that the "controller" had not
first tried to ascertain whether the rumours brought by our
messenger Sigau had not perhaps been exaggerated, which is
often the case. But the "controller" never consulted us in any
of these matters, nor could we, forming part of an expedition
so essentially supported by government, in any way oppose the
decision of the man who had been appointed as our guide. Ob-
jection on our part would moreover have been quite futile, for
the Kajans were under the direct orders of the "controller", and
his instructions were that they must return. Very reluctantly therefore and very much doubting both the justice and the necessity
of this decision, we had to give up all thought of penetrating
further into the Mahakkam territory. Shortly after, Nieuwenhuis
made a sojourn of several months duration with the Kajans of
the Měndalem, and learned that there were no really great
obstacles to a journey further inland; and two year afterwards, having
occasion to go again into these same parts, it was proved that
the population put no difficulties whatever in the way, and he
even managed to descend the Mahakkam right down to its mouth
and by this route to reach the east coast of Borneo below Samarinda.

July 16. We returned to the Boelit by the same way we had come, and
travelling slowly I had the opportunity of verifying my previous
observations, and, where necessary, to complete them. We spent
the night in the bivouac at Nᵃ Boewang.

July 17. The next day we had a difficult tramp before us, on account
of the very high state of the river. Our destination was Nᵃ Banjoe,
and in most places we had to wade up to our waists in water,
but in spite of that I reached Nanga Banjoe at 10.30 a. m.

July 18. Next day we arrived at the pangkalan Mahakkam at 10 a. m.,
and here ended the land-route for us, just at the right time, as
far as my personal comfort was concerned, for my shoes were

Sᵗᵉ KÉRVAU.

by this time reduced to a few rags kept together with rattan strings; this had caused me a good deal of discomfort during the last days. The boats were launched, caulked, and loaded, and on the 19th of July, at 7 a. m., we commenced the pleasant termination to our excursion in the Upper Kapoewas district, viz., the descent of the mountain streams. Favoured by beautiful weather, we glided along the various currents with marvellous rapidity, and at noon we found ourselves once more at N' Boelit.

One day of 10 hours, not counting the rests, brought us to July 20. the mouth of the Boengan. Without mishap we passed the gocroeng Bakang, where once more all the sampans had to be unloaded; one by one the empty boats with only three men in them shot vigorously down the falls. A glorious, never to be forgotten sight. All our attendants, anxious to get home, worked with unremitting zeal, so that in little more than an hour's time, all our boats had passed the falls, and were reloaded and ready to start. Not quite so fortunate were we at the goëroeng Matap, just below N' Poenan. This is a short fall, and looks comparatively harmless, so that I remained in my boat, and was down at the bottom in no time and without any bumps. But just behind me came another boat freighted with rice and two karongs of rock-specimens belonging to my collection, under the care of the chief Séniang. A false movement caused the boat to strike a rock and capsize, and the karongs disappeared in the fall. I promised at once a reward of ten ringgit for each karong with specimens, which should be recovered. A few anxious hours ensued; it seemed a truly hopeless case. But the Kajans worked wonders, and, if I had not been so anxious, it would have been a real treat to watch their feats in diving and swimming. At last, after many fruitless efforts, they succeeded in rescuing the heavy karongs from the deep pools immediately below the fall. We reached N' Boengan at 5 p. m.

At 6.30 a. m. we resumed our journey and passed the goe- July 21. roeng Délapan favoured by fine weather and low water. But little time was lost, and nothing went amiss.

July 22. N° Kéryau was reached as early as 2.55 p. m., and at 6.30 we once more set eyes upon the Upper Kapoewas plain. We stayed the night over at Poelau Tëngkidoe. On the up-journey I had left at different stations, cases and karongs belonging to my collection, which I called for in passing. Many of them were in a very bad condition, some had suffered from the rain, whilst others, notwithstanding my precautions, had been injured by the rising water, and were simply saturated. The boxes left behind at Poelau Balang had been opened by some inquisitive individual, and the contents all tumbled about. Fortunately no specimens had been taken nor any of the labels destroyed.

Thanks to a system of duplex numbering and labelling, I was able to put the whole collection in order again, without losing one single specimen.

If I had continued my journey into the Mahakkam-land, and had thus been prevented from attending to these collections as soon as I did, I might possibly have found them altogether worthless; as the numbers and labels, inside the damp cases and karongs would have been entirely destroyed by the united efforts of mould and insects.

At 5 p. m. we reached Poetoes Sibau, where we delayed one day to load the sampans with my spoil. On the 24th of July I took leave of NIEUWENHUIS who was going to spend a few months with the Kajans of the Mendalem. I had to make my way back to Sémitau in a rowing boat, which was slow work, the water being low. I left on the 24th at 7 a. m. and, rowing by day and drifting by night, I reached Boenoet next day at 12 o'clock. On the morning of the 26th I had got as far as Oedjoeng Said, and proceeded that day till just past Djongkong when I met the steam-boat which had been sent from Sémitau to meet me. On the 27th of July at 10 a. m. I safely landed with all my collections at the station of the expedition at Sémitau.

GEOLOGICAL STRUCTURE OF THE UPPER KAPOEWAS TERRITORY.

Orographically the Upper Kapoewas territory may be defined as consisting of two mountain ranges, trending almost due east and west, and separated by a strip of lower lying ground, the Upper Kapoewas low land. To the east and nearer to the sources of the river this low ground gradually rises, passes into hilly ground and finally merges into the mountain-land. In order to simplify the general description I shall distinguish these strips of ground as the northern-, the southern-, and the middle-strip. And this division acquires more significance as we find that both orographically and geologically these three strips include differently constructed territories.

The northern strip is a typical mountain-chain, embracing all the ground to the north of the Kapoewas plain and extending some distance beyond the boundary of Sarawak. This division I have called the Upper Kapoewas mountain-chain. Up to the point where the Upper Kapoewas plain is absolutely flat and covered with river deposits, the southern boundary of these mountains is very clearly defined, and forms the dividing line between the plain and the northern highlands. But further up-stream where the Kapoewas plain begins to rise, and gradually passes into a hilly district, the southern boundary of the Upper Kapoewas mountain-chain is much less clearly marked; here it gradually passes away into the more southern hilly districts of different geological structure. But even here the Upper Kapoewas-chain can still be recognised as that portion of the mountain-land which is most sharply accentuated; where the ridges are sharpest and the valleys deepest, and where especially the east to west trend of the ranges and crests has been most clearly preserved. The average height of the ridges in the Upper Kapoewas mountains is, for the basin of the Embaloeh, 900 metres; for the area of the sources of the Kapoewas, 1300 metres.

Posewitz's[1]) remarks as to the probable structure of this grand mountain-land, then unknown, are instructive. He says: „Vom östlichen Grenzgebirge Serawak's, östlich vom G. Saribusaratus, haben wir keine Kenntnisse. Das Nichtvorkommen des so verbreiteten Seifengoldes in diesen Gegenden Serawak's spricht aber indirekt dafür, dass die Verbreitung der „Gebirgsformation" hier keine grosse sein mag; dass vielleicht nur wenige kleine Gebirgsinseln daselbst vorkommen mögen".

One example out of many of the danger of drawing conclusions concerning the probable build of unknown districts, in works of a compilatory character.

On the whole the middle strip lies lower than the neighbouring territories. It may be divided into an eastern and a western half. The eastern half embraces the hilly district which towards the north joins the Upper Kapoewas mountain-range. The prevailing trend of the hilly ranges and ridges here continues to be east to west; but it is distinguished from the former district, by the lesser height, the more complicated build, and the more variable shape of its hills and valleys. The nature of this territory, may best be studied in the Boengan valley, and I therefore call the whole of this district the Boengan hills. The western half includes the real Upper Kapoewas low-land, an almost absolutely flat territory, with only a few scattered hills. Its average height above sea-level does not exceed 37 metres. Westwards this territory passes gradually into the low lands of the lake-district.

The southern strip, like the northern, is on the whole higher than the middle division. This division, which geologically and orographically forms one whole, has, like the others, its greater dimensions in an east to west direction. Nevertheless one fails to discern an east to west trend of the hilly ranges and a corresponding east to west direction of a great many of the valleys. On the contrary the shape of the mountains is most variable. Sometimes, as in the western half, the table-mountain-type pre-

· · ·

1) Th. Posewitz, 45, p. 119.

vails, a strongly eroded plateau-land, cut into ridges by erosion; sometimes, as in the eastern half, the mountains are more isolated, often most imposing and grotesque in shape and connected by lower lying hills. These are the volcanic Müller-mountains.

Probably the middle division, the Upper Kapoewas plain, is in reality a one-sided sunken territory (see sections AA' and BB'). The northern boundary of this sunken block is marked by the abrupt incline with which the Upper Kapoewas mountain-chain rises from the plain. One or more dislocations of the same type limit this middle division on its south side, but in most cases this boundary line is covered by volcanic material which no doubt has partly made its way to the surface of the earth through these lines of faulting.

Concerning the geological structure of these three divisions of the Upper Kapoewas territory the following should be noticed:

1. The Upper Kapoewas mountain-chain.

I examined this mountain-range in the river-basin of the Embaloeh. It consists there of a system of alternating strata of phyllitic clay-slate, drawing-slate, sandstone, quartzite, and graywacke-slate, much crumpled, folded and sheared by mountain pressure. The average trend of the formation is E.—W. to E. 5° S.—W. 5° N., the strata being tilted at a very high angle and often vertical. This complex is characterized by the number of larger and smaller quartz-veins which intersect the rock in all directions. It is evident from the rocks collected by BÜTTIKOFER in June 1894, in the river-basin of the Sibau, near N* Mĕnjakan and elsewhere, that the rocks further east in the Upper Kapoewas-range belong to the same formation. All the specimens collected there belong to the same type as those found by me in the valley of the Embaloeh (comp. Chapter VII). In the valley of the Sibau the strata are also as a rule vertical and much plicated, as I could see distinctly on the photographs taken by BÜTTIKOFER

of some rock-groups in the Sibau valley. From all that BÜTTI-
KOFER has told me, after he had rowed far up the Sibau
into the mountains, I have not the slightest doubt but that the
build of the Upper Kapoewas mountains in the river-basin of
the Sibau is, in every respect, identical with that in the Embaloeh valley.

Near the sources of the Kapoewas I have not personally
reached these mountains, but I gather from the character of
the boulders in the Kapoewas valley above N⁸ Boengan that
the characteristic rocks of the Upper Kapoewas mountain-chain
will be found *in situ* at no great distance up-stream from the
furthest point reached by me on the Kapoewas.

Nowhere in the system of strata of the Upper Kapoewas
mountains, have I met with any fossils. From the petrographical
character of the rocks, and the general build of the mountains,
I feel justified in drawing the conclusion that they belong to an
old formation, and that they probably are older than the cherts
and hornstones of the Boengan-hills, which are of pre-Cretaceous,
presumably of Jurassic age.

This formation, the old slate-formation, I consider to be the
oldest of the hitherto known sedimentary formations of Central
Borneo, with the sole exception of the contact-metamorphic
altered slates which I found afterwards along the Samba-river in
South Borneo.

It is probably of the same age as the oldest sediments which
in Sambas occupy a vast territory, the so-called old slate-formation of the Dutch mining-engineers.

Now although this formation is decidedly the predominating
formation in the Upper Kapoewas mountain-range, and the only
one found there as solid rock, there are a few facts which lead
us to conclude that other formations also play a part in this
mountain-range. In the first place I found in the valley of the
Embaloeh and especially in the bed of its right side tributary,
the Tĕkĕlan, several water-worn pieces of a kind of graywacke
with *Nummulites* and *Orbitoïdes*, which prove that there must

be rocks of eogene [1]) age in the Upper Kapoewas mountain range. BÜTTIKOFER found pebbles of an identical rock in the boulder-banks of the Sibau. On page 162, I discussed the appearance of these rocks containing *Nummulites*, and suggested the possibility that subsequent examinations would prove that the graywacke with *Nummulites* forms part of the system of the contorted strata of the Upper Kapoewas mountains. The petrographical relationship of the *Nummulite*-rocks and the other rocks of the Upper Kapoewas mountains, would certainly justify this conclusion, but on general geological grounds it seems to me, for the present, highly improbable. If however this should prove to be the case, then the rocks of the Upper Kapoewas range must have been folded after the eogene age, and consequently the whole or the greater part of this range must have originated after the eogene period, and later on have become denuded and eroded again into the condition in which we now find it. This idea would be in conflict with all hitherto acknowledged facts concerning eogene deposits in the Dutch East Indian Archipelago. True, the strata of this age have here and there, locally, been much disturbed, but on the whole they have not so far been seriously subjected to mountain-forming pressure. The appearance of real graywacke and clay-slate of the type of drawing-slate is unusual in eogene deposits and would be unique in the Indian Archipelago. This argument however loses in importance when we consider that in folded mountains in other parts of the world similar rocks of tertiary times are not unknown; for example the Flysch-schiefer.

In the second place BÜTTIKOFER found, in the Sibau-river, a small boulder of antimony. This find is of some importance because for years it has been an open question and a subject of investigation with the Dutch mining-engineers, whether antimony, does or does not exist in Dutch Borneo. True the ore was brought by the natives from different parts of Borneo, as

1) Eogene = old-tertiary, comprising the eocene and the oligocene formations.

for instance, from the little river Kéla (? perhaps the Ella) a tributary of the Mělawi[1]) and from the Ibau rivulet in the river-basin of the Pinoh[2]) and reported as actually found there; but further investigation generally proved the doubtful value ot such statements and led to the conclusion that the ore was derived from elsewhere, which could easily happen, it being a very common ore in the neighbouring province of Sarawak. Some natives brought me a piece of antimony, said to have been found in the river-basin of the Sělimbau-river.

The find of Büttikofer, is, as far as I have been able to trace, the only case in which the appearance of antimony in West-Borneo, although only as pebbles in a river-bed, has been confirmed with certainty by a trustworthy and competent person. These finds, however, would hardly justify speculating on the possible existence in this district of limestone of carboniferous age, which in Sarawak seems to be invariably connected with the appearance of lodes of antimony. The antimony needs not necessarily occur here in an exactly similar formation as in Sarawak, and in the second place our knowledge of the geological structure of the districts in the river-basin of the Batang-Loepar and the Batang-Rědjang in which the antimony is found, leaves yet much to be desired.

There seems to be a total absence of granite and other plutonic rocks in the region of the old slate-formation; neither from the Embaloeh, nor from the Sibau, nor from the sources of the Kapoewas above N⁴ Boengan, do I know of one single piece of granite. Dykes of igneous rocks, are also extremely scarce. Possibly the boulders of diabase which I found in small quantities below the goeroeng Moenhoet, in the upper part of the Kapoewas, may have originated from such dykes.

1) R. Everwijn, *11*, p. 31
2) See C. J. van Schelle, *53*, p. 81.

Pl. XXX.

THE BAKANG-RAPIDS.

2. The Upper Kapoewas lowlands, and the Boengan hills.

This territory is composed of:

1. A system of clay-slates alternating with diabase, diabase-porphyry, amygdaloidal diabase and diabase-tuff, with intercalated strata of chert, jasper and hornstone containing numerous Radiolaria.

2. Greywacke-sandstone, arkose, greywacke, conglomerate and sandstone in thick beds. In the greywacke-sandstone tiny coal-seams occur occasionally, and in one place unquestionable and clearly defined tests of *Orbitolina concava* Lam. were found.

The rocks of the first group are very strongly contorted and folded, both strike and dip changing constantly. All are in a very great measure altered by dynamo-metamorphism; and yet their position is not so intensely modified by mountain-pressure as is the case in the old-slate formation: A perfectly isoclinal structure which is of so frequent occurrence in the last-named formation, has, to the best of my knowledge, not been reached here anywhere. The diabases and diabase-tuffs have all become more or less schistose through mountain-pressure, and are internally crumbled and much decomposed. They are also in great measure silicified, so that in some places they have passed into gray or greenish chert. They are therefore both macroscopically and microscopically very variable rocks, which I have all included in the name of Poelau-Mélaioe type. The clay-slate is in many places not only affected by orogenetic forces but also by contact with intrusive sheets and dykes of diabase, and then altered into a kind of hornfels. This hornfels and the silicified strata of the Poelau-Mélaioe type form, almost without exception, in the Upper Kapoewas, the Boengan, and the Boelit, the obstacles which create the rapids and waterfalls. Chert, jasper, and hornstone with Radiolaria, equally folded and contorted, lie conformable between these layers. It appears from HINDE's investigations, that these rocks are of pre-Cretaceous perhaps of Jurassic age. Although they are easily recognizable and are generally very conspicuous in the field, yet their complicated structure would not permit us to decide with certainty whether these cherts occupy one, or more than one, horizon in this complex of layers of the Danau formation.

Because of the frequent appearance of Radiolaria, this series of rocks must be looked upon as a deep-sea deposit; it forms the eastern continuation of the corresponding formation, of which, in the region of the great lakes, a great part of the rock *in situ* is composed (see page 92).

The second group of rocks, the greywacke-sandstone, arkose, etc., on account of the appearance of *Orbitolina concava* Lam. must necessarily be of Cretaceous, probably of Cenomanian age. In contradistinction to the first group, this is evidently a deposit originating in the neighbourhood of a coast. The deposits are coarse, the beds thick, and the coal-seams in the sandstone betray the nearness of a coast-line. In other parts of West Borneo, viz., in the Sěběroewang valley, *Orbitolina concava* also occurs (as we shall see later on) in similar deposits, which also must be regarded as coast-deposits. The strata of this group of rocks, although for the greater part steeply inclined, have evidently been much less under the influence of mountain-pressure than the others. Probably the activity of the mountain-building forces was already in operation when these Cretaceous sediments were deposited in the neighbourhood of the rising coast, and as the folding-process continued they have occasionally become inclosed between the pre-Cretaceous sediments, in which they now form troughs.

Whether the almost horizontal beds of greywacke and arkose, near and on the boundary line between West- and East-Borneo, belong to this same formation, cannot be definitely asserted, because no fossils have been found in them. Petrographically there is no occasion to separate these horizontal sandstones from the arkose and greywacke-sandstones folded in between the pre-Cretaceous strata. It is merely on account of their horizontal, almost undisturbed, position, that I have provisionally separated these strata, as presumably Tertiary, from the Cretaceous formations.

In the Danau formation, granite as intrusive rock is exposed here in various places, another sign of its great resemblance in structure to that of the lake-district. I found granite along the banks of the Kapoewas above Poelau Tĕngkidoe; in the Boelit a tonalite massive was discovered not far above pangkalan Mahakkam, and the great quantities of granite blocks transported by the Mĕdjoewai justify my belief that there must be a granite region somewhere in the basin of this river. In some places, more especially on the Boelit, the granite has been under the influence of mountain-pressure, and the tonalite for instance has there been altered into a schistose rock resembling gneiss or amphibolite. We gather from this that the folding of the rocks was not completed when the intrusion of the granite took place, and that the mountain-building forces were at that time still at work in this region. Gabbro, norite and allied rocks such as harzburgite and others, mostly altered into serpentine, are found frequently and in great variety in this region, especially in the basin of the Kĕryau and from there up-stream, along the banks of the Kapoewas, into the Mĕdjoewai district. Wherever these rocks were found *in situ* they were generally altered by dynamo-metamorphism, and more or less schistose. Most of the varieties, however, were found amongst the boulders in the rivers, and nothing further can therefore be said as to the manner of their occurrence in this region.

3. The Müller-mountains.

Just as the formation of the Boengan-mountains is the eastern continuation of that of the Upper Kapoewas lowland and of the lake district, so the volcanic Müller-mountains on the Kĕryau, and on the Upper Boengan and the Boelit, form the eastward extension of the mountains of the same name, with which we became acquainted in the river basin of the Mandai (p. 65).

There are however important differences, both orographical and geological, between the western and the eastern portion of this range. On the Mandai it consists of a succession of table-mountains, an eroded plateau-region, built up of horizontal tuff-beds, rich in silicified wood. These tuffs are composed of inter-mediate and basic rocks, with a predominance of varieties of andesite, especially hypersthene-andesite and basalt. In one place, an offshoot of this territory reaches into the Kapoewas plain near P. Laap, where a basalt dyke (perhaps a flow), originating in the Mandai district, intersects the Kapoewas. On the Kĕryau, at any rate in its middle course, this same type of volcanic table-land prevails, but there is a marked absence of some of the basic eruptive rocks, especially of basalt. The examination of this district however was imperfect (see p. 179 et seq.).

Eastward of the Kĕryau district all traces of table-land have disappeared. We see before us a strongly marked, very pictu-resque mountain-land, in which some peaks of considerable height stand out boldly and with striking individuality. On the other side of the watershed, in the Mahakkam district, the Müller-mountains preserve this latter characteristic. There are no basic rocks, no basalt, and even the intermediate rocks are overshad-owed in number, and in diversity of form, by many varieties of acid rocks belonging to the groups of rhyolite and dacite, which here have attained a very strong and multifarious development. It seems that the framework, the older part of the Müller-mountains, consists here of andesite (especially mica-andesite), while the rhyolite and dacite were erupted later, and now form principally the higher peaks. With regard to the age of the volcanic rocks we can only say, that they must be later than Eocene. The enormous development of the volcanic territory points to a long continued period of volcanic activity, while the extensive encroachments made by erosion even in the most recent parts of this volcanic formation, justify the theory that volcanic activity must have ceased long since. My observations moreover, testify to the existence, in principle, of the present system of valleys at the time when these volcanoes were active, by which I do not mean to state that the volcanic activity commenced when the formation of the valleys had advanced almost to the stage which it has now reached, but which certainly does prove that volcanic activity had not yet ceased when their formation was initiated.

We must not omit to mention that the andesites are all of an ancient type, and they would, if examined under the microscope alone, perhaps be determined differently; thus the mica-andesite would for instance in that case perhaps be called kersantite. If from these indications we are justified in drawing any conclusion as to the age of this portion of the Müller-mountains, it can only be this, that volcanic operations commenced here in the late Cretaceous or early Tertiary, but that they have not continued into the present period. [1]).

Volcanic after products, such as gas exhalations, springs of warm water or water containing carbonic acid, I have not found any where, nor could the natives give me any information about them. To my knowledge there are no traditions among any of the Dyak tribes which refer to the volcanic activity of any of these mountains.

---

[1]) Perhaps an examination of the pieces of silicified wood imbedded in the volcanic tuff of the Mandai-river may throw more positive light upon this subject. Up to now this beautiful material has not been studied.

CHAPTER IX.

THE SĔBĔROEWANG AND THE EMBAHOE[1]).

The Sĕbĕroewang and the mountains where it rises, I visited twice, in the first instance from March 25 to April 3, and again from August 7 to 19, 1894. On the second occasion I crossed the mountains to the Embahoe basin and made my way down that river, back to the Kapoewas. The place where the Sĕbĕroewang or Sĕbĕroewang bĕsar[2]) empties in to the Kapoewas is 20 kilometres distant from Sĕmitau, and can be reached from there by sampan in three hours time. In its lower part the Sĕbĕroewang is 40 metres wide, and pursues a very meander-ing course, but when the water is high the distance may be considerably shortened by going through some pintas. The rela-tive state of the Sĕbĕroewang has a marked influence upon the rate at which one can row up-stream. The most favorable con-dition is when the Kapoewas is high and the Sĕbĕroewang low, for then the water in the latter river is pushed up by the water of the Kapoewas, so that to a considerable distance beyond the confluence the current in the Sĕbĕroewang remains almost imperceptible. When, on the other hand, the Kapoewas is low and the Sĕbĕroewang high, the current in the lower course is so strong that rowing up-stream is most difficult. The banks are at first low and marshy, and only in a few places, as indi-

1) For this chapter consult Maps III, IV, and VIII, and the sections KK' and LL' on page 19 of Atlas.

2) Sĕbĕroewang bĕsar, in contradistinction to Sĕbĕroewang Sĕni, a far less important stream, which also flows into the Kapoewas a little higher up on the left side.

cated on map VII, the solid rock appears; it consists of clay and sandstone with particles of coal, and occasionally of fine conglomerate. At Kwala¹) Pĕmindoek, marly limestone beds with hard concretions are interstratified with the clay. The hilly territory at some little distance from the banks, now and again sends out spurs close to the water's edge, as, for instance, at Sĕdjiram, a Dyak settlement where a R. C. Mission is established. On a small hillock just below the mouth of the Sĕdjiram, a tributary of the Sĕbĕroewang, stands a little church with a house and adjoining buildings. Two priests of the Jesuit order were stationed there, Fathers LOOYMANS and MULDER. They have

Fig. 53. DYAK CHILDREN FROM THE SĔBĔROEWANG.

baptized a goodly number of Dyaks, particularly children, and these little ones are frequently lodged and educated by them in the mission buildings. Fig. 53 shows a group of these juvenile

1) Kwala = Nanga (Mal.), mouth of a stream.

Christians, fair specimens of the merry, roguish little faces so often met with in the Dyak houses. The priests confine their missionary labours to the heathen Dyaks, they will not interfere with the Mahomedans or with the Chinese. Three times I visited Sĕdjiram and each time the priests received me with the greatest hospitality and kindness, each time also it cost me an effort to bid goodbye to this small outpost of European civilisation.

Under favorable conditions one can get in six hours to Sĕdjiram from Kwala Sĕbĕroewang; but if the current is very strong it takes much longer, it is then hardly possible to reach the settlement in one day. There is, however, a marshy Dyak footpath by which one can go from Sĕdjiram to Sĕmitau in one good day's march.

On the right shore of the Sĕbĕroewang, just above Sĕdjiram, rocks of serpentinized tuff-breccia make their appearance, and I noticed this same formation $^1/_4$ kilometre distant from the mission-house in the brook Lĕmiang a [right branch of the Sĕdjiram. It is there exposed in thick beds striking E.S.E.— W.N.W. and dipping 78° to the north. The mission buildings are erected on a foundation of grey fractured quartzite and chert with limonite veins, interstratified with micaceous quartzite with quartz and calcite veins; all these strata dip S.S.W. More northward the same kind of rocks reappear in the bed of the Sĕdjiram, but there they dip 75° to the N.N.E. These rocks form the wings of an open anticline. Upon the grey quartzite lie strata of tuff-breccia, and upon these, beds of red fractured chert and hornstone with Radiolaria, which strata also appear on the right bank of the Sĕbĕroewang below Sĕdjiram, dipping 60° to the S.S.E.

The position of the strata, as indicated on map VIII, shows that they are folded, and strike W.N.W.—E.S.E., and that the tuff-breccia lies conformably between the strata of quartzite and chert with Radiolaria. This system of strata, which will be described more fully later on, is of pre-Cretaceous age, and it is the oldest formation known in the Sĕbĕroewang valley. Above Sĕdjiram

the river basin is cut in a younger formation for which we may claim a Cretaceous age. In this latter formation three horizons may be distinguished (see section KK') 1. the oldest or clay series, 2. the middle or marl series, and 3. the upper or sandstone series.

There is scarcely any virgin forest left on the banks of the Sĕbĕroewang, fresh made ladangs and rank jungle growing on old ladangs now occupy its place. When cutting down the forest however, the tapangs or bee trees have generally been spared and a few magnificent specimens may be noticed among the younger wood, (see fig. 54). Here the river has no natural obstructions such as are caused by rocks or rapids, but the natives have in this comparatively well populated district, wilfully created many of them by their bad habit when making ladangs, of throwing the trees nearest the edge into the water. This evil practice seems to be the fashion in Borneo, for the Malays as well as Dyaks do it, and many a splendid piece of navigable water, has thus been radically spoiled. On the whole the banks are low, and rock

Fig. 54. TAPANG TREE.

in situ is but rarely seen. On the banks above Sédjiram, rocks of the oldest series viz., clay-stone may be seen, which in some places consist almost of pure kaolin. About 4 kilometres below Sajor, alternating layers of brownish marl and thick beds of grey, marly limestone, overlie the clay-stone. In this, the limestone or marl series, which unfortunately was nearly always submerged when I was there, I found a few fragments of shells, spines of a *Cidaris*, and some undeterminable shell impressions. Presumably it is in these beds that the cretaceous fossils described by Geinitz[1]) have been found. The average strike here is E. 20° S. corresponding with the general direction of the Sĕbĕroewang valley, and the dip is steep to the south. As a rule, the strata dip 45°—60° to the south, but they often stand vertical or nearly so, and they have even been found to dip in the opposite direction, i. e. to the north, as is the case above the kampong Sajor. This, however, is only a local occurrence due to folding, and the general dip of the strata in the Sĕbĕroewang valley is undoubtedly to the south-south-west.

At Nᵃ Sajor where there is a house with 3 pintas at about 160 metres distance from the river, I made an excursion inland in a north-eastern direction, as far as the house Soengai-Aping, in the vicinity of the house Sajor, about 4 kilometres from Nᵃ Sajor. In the ladangs and in the forests there are several rock-groups which clearly form part of two parallel beds of a kind of tuff-breccia, identical with that which, as already stated, is found at Sĕdjiram on the right shore of the Sĕbĕroewang. This then is again the pre-Cretaceous formation which, at some distance from the Sĕbĕroewang in a northern direction, constantly skirts the cretaceous rocks. The thick beds of breccia stand vertical and trend W. 30° N.—E. 30° S. They contain pieces of diabase, clay-slate, quartz, and limestone, consolidated by a kind of cement which might be called chloritized and serpentinized diabase-tuff. The pieces of diabase belong for the greater part to a glassy type, a kind of vitrophyre, microscopically not to be distinguished from andesite. I shall call this breccia, which is splendidly developed and exposed at Mt. Rajoen, the Rajoen breccia.

Above Sajor, sandy clay-stone, and clayey sandstone of the upper series prevail, and in many places coal-seams or thin

---

1) H. B. Geinitz *13*, p. 205.

layers of coal occur. The beds are well exposed in the Kator river, a right branch of the Sěběroewang, at no great distance from its mouth. The karangans which near Sědjiram contain only sand or fine gravel, are here formed of pebbles as large as a fist. A very remarkable feature amongst them is the great abundance of muscovite-granite and pegmatite. Near Tandjong Těběroe a fault, cutting right through the river, brings the lower series once more to the surface, and near Kwala Měndjalin it is overlaid, along the left shore, by coarse, marly sandstone with *Orbitolina concava* Lam. These fossils are found in much greater abundance a little further up-stream on the right shore, where, for a short distance, the rocky banks, 3 or 4 metres high, are entirely composed of beds of *Orbitolina*-marl, striking E. 25° N.— W. 25° S. and dipping 32° to south-south-east. Grey sandstone and clay sandstone of the upper series overlie this marl.

A little higher up on the left shore, the bed of a small tributary, the Běkoewan, cuts through the sandstone stage, thick beds of which alternate with thinner layers containing numerous coal-seams. These strata dip steeply S. 30° W. or they are absolutely vertical. About 3 kilometres further up-stream, near Batoe Běnawah, the layers of this same system dip somewhat less to the south, here the clayey sandstone contains fragments of shells and a few tests of *Orbitolina*. Then, by folding, the marly-stage is again brought to the surface in two distinct beds, one having a steep incline to the N. N. E., the other to the S. S. W. The river has cut almost straight through the anticlinal axis of the fold, so that both portions appear at intervals along the shore. I found the best preserved tests of *Orbitolina* in the sandy marl opposite the mouth of the Měnijin, a small brooklet which empties into the Sěběroewang, 3½ kilometres above the Bělikai. Here, the marl is rich in vegetable remains and thin coal-seams. About 200 metres below Nᵃ Gaman a remarkable section is exposed (section D on map VIII). Between the sand-beds of the upper series, dipping 45° to the south, lies a thick bed (0.40 metre) of coarse conglomerate, containing numerous boulders of

granite, some having the size of a man's head. The pebbles in
the banks of the Sĕbĕroewang are here displayed in great variety.
I collected specimens of amphibole-granite, muscovite-granite,
amongst which was a beautiful variety with red garnets, muscovite-pegmatite, quartz-diorite, coarse grained diorite, quartz-porphyrite, amphibole-porphyrite, quartz-amphibole-porphyrite, chert
and silicified tuff with Radiolaria, quartz, and hornstone, and I
noticed that the greater part of the crystalline rocks was brought
down by the Bĕlikai, a left branch of the Sĕbĕroewang, which
rises on Mt. Oejan.

The pangkalan Pyang lies $2^1/_2$ kilometres above Kwala Mĕnijin.
Here the Sĕbĕroewang ceases to be navigable for loaded sampans.
Close to the pangkalan lives a Chinese merchant. From Kwala
Sĕbĕroewang the pangkalan can be reached in two days, provided
that the water is high, the sampan light, and the traveller can
dispose of a crew in a fit condition for smart rowing or poling, as the
case may be; otherwise it is a question of at least three days.

From pangkalan Pyang mountain tracks lead in different directions through the hills between the rivers Sĕbĕroewang, Soehaid,
Embahoe and Silat; everywhere Dyak houses stand scattered
about. One of these tracks leads through the Sĕbĕroewang valley
some distance from the right shore of the river, to Sajor, and
from there northward to Soehaid.

At Pyang a splendid view can be obtained of the hills of the

Mt. Mĕrangat.   Mt. Biroe.   Mt. Rajoen.
365.   500.

Fig. 55. THE HILLS NEAR THE SOURCES OF THE SĔBĔROEWANG, SEEN FROM PYANG.

Upper Sĕbĕroewang, conspicuous amongst which are Mt. Rajoen
(500 metres), with its precipitous cliffs and delicate outlines,
and the lofty, thickly wooded, Mt. Oejan (900 metres) On the
10th of August I visited this latter mountain starting from Pyang.

I took a path which led up-stream sometimes along the banks, sometimes through the river or over it by tree-trunks, for a distance of 4 kilometres until I came to N° Ajoe. Then I followed the Ajoe brook up-stream; no solid rock appeared ·until the place where the brook Sĕpan flows into the Ajoe over banks of clay-stone, slightly sloping southward. Up to this point the path had run, mostly in a south-easterly direction, but now it verged to the southwest and leaving the Ajoe, brought me to the western extremity of Mt. Oejan. The Sĕpan coming down from that mountain flows past Roemah Bĕlôh ¹) two hours distance from Pyang, which lies at the foot of Mt. Oejan on the west side.

The solid rock in the bed of this streamlet is clay-stone alternating with sandy clay-stone, the strike E.-W. and the dip $15°—20°$ to the south. Loose pieces of a fine breccia and a conglomerate, which higher up forms the solid rock, lie scattered about in the stream. The beds of conglomerate dip slightly to the south and consist of boulders, moderately water-worn and of different sizes, but as a rule not smaller than a man's head. They are fragments of widely different rocks, consolidated by a quartzose cement. This conglomerate resembles that which I found in the Sĕbĕroewang, just above Kwala Gaman, interstratified with beds of sandstone.

It seems probable that the plutonic crystalline rocks, muscovite-granite, pegmatite, diorite, etc., which I found in such quantities in the boulder-banks of the Sĕbĕroewang have originated from this conglomerate. This supposition is supported by the fact that in the Sĕbĕroewang the greater part of these stones have been deposited by the Bĕlikai, and that this stream rises on Mt. Oejan. The largest pebbles in this conglomerate are, for the most part, different varieties of muscovite-granite, pegmatite, graphic granite, syenite, diorite and occasionally limestone. From here a steep ascent through dense wood leads to a small ladang

---

1) Roemah Bĕlôh is a small Dyak house, not marked on the topographical map and according to my estimation about 2 kilometres to the east of the house Soengai Roesak. The Sĕpan is probably the same stream which on the map is called S. Roesak.

400 metres high, the highest ladang on the slopes of the Oejan. Then follows another stretch of dense wood, and being told by a Dyak who accompanied me that the summit was entirely covered with this thick growth, and that there was no view to be had from the top, I decided not to continue the ascent. The solid rock was still the same coarse conglomerate, just mentioned, the beds dipping slightly to the south. The soil was, however, overstrewn with loose fragments of quartz-amphibole-porphyrite, and about 100 metres higher up in the wood, this same porphyrite formed the rock *in situ*. Although I did not climb to the top, I have no doubt whatever that it consists entirely of this porphyrite.

This high ladang offers a most magnificent view. The wide valley of the Sĕbĕroewang opens towards the west, bounded on the south by the Sĕraboen range of hills and the other ranges parallel with it which separate the Sĕbĕroewang from the Silat. They all trend E.—W. or E.S.E.—W.N.W. To the north, the valley is separated from the plains of Soehaid and Sĕlimbau by the Biroe range trending E. 25° S.—W. 25° N. In this range of hills, Mt. Mĕrangat, to the west of Mt. Biroe, is conspicuous by its large white patches, probably of naked rocks. Both Mt. Oejan and Mt. Rajoen, form part of ranges of hills trending E.S.E.— W.N.W., which divide the extensive valley at the head-waters of the Sĕbĕroewang into strips running in the same direction. They only send out a few unimportant spurs in a more westerly direction. The trip from Pyang to Mt. Oejan can be made in one day.

The track which keeps up the communication between the river basins of the Sĕbĕroewang and the Embahoe, begins at Pyang. At first it runs in an E.N.E. direction straight to Mt. Rajoen, sometimes closely following the meandering course of the Sĕbĕroewang, sometimes leaving the river at a considerable distance. The soil is sandy, the path leads through old ladangs, and no solid rock is to be seen.

About 2¹/₄ kilometres from Pyang we crossed the bed of the

Bĕdoengan, a rather important feeder of the Sĕbĕroewang. Pebbles of sandstone, whet-slate, clayey sandstone, and diabase, cover the bed of the stream. Presently the path turned to the north and led across some sloping ground to the west side of the Rajoen. This slope we may take to be a continuation of the western spur of that mountain. The rock *in situ* is exposed in a few solitary blocks, showing the same formation as found on the mountain itself, viz., the Rajoen breccia just mentioned, which here contains a good deal of limestone.

Soon we struck the Bĕdoengan again, near the deserted house of that name on the right shore. In the bed of the Bĕdoengan, and particularly in the cliffs which line the right shore, the solid rock is frequently exposed, and I was able to examine the section, given in map VIII, section E. The bed of the river cuts through thinly laminated diabase-tuff (9), many of the diabase fragments bearing an andesitic character. Some of these layers of tuff contain tests of Radiolaria, besides a few Foraminifera. The Radiolaria have been still better preserved in a greenish-grey, silicified tuff (10), overlying the former, in which the tests have been infiltrated with some chloritic mineral. The next layer is chert (11), possibly very fine silicified tuff, followed by fine tuff-breccia containing fragments of this chert (12), then again tuff, and finally fine tuff-breccia (13 and 14). HINDE, who has examined the fossils in these beds, comes to the conclusion that this system must be of about the same age as the chert and jasper of pre-Cretaceous age familiar to us from the lake district and from the Upper Kapoewas, although the resemblance of the fauna of these two localities is not so close as to enable him to assert positively that the tuffs with Radiolaria from here, belong precisely to the same "horizon" as the cherts with Radiolaria from the Upper-Kapoewas. The entire system dips about 40° to the north-north-west; but dislocations occur here and there which materially modify both trend and dip.

All these rocks are extensively jointed and break easily in parallelopiped fragments. The joint-planes show invariably the well known aspect of a broken piece of serpentine, so that, at first sight, one is apt to mistake the rock for a serpentine. But as a matter of fact this mineral, the result of decomposition, is only a comparatively insignificant constituent, only the joint-planes of the rock being infiltrated with it.

The house Loengang Bĕdoengan lies on the north-west slope of the Rajoen, about $^3/_4$ kilometre higher up-stream on the left of the Bĕdoengan, which is there only a small mountain stream. On the 29th of March we made our bivouac near this house [1]), on the outskirts of the forest. The view from here was most beautiful; on the one side lay the house with the Biroe hills rising behind it on the northern horizon (see plate XXXIV), on the other side the luxurious rice fields which cover the hills opposite. The view of Mt. Rajoen, on the other hand, with its precipitous slope terminating abruptly on the north-west side, is equally beautiful, when seen from the ladangs on the opposite hill. From there, facing Mt. Rajoen, the house Loengang Bĕdoengan and the newly laid out ladangs with huge boulders laying scattered about, make a handsome foreground (see plate XXXV). The little clump of high trees on the summit of Mt. Rajoen, is the only remaining bit of virgin forest, all the rest has been cleared to make ladangs.

From here I ascended the western top of Mt. Rajoen, 400 metres high, following a very steep and slippery track. The solid rock is frequently laid bare on the slope, and I was able to get the section E, given in map VIII. The base of the mountain is composed of fine tuff-breccia (1). Then follow several beds of very coarse tuff-breccia (2—4) containing so many fragments of limestone that the rock may be said to be composed of limestone blocks consolidated by tuff cement. The blocks which lie dispersed over the ladang at the foot of Mt. Rajoen are of this same coarse breccia. Although the limestone is not

---

1) This house has only lately been built, and it is not yet marked on the topographical map. The house there marked 'Bĕdoengan' is deserted and the inmates have gone to the new house on the other side of the river.

Pl. XXXII.

STALACTITES ON THE LIANG-BOEBOEK.

altogether devoid of structure, I have failed to discover in it any determinable organic remains. On the breccia lies chert and fine silicified tuff (5. 6. 7), and near the top fine tuff-breccia (8) again occurs, similar to the breccia exposed in the upper strata in the bed of the Bĕdoengan.

The whole of this system of strata dips slightly south or south-east. I found that the whole of the range, about 4 kilometres long, known as Mt. Rajoen, had once been cultivated ground, and that only here and there on the precipitous slopes and on the highest peak (500 metres) patches of virgin forest had been left standing. Much land has been cultivated in this fairly well populated district, and the great charm of the view from Mt. Rajoen lies principally in the harmonious blending of contrasting tints; the fresh green rice fields, the monotonous, dull grey-green of the secondary growth on the old ladangs, and the characteristic dark foliage of the primeval forest, amongst which a few giant trunks display their white glistening bark. The landscape cannot be called grand, but it is charming on account of its great variety. As far as the eye reaches, range upon range of hills follow in close succession one upon another, all trending E.S.E.—W.N.W., not however as in the Upper Embaloeh valley, all built after the same pattern, but each with its individual characteristics, and this points *a priori*, to differences either in the geological structure, or in the position of the strata.

Nearest to us, looking south, was the Oejan mountain range, on the other side of the head-waters of the Sĕbĕroewang and Tĕpoewai. This range, as we saw above, does not extend any further to the west, but almost entirely closes the Sĕbĕroewang valley on the south-east side. Following Mt. Oejan (900 metres) to the east is Mt. Mĕrakai (719 metres) (see maps III and IV), which curves to the north-east and joins the Pyaboeng mountains (1130 metres). These latter are isolated mountains, and totally different in shape to any of the others; they are the loftiest in this district. Their principal components are igneous rocks, such as amphibole-porphyrite and quartz-amphibole-porphyrite.

17

Then to the south comes the Séraboen range, which, as far as one can see, appears to be uniform in height, and from our stand-point looked like a giant's causeway. It much resembles the renowned Magalies mountain, north of Pretoria, a range of hills, composed chiefly of quartzite, which stretch away for miles at apparently the same height. The gaps in this wall, in the Transvaal characteristically called "poorten," are not lacking in the Séraboen range. The Silat, for instance, breaks through these hills with a typical "poort" (gate). The Séraboen range (see Atlas map IV) is supposed to commence at the Kapoewas river¹), and Mt. Sĕtoenggoel (355 metres), composed of sandstone with a slight southward dip, may be called its first outpost.

Then follows Mt. Sébalang (507 metres), separated from the former hill by a strip of lower ground and to the east of this, the Séraboen range bifurcates into two parallel branches running in an E. S. E. direction. The most northerly branch, the Séraboen range proper, continues with undisturbed regularity over a distance of 100 kilometres, its height being hardly ever under 400, and never exceeding 500 metres, except of course at the "poorten." Mt. Bĕranak, at the head-waters of the Gilang, is pointed out as the eastern termination of the Séraboen range, because it there loses its peculiarly uniform character and the mountain ridges become considerably higher. Mt. Bĕranak (540 metres) is the only peak in this range which exceeds the 500 metres level. The Seraboen range here amalgamates with the mountains which might be called the southern outposts of the Madi plateau. The width of the Séraboen mountains, at least where they exceed the 100 metres level, is as a rule less than a kilometre. The southern branch, also called the Mérdja range, after its highest peak Mt. Mĕrdja (690 metres), is much less regular, although closely connected with the Séraboen range, which is best proved by the fact that the valley of the Djitan river, which separates the two ranges, forms a high-plain above the 100

---

1) It is uncertain whether these hills extend any further on the right shore of the Kapoewas.

metres level. The valley of the Dangkang and the Ĕntibah separates the Mĕrdja from the Sagoe range of hills, which is the
direct continuation of the Pĕnai mountains already familiar to
us. Although lower, these hills are no less remarkable than
the Sĕraboen mountains. Like these they rise abruptly from the
plain in a narrow ridge scarcely half a kilometre wide. They
extend over a distance of 68 kilometres, and are very uniform in
height, the crest never exceeding 250 metres. I noticed in it four
typical "poorten." The Silat river breaks through two of them, the
Ĕntibah through the third, and the Sĕlimoe through the fourth.
In the same way as the Sĕbĕroewang bifurcates to the east
of Mt. Sĕbalang, so the Pĕnai range splits into two· branches
when approaching the Kapoewas. From the top of Mt. Kĕlam
this range looks like a tremendously high wall, stretching right
across the plain and curving round as it approches the Kapoewas.
This curved part, the southern branch, is 44 kilometres long,
but not nearly so regular in structure as the northern branch.
This latter, called the Sagoe range, touches the Sĕraboen range
near Mt. Tangga, and from there, extending eastward, it gradually
loses its peculiar orographical individuality. The striking similarity
of these mountains necessarily leads one to presuppose uniformity
of geological structure, and it is therefore quite allowable to
judge of the general structure of the mass by examining the
most westerly spurs, the Sĕtoenggoel and the Pĕnai mountains,
which are both composed of sandstone. I even go so far as to
express my firm conviction that subsequent investigations will
prove that these remarkable hills originally formed a plateau district
built up of beds of sandstone gently dipping to the south and
now dislocated by faults striking W.N.W—E.S.E., as represented
in section KK'[1]).

To the west, in the direction of the Sĕbĕroewang valley, one

---

[1]) Moreover it is a known fact that on Mt. Sagoe, sandstone is found. In the Geological
Museum at Leiden there is a piece of sandstone from Mt. Sagoe, collected by VAN SCHELLE
in 1879, corresponding in every respect to the sandstone of Mt. Sĕtoenggoel.

can see distinctly that Mt. Měrangat is the connecting link between the Rajoen hills and the hills near Sajor where I found tuff-breccia (see page 250) containing fragments of limestone, trending E.S.E.—W.N.W., and exactly corresponding to the formation of Mt. Rajoen. To the north the view is limited by Mt. Biroe close by, and beyond it only a few peaks are visible, amongst them Mt. Ampan N. 42° E., and slightly more to the front, Mt. Bekajoe N. 45° E., both situated on the Embahoe, at a short distance from N° Boeloeng.

From Pyang the top of Mt. Rajoen can be reached in $2^1/_2$ hours, provided there are no detentions by the way, so that with Pyang as the starting point, the trip can be accomplished in one day. Following the footpath the distance from Pyang to the house Loengang Bědoengan is about $5^1/_2$ kilometres.

August 11.   On August 11 I went from Loengang Bědoengan across the divide separating the Bědoengan from the Gaäng, thus passing from the river basin of the Sěběroewang into that of the Embahoe. The divide is not far distant from the house Bědoengan, and only 200 metres high. From this point, facing east, one gets a splendid view of the shallow valley of the Gaäng, gay with an endless variety of shades of green. Mt. Pyaboeng is only partly visible, the greater portion being hidden behind the eastern extremities of Mt. Rajoen. From the divide, a steep path brought us quickly to the bed of the Gaäng, about 75 metres lower down the pass. It is here only a little mountain stream, the solid rock in its bed being the familiar Rajoen breccia with fragments of limestone imbedded in it. The path goes up and down through the valley of the Gaäng, through low brushwood and over clayey and sandy ground; the solid rock being only occasionally exposed as shown, on map VIII. From N° Běnoewang, 4 kilometres from the divide, I ascended the eastern summit of Mt. Rajoen (454 metres) called Batoe Raoeng. Halfway up we passed through a natural grotto under overhanging rocks of very coarse Rajoen breccia. The mountain seems to be entirely composed of this rock, but nearer the top its grain becomes much finer. The beds are vertical and strike east and west. The view from the top is simply glorious. The Oejan seen from here in its widest proportions is strikingly beautiful. The higher part, viz., all that lies between 600 and 900 metres, appears like a solitary cone resting on a broad pedestal. This cone is probably entirely composed of amphibole-porphyrite. The semicircle of peaks, connecting Mt. Oejan with Mt. Pyaboeng, is very conspicuous, the natives call it the Bĕloewan mountains.

In the Gaäng near Nᵃ Bĕnoewang completely rounded rocks of marly limestone become visible above the water. Petrographi-cally speaking, this limestone, which plays a conspicuous part more eastwards in the valleys of the Gaäng and the Tépoewai, is closely allied to the marly limestone in the valley of the Sĕbĕ-roewang, and I believe both to be of Cretaceous age.

At Nᵃ Bĕnoewang the Gaäng becomes navigable for small August 12. sampans, but as I was unable to procure a boat at the house close by, on the right bank of the Bĕnoewang, I continued on foot as far as the house Kyaja on the left bank of the Gaäng. Five times in succession we had to wade some distance through the river, and yet oftener through some of the small side streams, which were all much swollen through recent heavy rains. I found clay and loam all the way, and solid rock was hardly ever visible (see map VIII). At Kyaja a small Malay kam-pong 4 kilometres from Nᵃ Bĕnoewang, I managed to secure a few small sampans in which next day I further descended the river.

The high state of the water was most unfavorable to my observations, for the solid rock, which, at the best of times, is difficult to trace in the shallow bed, was now almost entirely submerged. We saw nothing but granite-breccia and marly limestones of the same type as those found higher up the river. About 3½ kilometres below Kyaja, the Gaäng unites with the Tĕpoewai, this latter river rises on the eastern slope of Mt. Oejan. At the confluence the current of the two rivers is almost diametrically opposed, causing a momentary violent disturbance in the water, and a local decrease in the velocity and power of transport of the stream current. This accounts for the existence of a large boulder-island just below the confluence in the Tĕpoewai, the name by which the joint river is known for the rest of the way down-stream.

This boulder-island is connected by a shallow with a karangan at the right bank. In the karangan I also found, besides marl, granite-breccia, quartzite, and Rajoen breccia, several fragments of amphibole-porphyrite, which had evidently been transported by the Tĕpoewai [1]).

Just below the boulder-island an exquisitely shaped rock rises above the surface, but the natives begged me not to knock any pieces from it. In several other places down the river the same superstitious restriction still exists. I have marked the places on the map, with the name given by the natives to such rocks, „batoe pomali", i. e. haunted rock. The curious part of it is that the population, although now for the greater part turned Moham-medan, still revere the places which· once they held sacred.

From N⁴ Tĕpoewai down-stream one sees now and then a boulder-bank or a small island above the surface, but on the whole the water runs smooth, as it does in the Gaäng. I wanted to climb up the Pyaboeng, and for that purpose called at the house Antoek Oeloe [2]) which from Kyaja can be reached by sampan in $2^1/_2$ hours, the distance being about 10 kilometres.

The house is on the right bank of the Tĕpoewai, at the foot of the Pyaboeng mountains. The inhabitants are called Malays, but they are really Dyaks converted to Mohammed-anism. In Borneo these are always called Malays, and hence "to turn Mohammedan" is very characteristically called "Massok Malaioe". Of all the districts which I have visited in Borneo,

---

1) Besides these, there were fragments of quartz-biotite-porphyrite with large quartz bi-pyramids and rhyolitic habitus.

2) The inhabitants told me that the name of the house was Antoek, but the real kampong Antoek is about $1^1/_2$ kilometres further down-stream; I have therefore called this house Antoek-Oeloe.

it is here, in the head-waters of the Embahoe and Silat rivers, that Islamism has made most progress, and its influence upon the Dyak population is to my mind, without any exception, most pernicious. The terrible neglect of their homes, bears witness to their exceeding dirtiness and laziness. The difference between the Malay and Dyak kampongs is very striking; amongst the Malays I looked in vain for the friendly and even cordial reception which I had always received in the Dyak houses. Deceitfulness and avidity have totally supplanted the true Dyak virtues of hospitality, honesty, and trustworthiness. I was also struck with the absence of all articles of home Industry among the Malay-Dyaks. On the Sĕbĕroewang, where, besides the previously established Sĕbĕroewang Dyaks, a number of Kantoek Dyaks have settled, I found amongst the latter the most tasteful badjoes[1]) and koemboes[2]), and a large assortment of exquisite and dainty wood carvings. But there is no more of this work when once they become converts to Islamism. It is much to be regretted that the Dutch Government indirectly favors the propagation of Mohammedanism in so far that by preference Malays are always appointed as kampong chiefs; the Dyaks naturally infer from this that the Dutch prefer Mohammedanism to the Dyak ideas of religion; and I have often had much difficulty in convincing them of the contrary, and that the Dutch have no special predilection for Islamism.

In the house Antoek, a dirty hovel, all the available men were absent, working in the ladangs on the mountain slopes. So I sent for the kampong chief of the house Madoeng close to Mt. Pyaboeng. He also was unable to provide me with carriers, but he offered to act himself as guide up the mountain (Fig. 56).

The next day I set out, with some of my rowers, to carry August 13. the baggage to the house Madoeng, situated half an hour's distance at the confluence of the brook Madoeng and the Boeloe, a small mountain stream rising on the Pyaboeng. Many

1) Badjoe, coat of home-made tissue (see Pl. XX).
2) Koemboe, a kind of plaid of home-made tissue.

of the people here suffer from goitre, a very prevalent disease on the Gaäng and the Tĕpoewai, not only among adults but also among children.

Fig. 56. MOUNT PYABOENG.

We commenced the ascent by one of the north-western spurs of the mountain, which consisted of clay-slate and granite-breccia. At an altitude of 350 metres we came to a small ladang house from where I sent my three carriers back to Antoek to fetch the remainder of the luggage, which had to be left behind in the morning.

August 14   We went through the valley of the brook Tĕlĕpan, and then the actual ascent began. The rock *in situ* in the brook is whet-slate, with well defined strata dipping at an angle of 78° to the south. Large fragments of amphibole-porphyrite, porphyrite-tuff, porphyrite-tuff-breccia and chert lie scattered about in the stream. We somehow clambered up the precipitous height, always keeping to the south; with tremendous exertion we reached the top of an almost perpendicular wall of crumbling rock, 65 metres high, with only here and there an occasional stump or root to hold on by, and at the end of two hour's toil we stood on the top of the ridge, from where the highest peak is within easy reach. During the ascent, I noticed that this part of the Pyaboeng mountains consists of successive strata of diabase-porphyrite, tuff, tuff-breccia and chert, all dipping at an angle

Pl. XXXIII.

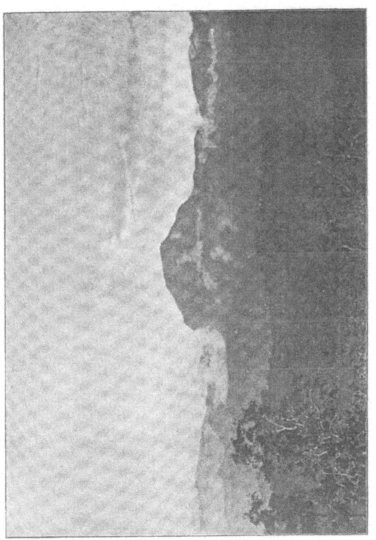

BOEKIT TĚMATA.

of about 55° to the south. They brought to my mind again the
rocks of Mt. Sĕmbĕroewang near Danau Loewar, and together
with these I include them in the pre-Cretaceous Danau formation. In another hour we reached the highest point (1150 metres)
following the ridge eastward. It rises in a very steep pointed
peak above the ridge consisting of amphibole-porphyrite. The
forest on the Pyaboeng is again of the marshy jungle type;
conspicuous are the colossal Nepenthes-plants, with cups 30 centimetres long and 6 wide. My expectations were only partially
gratified when I reached the top of Mt. Pyaboeng. I had hoped
for a good view of the mountain district extending towards the
east; but the weather had changed, clouds and mist enveloped the
mountain, and only now and then could I catch a glimpse of the
scenery around. This much however I could see, that a semi-
circular range of hills, known as the Bĕloewan hills, stretches from
Mt. Oejan to Mt. Pyaboeng, turning its concave side to the south-
east, and there joins an eastern spur of this last named mountain.
Within this semicircle are the head-waters of the Tĕpoewai. Between the mass of Mt. Pyaboeng and the Bĕloewan hills is a
deep narrow ravine where the Pĕnĕsah, one of the principal
feeders of the Tĕpoewai, has its source. From this point of view
also the Sĕraboeng range is very remarkable, stretching far away
to the south-east like a gigantic wall, with hardly varying height
and width. The Madi mountains and the whole of the hilly district
east of Mt. Bojan remained hidden behind the clouds. I returned
in the afternoon and discovered a very usable mountain path at
a short distance from the place where we made our toilsome
ascent. We came down quite easily, and reached the little ladang
house before nightfall. I cannot help thinking that the guide
showed us the dangerously steep ascent purposely, in order
to discourage me or my men, and so to earn his promised
reward more easily. It is only among the Malays or Malay-
Dyaks that I ever experienced any chicanery of this sort, such
a thing would never have happened among real Dyaks unless
they were already contaminated by Malay influence. It is al-

together contrary to the ruling moral principles of the Dyaks.

August 15. After spending one more night in the little ladang house, I returned to Antoek, and in the morning I was enjoying the beautiful panorama which displayed itself before me, as I descended the ladangs on the north-western spur of the Pyaboeng above Roemah Madang. The landscape, bathed in the early morning light, was most entrancing, especially towards the north across the valley of the Embahoe, which is bordered by fresh, pale-green ladangs. In the centre of this valley stands Mt. Ampan, looking in the distance like some colossal monument of stone, its shape and isolated position reminding me of Mt. Kĕlam near Sintang.

We reached Antoek Oeloe at 10 a. m., and with a good deal of difficulty I procured from the ungracious inhabitants, two little sampans in which I descended the river. Some of my coolies had to go on foot alongside the water or through it, because there was not room for them all in our two small boats. About 400 metres below Antoek, I suddenly saw a solitary rock, composed of breccia, standing in the middle of the stream. A little lower down on the left bank is a house and over against it a boulder-bank composed principally of boulders of fine Rajoen breccia and decomposed amphibole-porphyrite. A group of rocks lower down on the opposite side, also deserves special mention, because they are composed of liparite. Farther on, breccia and limestone (see map VIII) are the only rocks occasionally visible in the river-bed as far down as Nᵃ Tĕpoewai, where I stayed in the dirty, ugly house of Oeti, the kind-hearted chief of the kampong. Nᵃ Tĕpoewai is a rather large and exceedingly unclean Malay kampong, the houses of which lie hidden amongst palm trees.

August 16. So far the passage had presented but very few difficulties, as might be expected from the geological structure of the ground. Judging by the nature and position of the rocks which here and there protruded above the surface or along the bank, I feel convinced that the bottom rock in the valleys of the Gaäng and Tĕpoewai consists of marly limestone with strata striking east and west, and dipping steeply to the south. Overlying this limestone, and with the same strike, occur claystone and breccia, the breccia being principally composed of granite fragments. I think it probable that this entire system belongs to the middle or the upper division of the Cretaceous formation, as found in the Sĕbĕroewang valley. Both the Gaäng and the Tĕpoewai therefore run parallel with the prevailing strike of the rocks, and consequently no serious obstructions were likely to occur. But on reaching the Embahoe, ¹/₄ kilometre below the house of Oeti, the aspect of things was bound to change, for the flow of the current now became N.—S., and lower down, the river cuts a transverse valley through the hills which still trend E.—W. As might be expected, the rocks which gradually came to view below Nᵃ Tĕpoewai presented also a much more varied appearance. From the sketch on map VIII and the section LL', it will be seen that the rocks are folded and trend E.—W. or thereabout, and that they dip steeply to north or south or sometimes even they are vertical.

The chief components of this system of folded strata are fine grained breccia, much akin to the Rajoen breccia, silicified and serpentinized diabase-tuff, closely resembling the Poelau Mĕlaioe type, clay-stone, sandstone and quartzite, which to my mind belong to the Danau formation. Moreover, amphibole-porphyrite frequently occurs, and the numerous dams of rocks running across the stream, which impede the water in its downward course, are almost without exception composed of this igneous rock. These dams are most likely dykes, although, judging from their position, one might also take them to be intrusive sheets intercalated between the beds of the above named system of pre-Cretaceous rocks. For the ridges of amphibole-porphyrite rocks which cause the rapids, are generally grouped in a due east to west direction in the bed of the river, running parallel therefore with the trend of the hills, as, for instance, near the Riam Mĕlalang, the Goeroeng Boewang and the Goeroeng Balik. Highly interesting is the appearance of a dyke of amphibole-andesite, on the left bank, 100 metres below the Pĕmali rivulet, but it is only exposed for a very short distance. This dyke cuts through silicified tuff-breccia, near the contact with the breccia the andesite is highly glassy, and might almost pass for obsidian. Farther from the contact, the rock becomes gradually less vitreous, and at a distance of 3 metres, it has become ordinary andesite with micro-lithic groundmass. In some places also a few quartz-porphyrite rocks protrude out of the water, which in all probability form part of dykes. To the north of the Laki rivulet, the tuff-breccia along the shore gives place to white quartzite and quartz-sand-stone, the strata of which stand vertical, like those of the tuff-breccia, and they are, moreover, plicated and intersected by numerous dykes of amphibole-porphyrite and quartz-porphyrite.

Between the Riam Pajang and the Riam Toentoen, the river cuts through a small boss of granitite and amphibole-granitite. Through pressure the granitite has become very schistose and, seen with the naked eye, it looks in several places exactly like gneiss. Sometimes this granitite is injected by numerous veins of muscovite-pegmatite. In this case the pressure has converted the granitite into a rock which, macroscopically, looks as if it were composed of alternate bands of fine biotite- and coarse mus-covite-gneiss. Near the Riam Ambang the granitite is followed by protogine-granite, which has likewise become gneissic through pressure.

Below Goeroeng Toentoen granite entirely disappears, but in several places, as indicated on the maps, dykes of amphibole-porphyrite were found. This amphibole-porphyrite contains in many places a good deal of biotite and quartz. The habitus is andesitic, as is the case in all the different parts of the Embahoe valley which I have explored.

Just below the kampong Oelak I came upon the most serious obstacle to communication by water in the Embahoe. It is the renowned Goeroeng Oelak, where every year many boats come to grief. For a distance of 100 metres the bed of the Embahoe is much narrowed, and obstructed in a grotesque manner by masses of amphibole-porphyrite of andesitic habitus, the rock shows at the same time a separation into horizontal planes, and a roughly vertical, columnar structure. The height of Goeroeng Oelak is about one metre, and it is necessary entirely to unload the boats and to carry the luggage for some distance along the bank till the riam is passed. Considering the enormous area over which the porphyrite extends here, I am inclined to believe that this is a flow of andesitic porphyrite, through which the Embahoe has been forced to cut its bed. Below Goeroeng Oelak, on the left bank, hard, white, arkosic sandstone, rich in kaolin, constitutes the principal rock; it alternates with beds of a kind of conglomerate, till below Goeroeng Soenai. In the rocks in this goeroeng, thick undisturbed beds of white sandstone, rich in kaolin, with a gentle dip to the north, are well exposed. This rock belongs no doubt to a younger system, of possibly Tertiary age, deposited after the pre-Cretaceous strata had been folded and the granite had become gneissic through pressure. My limited observations however did not permit me to state with certainty whether this sandstone is younger or older than the porphyrite. In the Goeroeng Pĕlai, the last and most northerly of the rapids in the Embahoe, the rocks are again composed of andesitic biotite-amphibole-porphyrite. I visited one of the Chinese gold-diggings here. These diggings are often called parits; 'parit' means the cutting or channel where the gold is washed. The ground which they were digging out and washing at the time, contained a small quantity of fine gold dust, it consisted of yellowish grey, clayey sand containing large boulders of biotite-amphibole-porphyrite of andesitic habitus belonging to the same type as that exposed at Goeroeng Pĕlai. Small hillocks of amphibole-porphyrite stand out of the undulating ground which slopes gently towards the river. Evidently the soil, in which the digging is done, chiefly consists of the unremoved decomposed porphyritic rock [1]) itself.

---

1) The manner of working these mines is the same here as everywhere else in Borneo. It might be called an extremely primitive sluice-system. Of the many books written upon this subject I will only mention that by C. J. VAN SCHELLE. *52* I, p. 263.

According to the Chinese who were at work in this 'parit' the gold return is very small, but one cannot attach any import- ance to their statements. There are many gold-diggings along the Embahoe, those that I saw being worked in August 1894, have been marked on the map. But no doubt there are several others, and as I learned afterwards from the chief of the kampong at Nᵃ Boeloeng and also at Sĕmitau, that the total gold pro- duction per annum on the Embahoe from these diggings is pretty considerable, it might be well worth while to have the occurrence of the gold in the basin of the Embahoe reported on by a capable mining-engineer. Below Goeroeng Pĕlai, friable white sandstone and fine conglomerate again prevail.

At 5 p. m. I arrived at Nᵃ Boeloeng, a Malay kampong, the chief of it being at the same time the acknowledged head of the neighbouring kampongs. The distance from Nᵃ Tĕpoewai to this place is 20 kilometres, and in spite of the many delays we had accomplished it in 10 hours. Just below the place where the Boeloeng empties into the Embahoe, there is on the left side a large boulder-bed, the only karangan of any importance on the Embahoe, between Nᵃ Tĕpoewai and this house. The pebbles found here belong to the type which higher up-stream I had found *in situ*.

August 17. On August 17ᵗʰ at 9.30 a. m., I set out for Mt. Ampan with the kampong chief as my guide. From Nᵃ Boeloeng we went first a little way down-stream to the western side of the mountain, and landed on the right bank, where Mt. Ampan rose up before us, as illustrated in Plate XXXVI. It is a huge mass of rock, in outline not unlike the crude shape which children usually give to mountains. In many ways this rock resembles Mt. Kĕlam, but it is probably still steeper. The western slope, which alone is scalable, dips at an angle of about 42°. We now made for the foot of Mt. Ampan, picking our way over low hilly ground for a distance of ³/₄ kilometre. The lower part of the mountain is laid out in ladangs, where large fragments of porphyritic am- phibole-andesite lie scattered about. When the ladangs are passed, the ascent becomes much steeper, and the path leads up to an altitude of 305 metres through thick wood [1]) to a natural grotto, spacious enough, but of very little depth. A little higher up the wood becomes thinner, and soon gives way altogether to a growth of low ferns alternating with smooth, perfectly bare, rock-faces, which are most difficult to climb. I reached the top at 9.15 and found it covered with low brushwood, conspicuous amongst which are the Nepenthes with their enormous cups. More than once the mountain has been burnt, either partially or entirely bare, and this accounts for the scantiness of vegetation on and near the top. The kampong chief assured me that the worst of these fires had been in 1849 and in 1871, and that for years afterwards the mountain had presented the appearance of a gigantic naked mass of stone. During the ascent I noticed that the mountain is built up entirely of a kind of porphyritic amphibole-andesite [2]).

The panorama from the top of Mt. Ampan is most beautiful, and as it is the highest (545 metres), steepest, and smallest in circumference of all the surrounding hills, one gets from it a bird's eye view of the country, as if seen from a high tower or from a balloon. To the north stretches the wide plain of the Upper Kapoewas and the Great Lake district, a few isolated hills in the foreground attracting special attention, viz., Mt. Sĕnarah near Djonkong, N. 10° W. and Mt. Tang N. 10° E. Along the Embahoe the Kapoewas plain is bounded on the south side, not far from Karangan Pandjang, by low ranges of hills, their precipitous slopes facing north, their gentle inclines southward. The trend of these hills is on the east of the Embahoe, E.—W., west of that river, E. 20° N—W. 20° S. The latter portion, being an elongated, steep mountain ridge, is called Mt. Raja. Looking to the east and to the west, one can see nothing but hilly ground grad-

1) Here as on the water-parting between the Kapoewas and the Mahakkam, my native companions cut little bits of wood from a certain tree; these they carried with them as a safeguard against the evil influence of the mountain spirits.

2) A piece of this rock is in the Geological Museum at Leiden, in the collection made by VAN SCHELLE, who visited Mt. Ampan in 1879.

ually rising towards the east in the direction of the Madi mountains. In the more immediate vicinity, there are a few mountains
which both in form and character bear a close resemblance to
Mt. Ampan, and probably they are composed, like the latter, of
andesite or andesitic-porphyrite. Amongst these I reckon Batoe
Boetak (342 metres) E. 43° S., Mt. Embahoe (330 metres) E.23° N.,
and on the other side of the Embahoe, Mt. Toenggoel (445 metres)
W. 3° N. and Mt. Sĕnai (441 metres) W. 14° S. To the south,
the low hilly district of the Upper Embahoe and the Litoek
stretches, like a wide valley, between the mountains of the Middle-
Embahoe on the one side and the Pyaboeng and Bĕloewan mountains on the other.

Returning to N° Boeloeng in the afternoon we found the
boats, which had been left behind at the pangkalan Pyang in
the Oeloe Sĕbĕroewang, ready waiting for us, so we continued
our journey rowing down the Embahoe.

From N° Boeloeng downwards, the banks on the whole are
low and sandy; the little sandstone visible, is very light in
colour, with beds dipping gently to the north. A small distance
down-stream from Ririt, a remarkable rock, projecting from the
right shore, juts out far into the river. It is called Batoe
Raoeng, and consists of sandstone. Further down the river, the
only other rocks of any importance are situated about $2^3/_4$ kilometres below the kampong Ririt, where the river breaks
through the range of hills called Goenoeng Raja. Rapids of no
great importance are formed here, and along the banks thickly
wooded, precipitous, crags project into the stream. They are composed of white quartz-sandstone, striking E. 25° N.—W. 25° S.
and dipping 45°—60° to the N.N.W. In the evening I arrived at the
kampong Karangan Pandjang [1]), which is certainly long enough,
but where, strange to say, no boulder-bed is to be seen. Almost
the entire length of the kampong is lined with dourian-trees,
laden with ripe fruit and filling the air with their balsamic odour.

―――― ――――

1) Karangan Pandjang = longboulder-bed.

Pl. XXXIV.

Roemah Loengan Bĕdoengan.

As we walked under these trees by the riverside we were painfully made aware that the golden rule, "high trees bear small
fruit", does not apply to the tropics.

On the 18th of August at 7. a m. I left Karangan Pandjang. August 18.
Below the kampong the river becomes more sinuous in its
course, the shores are flat, and the surrounding territory flat
or slightly undulated. Here and there however at the bends
the solid rock becomes visible. The strata of friable grey
sandstone, probably of Tertiary age, lie, below Karangan Pandjang, in a nearly flat trough-shaped position. At first they
dipped slightly to the north, but as we advanced northward the
inclination became imperceptible, and opposite the kampong
Sasan, on the right shore, I saw that the sandstone was overlain by finely stratified clay with coal-seams which dip about
2°—4° to the south.

I was told that in a small tributary of the Embahoe in the
neighbourhood of Sasan fairly thick seams of coal are found in the
clay-stone, of the same quality as the coal struck on the Sélimbau-
river and which, according to EVERWIJN's description, belongs to
this same formation. I did not visit these coal beds at Sasan,
because I was not aware of their existence at that time, it was
only afterwards when at Sěmitau that I was told about them.
Six kilometres down-stream from Sasan, close to Poelau Badoeng,
where the river makes a sharp bend near a steep rock face, I
found shell impressions in a clayey sandstone. The strata there dip
about 4° to the south. The last rocks along the banks of the Embahoe. are found on the left shore, 400 metres below Nᵃ Taman, they
consist of biotite-andesite containing amphibole. EVERWIJN [1]), who
in 1853 went up the Embahoe to this point, justly remarks
that these rocks, which he calls merely an igneous porphyritic
rock, resemble in every respect the rock of which Mt Sénarah,

1) EVERWIJN II, p. 67. He calls the Embahoe, Embocan or Emboan. He was told by
the natives that coal was to be found a two day's journey from the Kampong Rassan. His
statement that the Embahoe ceases to be navigable for small open boats as low down as
Rassan is incorrect. He calls Mt. Sěnarah, Mt. Sindoro.

6 kilometres further on, is composed. I have compared the specimens I collected on the Embahoe with the specimens in the National Museum in Leiden brought by van Schelle in 1874 from Mt. Sĕnarah, and I found that they are identically the same, viz., biotite-andesite containing amphibole. Everwijn speaks of dykes of porphyry, and I agree with him that this is in all probability an andesite dyke, connected with the andesite Mt. Sĕnarah.

Below this point no more rock *in situ* appeared, for we had now reached the Kapoewas plain once more. The river here describes a very sinuous course and forms many danaus, amongst which Danau Kota Bĕharoe and Danau Parak are the principal ones. I was able to shorten the way considerably by making use of the pintas Entĕbinga, and at 12.30 we arrived at Djong-kong, where we stopped a couple of hours. From there going down the Kapoewas, we only just managed to work our way through the pintas Pyasa and Niboeng, which had nearly run dry. The last night of the excursion was spent on the Kapoewas in the neighbourhood of Sĕlimbau, and at 10 a.m. on August 19th I reached my headquarters at Sĕmitau.

SUMMARY OF THE GEOLOGY OF THE RIVER BASIN OF THE SĔBĔROEWANG AND THE EMBAHOE.

(See maps III and VIII)

The territory described in the preceding pages forms part of the hilly district which borders the Upper Kapoewas plain, and more particularly the Lake district, on the south side. The Sĕmitau hills (comp. page 26) form the oldest part, they are composed of crystalline schists, amphibolite, chlorite-schist, quartzite-slate, micaceous quartzite and silicified clay-slate. This oldest formation is continued on the other side of the Kapoewas, how far is still unknown. To the east, in the valley of the Embahoe, there is no trace of it, unless the gneissic granite and quartzite above

Oelak may be reckoned to belong to this formation. The same may also possibly be the case with the granite, which, according to VAN SCHELLE[1]), is found in the hills to the north of the Sěbě-roewang valley, near Sajor.

The trend of the hills and the strike of the strata within the area of the oldest formation are nearly due east and west. The strata are always much folded, and frequently intersected with quartz veins. Whether the silicified clay-slate, which I observed in different places in the hilly district of Sěmitau is of the same age, or whether it belongs to a more recent formation infolded in the crystalline slates, is as yet uncertain. On the sketch map this silicified clay-slate is not separated from the crystalline schists. Most probably there lies between this area and the Upper Ka-poewas plain a line of faulting, and so its continuation ought to be looked for at a considerable depth in a northern direction.

In a southern direction, against this belt of old crystalline rocks, there leans a very peculiar formation, chiefly composed of strata of diabase-tuff, alternating with banks of tuff-breccia, in which, besides fragments of diabase-tuff, pieces of limestone, diabase and porphyry are found. In this same complex occur now and again layers of chert and clay-slate. The tufas are conspi-cuous by the frequent occurrence of a vitrophyric variety of diabase, petrographically not to be distinguished from andesite. The tuff, as well as the chert, contains Radiolaria in great quan-tities and endless varieties. The tuff-fragments enclosed in the tuff-breccia contain the same kinds of Radiolaria, and in the lime-stone of this breccia Foraminifera occur. When examining the Radiolaria, HINDE came to the conclusion that these tufas are of pre-Cretaceous age and therefore must belong to the same for-mation as the cherts and jaspers with Radiolaria of the Lake

---

1) According to an unpublished note from C. J. VAN SCHELLE dated 1879, in the „Archief van het Rijksmuseum te Leiden." He found granite near Mt. Lěmbang Moeda, on the path from Sajor to Empěriang in Soehaid. This Mt. Lembang Moeda must be the same as Mt. Sěboeloeh (250 metres) of the topographical map, as it is the only hill which lies between these two places.

district and the Upper Kapoewas. Yet the resemblance is not sufficiently close to justify a positive assertion that they are all on precisely the same horizon. The Foraminifera in the limestone fragments of the tuff-breccia of Mt. Rajoen have also a pre-Cretaceous character. This pre-Cretaceous tuff-breccia forms a kind of continuous belt along the south side of the axis of the old Sěmitau formation. The strata are highly tilted and more or less folded. They dip generally at a considerable angle to the south, sometimes also they stand vertical. At Sědjiram this belt touches the Sěběroewang and along the banks of the river tuff-breccia and chert with Radiolaria make their appearance. The strike is W.N.W.—E.S.E. No other trace of the pre-Cretaceous formation is to be seen either higher up or lower down in the bed of the Sěběroewang, but it crops out to the north of the river wherever the hills slope down towards the valley. Near Mt. Rajoen the pre-Cretaceous rocks are exposed in a broken saddle, the north wing of which has a steep incline to the north, the southern, a gentle incline to the south. The sharp bend noticeable in the strata of this saddle (compare section B on map VIII) towards the east merges into a fault in consequence of which the pre-Cretaceous formation becomes exposed in two parallel strips, both dipping to the south, and between these, in the valley of the Těpoewai, marls and breccias of Cretaceous age are found. More to the east again, in the valley of the Embahoe, the pre-Cretaceous rocks are not only highly tilted, but also greatly folded and compressed. The rocks here begin to show very decided signs of relationship to the Poelaü Mělaioe type, familiar to us in the Lake district and in the Upper Kapoewas, and which we admitted to be a diabase-tuff highly altered by mountain pressure and subsequent infiltration with silica. It is a remarkable fact that neither the diabase nor the limestone found so abundantly as fragments in the tuff-breccia, were found *in situ*.

The Sěběroewang cuts its bed almost entirely through a more recent formation, which consists, as may easily be seen along the banks, of a system of strata in which, beginning from the bottom, the following divisions can be distinguished.

A. Variegated green or yellowish clay-stone, here and there passing into almost pure kaolin, rarely also clayey sandstone. (Lower or clay-series).

B. Marls or sandy marls with *Orbitolina concava*, Lam., and shell fragments; the sandy marls contain moreover particles of coal. (Middle or marl series).

C. Sandstone, clayey sandstone and clay-stone. (Upper or sandstone series). In contradistinction to the lower horizon, sandstones here predominate; they are generally of a greyish or greenish colour, and like the clays they sometimes contain quantities of undeterminable vegetable remains. Coal-seams occur in most places, but no coal-beds of any importance have been found. In the upper part of this series strata of coarse conglomerate occur now and then, alternating with the sandstone. This conglomerate, as mentioned above, principally contains boulders of muscovite-granite, pegmatite, syenite, diorite and limestone, none of which have I, with certainty, found anywhere as solid rock. The second series, marls with *Orbitolina concava*, Lam., is, stratigraphically speaking, the most interesting horizon in this system, and is much thinner than either of the other two. I estimate its thickness to be not more than 20 metres.

The strata of this formation, which, as a whole, we will call the Séběroewang formation, rest on or lean against the tuff-breccia formation which separates them from the hilly district of Sémitau. The general strike of the strata is W.N.W.—E.S.E. often passing into an E.—W. or W.S.W.—E.N.E. direction. It is worth noticing that in the marl series the strike was generally W.S.W.—E.N.E., while in the two others a W.N.W.—E.S.E. strike is by far the more usual. The general dip is either moderate or steep to the south, but because of an important strike fault, and through folding, the same horizon crops out more than once, a fact which is most evident in the marl series.

Further east, the Séběroewang formation again makes its appearance to the north of the tuff-breccia formation of Mt. Rajoen, in the valley of the Gäang and Tĕpoewai; here it rests on the tuff-breccia formation which is thrown down along a fault on the north side of Mt. Rajoen.

The rivers Sĕbĕroewang, Gaäng, and Tĕpoewai below N⁂ Gaäng, have all three cut their beds in the Cretaceous form‧ ation and their courses are parallel with the direction of the strike of the rocks. Thus they form longitudinal valleys in an area where the rocks are generally similar in their power of resisting the eroding influence of the water. It is owing to this peculiar circumstance that they are, so to speak, the only rivers in West Borneo which, in their upper courses, have no rapids or other special obstacles to navigation.

In attempting to fix the age of the Sĕbĕroewang formation, the marl horizon is the only one which can assist us, as it is the only one in which determinable fossil remains have been found.

EVERWIJN[1]), in 1856, was the first to discover in these marls several fossils which he took to be Nummulites, and from which he consequently deduced the age of the formation to be Eocene. In 1872, SCHNEIDER collected fresh fossils from these marls, which are described by FRITSCH[2]), as probably two new species of the *genus Patellina*. He leaves it open whether these rocks are of Eocene or of Cretaceous age.

VAN SCHELLE, in 1879, collected in the marls in the neighbourhood of Sajor a quantity of fossils, and amongst them not only the above named *Patellina* but also other shells which have been examined by GEINITZ[3]) and from which he feels

---

1) EVERWIJN *11*, page 25.

2) K. VON FRITSCH *12*, Palaeontographica. Suppl. bd. III, p. 144, 1877 and Jaarboek van het Mijnwezen 1879. I, p. 246.

3) II. B. GEINITZ *13*, p. 205, and also R. D. M. VERBEEK *61*, p. 41.

GEINITZ here informs us that the fossils of this formation are closely allied to the following Cretaceous (Senonian) species: *Natica Gentii*, Sow., *Natica lamellosa*, Roem., *Panopaea gurgites*, Brog., *P. mandibula*, Sow., *Goniomya designata*, Goldf., *Trigonia limbata*, d'Orb., *Lyonsia germari*, Gein., *Vola quadricostata* Sow., *Modiola capitata*, Zitt., *Gervillia solenoïdes*, Defr., *Hemiaster regulusanus*, d'Orb., *Hemiaster sublacunosus*, Gein., *Hemiaster*

justified in deducing a Lower Senonian age for this formation.

The examples of *Patellina* collected later by VAN SCHELLE were sent to the National Museum in Leiden and have been reexamined by MARTIN, who finally proved that the tests in question do not belong to *Patellina*, but to the genus *Orbitolina*; and that they may safely be determined as *Orbitolina concava* Lam. MARTIN [1]) comes to the conclusion that the marls of the Sĕbĕroewang, on account of the presence of *Orbitolina concava*, must be considered to be of Cenomanian age.

Although the general character of the Sĕbĕroewang formation points towards a shallow water deposit, possibly in a gulf or bay at a greater or shorter distance from the open sea, the deposits of its upper horizon could only have formed subaerially, or in the immediate vicinity of land. I am referring to the conglomerates which I detected in one locality on the Sĕbĕroewang not far from the brook Gaman, and again on the western slope of Mt. Oejan where the beds of conglomerate incline gently to the south. This coarse conglomerate contains many water-worn boulders of different kinds of rock, mostly granitic, the origin of which I am unable to trace. As already stated, I consider that these conglomerates alternate with strata of sandstone from the higher series of the Sĕbĕroewang formation and I am inclined to believe that in a southern direction they are overlain by the more recent, probably Tertiary, sandstones of the Sĕraboen-Sĕtoengoel range of hills. I must however admit that the conditions were not sufficiently favorable, and my stay was too short to enable me to give a positive opinion.

For the sake of completeness I should add that these conglo-

---

plebejus Novak; furthermore representatives of the genera *Phasianella*, *Avellana*, *Astarte*, *Spondylus*, *Lima*, *Arca* and *Ostrea*.

As mentioned above (page 250) I did not succeed in finding again the place where these fossils had first been discovered, which is much to be regretted, for the shells found by VAN SCHELLE and described by GEINITZ appear to have been lost. At any rate, notwithstanding a most careful search into this matter, I have not yet been able to trace their whereabouts.

1) K. MARTIN *32*, p. 209.

merates and the great variety of their boulders have already been noticed by CHAPER [1]) who visited this district in 1890.

. The most recent sedimentary formation in this area is sandstone with alternate layers of clay-sandstone, soft grey clay-stone and coal-beds which in many places are thick and pure enough to deserve to be prospected for possible mining purposes.

The less disturbed and almost horizontal position of the strata and the presence of coal-beds, are the only two constant characteristics which distinguish this formation from the older one. The sandstones in this younger formation are, moreover, generally of a pure white or grey colour, they are more friable and contain less clay than the cretaceous rocks. Occasionally, as on Mt. Sĕtoengoel, they pass into greywacke sandstone. This recent sandstone formation is met with on the northern slope of the valley east of the Sĕmitau-hills, and the strata are particularly well exposed on the banks of the Embahoe. They are either gently undulating or with a slight dip to the north. In the hilly district of Sĕmitau itself horizontal layers of this sandstone rest, here and there, unconformably on the older formation. I noticed this on the Embahoe near the riam Soerai, and the horizontal strata of sandstone found by VAN SCHELLE on the top of Mt. Sĕboeloeh (Lĕmbang Moeda),

---

1) M. CHAPER 5, p. 877. CHAPER speaking of these conglomerates says: "Les poudingues, souvent très durs et dont les éléments bien roulés atteignent parfois de fortes dimensions, contiennent une tres grande variété de roches (roches granitiques, porphyriques, quartz-diorite et même des calcaires fossilifères)."

CHAPER, who in 1890 undertook some geological and mining researches in a part of West-Borneo, rowed up the whole of the Sĕbĕroewang and visited Mt. Rajoen and Mt. Oejan, just as I did. He comes to the conclusion that besides the conglomerates only clay-stone and soft sandstones occur in the Sĕbĕroewang valley, that absolutely no fossils are found, and that therefore "en l'état actuel il est impossible d'arriver à une détermination, même approximative, de l'âge de ce puissant système de grès et de schistes". The marls with *Orbitolina*, which surely are conspicuous enough, he ignores altogether. He sees no difference between the Rajoen tuff-breccia which does not contain a particle of granite, and the Oejan conglomerate. With reference to the general structure of the Rajoen and the Oejan he observes: "La, les grès et les roches qui leur sont sous-jacentes, se redressent fortement et ont subi un métamorphisme assez intense qui les a transformés en quartzite, jaspe et diorite vert foncé, avec grénate, le tout ayant conservé ses plans de stratification". — As a contribution to the geological knowledge of Borneo, CHAPER's publication is of little value.

Pl. XXXV.

BOEKIT RAJOEN.

most likely belongs to the same formation. The extensive sandstone deposits in the Sĕraboen-Sĕtoengoel, Sagoe-Pĕnai and intervening ranges to the south of the Sĕmitau hills, I also place in
this formation. As far as we know these sandstone beds all dip
slightly to the south.

On the other side of the Kapoewas I include in this formation
the sandstones of Batoe Kĕling, those of Mt. Sĕtĕboe and Mt.
Sĕgĕrat, in the neighbourhood of the Kĕnĕpai, as well as the
recent sandstones of the Lake district, which I formerly called
Lĕmpai sandstone. Coal-beds are known to exist in the upper
strata of Mt. Lilin near the Sĕtoengoel, on the S. Sĕntabai,
near Danau Kĕnĕpai, on Mt. Sĕtĕboe, in the neighbourhood of
Sĕlimbau, and on the Embahoe near Sasan. Only those at Sĕlimbau, near Danau Kĕnĕpai, and near Mt. Sĕtĕboe are being
worked to this day.

On the lower Embahoe near Poelau Boedoeng, I found shell
impressions in clay belonging to this formation. On the authority
of van Dijk [1]) and Everwijn, this recent sandstone formation
is generally regarded as of Tertiary and more particularly of
Eocene age, but it is only fair to add that positive proof as to
the age of this formation, in the district now under consideration,
is as yet lacking. As a matter of fact we have no right to go
beyond the statement that the practically undisturbed position
and the character of the coal-beds (the coal is of a dark brownish
colour) lead us to accept this formation as Tertiary or more correctly Eocene. Hence it is marked on maps III and VIII as Tertiary.

Finally, Quaternary and modern deposits cover the greater
part of the surface of this vast area. They are principally
layers of sand, clay and gravel, forming a belt round the higher
ground, and in the valleys constituting the slopes leading down
to the river-beds. These older fluviatile deposits, (which I will
call Quaternary) are in many cases mixed with parts of the
weathered underlying rock, into which, in some instances, they

---

1) P. van Dijk 8, p. 138.

gradually pass. It is in these deposits that the Chinese in many places carry on their gold diggings.

In the immediate neighbourhood of the brooks and rivers which so frequently cut their beds in these older fluviatile deposits, a larger or smaller extent of the soil consists of recent fluviatile, i. e. true alluvial deposit.

Igneous rocks do not occur *in situ* in the river-basin of the Sĕbĕroewang or in the western division of the hilly district of Sĕmitau, although they are quite common in the river-basin of the Embahoe, where they appear frequently and in many varieties. This, added to the fact that the Embahoe forms a transverse, and the Sĕbĕroewang a longitudinal valley, satisfactorily explains the striking difference which exists, and which even a layman cannot fail to see, between the navigability of these two rivers.

The main peaks of the Bĕloewan mountains consist of amphibole-porphyrite and they furnish the immense quantity of porphyritic boulders which are being carried down by the Tĕpoewai. I visited the two remotest peaks of this curved range, the Oejan and the Pyaboeng. In the valley of the Embahoe, both granite and gneiss, as well as the pre-Cretaceous formation have been cut through by a multitude of dykes, the predominating strike of which is always east and west. Of all these different dyke rocks we would in the first place draw attention to a very beautiful micro-granitic quartz-porphyry with dacitic character, containing the most exquisite phenocrysts of quartz in the form of crystallized bipyramids, which, about 1 kilometre above Antoek, forms a dyke, near the boundary line between the pre-Cretaceous formation and the marl horizon of the Cretaceous formation.

There are some more dykes of quartz-porphyry, close to Goeroeng Boewang, but all the rest are of amphibole-porphyrite, or quartz-amphibole-porphyrite often containing biotite, and they bear a striking resemblance to the rocks of the Oejan and the Pyaboeng. This porphyrite without any exception, is andesitic in its character, which is particularly evident in the lower course of the Embahoe, for instance at Mt. Ampan, where it was most difficult to decide whether these rocks should properly be called porphyrite or andesite.

Geological age cannot help us in this matter, as the only fact ascertained about them is, that the porphyrite in question is more recent than the pre-Cretaceous and probably even more recent than the Cenomanian [1]).

All the rocks in the lower course of the Embahoe, where mica-andesite is also of frequent occurrence, have been classified more or less arbitrarily as andesitic rocks, while those higher up-stream are all called amphibole-porphyrite. By way of exception the dyke which, 3 kilometres below N[a] Tépoewai, breaks through the pre-Cretaceous formation, has been singled out, although situated within the area of porphyritic dykes, as being andesitic, because even under the microscope this rock possesses all the marked characteristics of a true andesite.

For the rest, therefore, it must be kept in mind that the types distinguished in map VIII as amphibole-porphyrite and amphibole-andesite, have many striking points in common, and that the first named type, seen only with the naked eye, has a decided andesitic character, whereas the ground-mass of the second type displays under the microscope somewhat porphyritic tendencies. I rather incline to the theory that these porphyrites are neither more nor less than modifications of andesitic rocks solidified at a certain depth, which if they had reached the surface would have become dacites or andesites. The profound denudation of which this entire area bears the traces has made these deep-seated varieties of andesite accessible to us.

The mountains and hills in the drainage-area of the Embahoe, mentioned in this chapter so far as they are composed of porphyrite and andesite, I consider to be really the most westerly outposts of the volcanic Müller-mountains.

---

1) In section LL.' I have accepted for the Oejan porphyrites a post-Cenomanian age.

CHAPTER X.

Right across Dutch Borneo from North to South.

A. The Boenoet, the Tĕbaoeng and the Mádi plateau.

My investigations in the upper course of the Kapoewas and its tributaries had convinced me that the trend of the principal lines of dislocation is there from east to west, and my travels in the territory of the Embahoe and the Sĕbĕroewang had shown me, that this same direction prevails not only through-out the Upper Kapoewas territory but also more southward. This in connection with what Schwaner relates about the mountains which form the watershed between the Mĕlawi and the Katingan in South Borneo, made me come to the conclusion that lines of dislocation trending from east to west control the build of the whole of Central Borneo. Therefore it was to be expected that by travelling in a west to east direction, as in the Upper Kapoewas district or in the Sĕbĕroewang river-basin, I should always meet with the same geological formation, or at any rate should find little variety in the structure of the country. But by tra-velling in a direction from north to south, I should cut right through all the formations, and this would be the only way of procuring a section from which, in connection with my previous investigations, the frame of the geological structure of Central Borneo might be deduced, in outline at any rate.

After thoughtful consideration, I decided to attempt to take my journey in a north to south direction right across Borneo to the Java Sea, following, as closely as possible, the 113th degree of longitude from Greenwich. I chose this route, in the first place, because it would lead me through a territory absolutely unexplored scientifically and for the greater part never set foot on by any European; in the second place because the journey would probably enable me to obtain, in connection with my observations in the Embaloeh valley, a consecutive section from the boundaries of Sarawak right across Dutch Borneo as far as the Java Sea; in the third place because it would most probably give me the opportunity of obtaining a more intimate knowledge of the highest mountain-land so far known, of Dutch Borneo, the region of Mt. Raja (2278 metres), and, last but not least, because the journey through South Borneo would partially run parallel with SCHWANER's route, and I should thus have the opportunity of comparing his observations with my own.

The greatest difficulty in this projected excursion seemed to me to be the question of food, for I knew that along the whole of the route, (and I calculated the journey to take fully two months) I should probably nowhere be able to procure provisions in anything like sufficient quantity. I knew that by crossing the Madi mountains from the river Boenoet, I should reach the Mélawi at a point where this river is still navigable for fairly large boats, and it seemed therefore advisable to send a few boats with food up this river on the chance of their reaching the place on the Mélawi about the same time that I expected to get there. At Sintang I engaged 29 coolies, — for the greater part Dyaks — at a daily pay of ƒ 0.80 or $ 0.50 and some tobacco, and of course I had to keep them in food, rice and salt. I had no trouble in finding my men, although they knew that some of them would have to come with me as far as Bandjermassin, and they would therefore be over three months away from home.

On the second of September I sent fourteen of these men with provisions, chiefly rice, salt, and dried fish, in three boats up the Mélawi, with instructions to deposit the provisions at N°. Gilang and then with two of the boats and a small quantity of rice to row up the Kĕrĕmoei, a right tributary of the Mĕlawi, as far as Kwala Panai and there await my arrival. I left on the 3rd of September with the remainder of the coolies in the little steamer of the Resident, the Karimata, for Sĕmitau, where I completed my equipment for the great undertaking. As I did not intend to go back to the house of the Borneo-expedition at Sĕmitau, I sent all my collections and baggage to Sintang, the assistant-resident SNELLEBRAND having kindly offered to take charge of them there for me. From Sĕmitau I proceeded, on September 5th, by the steam-launch Poenan as far as Boenoet, where I arrived the next day. The pangeran of Boenoet, a young opium-smoker, gave me a beautifully sealed letter which was to insure the co-operation in my undertaking of the Kampong-chiefs of Nᵃ Sĕbilit and of Lĕmatak. Afterwards I found however that these chiefs could not read and the royal epistle had not the slightest effect upon them. The pangeran also promised to follow me in a fast sailing boat and to accompany me on my journey into the mountains, which promise he, fortunately, did not keep. I left Boenoet at 11.30 a. m. in two boats, one small bidar and one open rowing-boat, called by the natives a *pĕlĕle*. The water in the Boenoet was low, and the passage between high mud-walls, which at the existing water-level, lined the river on either side was extremely tedious. The pintas Kadap fortunately was found to be navigable, so we cut off $2^1/_2$ kilometres and reached Nᵃ Toewan that evening, and spent the night on a sand-bank which swarmed with mosquitoes.

Sept. 7.   At 4 p. m. we passed the river Bojan, and that day we advanced another three kilometres as far as Landau, a small settlement of Kantoek-Dyaks. I watched them catch fish here with javelins, the upper part of which, the hook proper, as soon as it had stuck fast into the fish was detached from the rattanstick, and only remained connected with it by a thin rattan-rope. This arrangement was to prevent the breaking of the rattanstick through the sudden shock of the fish's rapid movements, should a particularly strong one be hooked. The handles of these javelins were beautifully carved and painted. In one of the houses of these Kantoeks there was a harmless idiot, whose object it evidently was to amuse everybody, and whom the inhabitants used as a kind of fool.

Up to this point the Boenoet meandered with an almost imper- Sept. 3. ceptible fall, through flat, alluvial, marshy regions, a portion of the Upper Kapoewas plain, but presently the hills to the south and south-east, began to come closer to the river and near Alak, 8¹/₂ kilometres above Nᵃ Bojan, the first solid rock is exposed on the right shore. It is composed of rough beds of friable sandstone, 0.40—0.70 centimetres thick, alternating with strata of clay-stone 0.05—0.15 centimetres thick. At 11 a. m. we reached the point where the Tĕbaoeng and the Soeroek meet; the two rivers combined are called the Boenoet. The Soeroek comes from the east. It has a length of about 80 kilometres and is navigable for canoes for a distance of 60 kilometres. From the point where, leaving the mountains, it enters the Kapoewas plain, its course is a remarkably winding one. It rises on the southern incline of the Maroeng mountains, very probably a portion of the volcanic Müller mountains. I look for important results to be derived from geological researches in the Soeroek valley, more especially for the determination of the boundaries of the area of volcanic tuffs, and of the connection between the Müller mountains and the Madi plateau.

I turned southward up the Tĕbaoeng, here 40 metres in width. The banks consist in some places only of sand and gravel, more generally however of alternate layers of sand, clay, gravel and strata of compressed vegetable remains. The sand-banks along the bends up to the point of meeting with the Soeroek also contain gravel in their upper parts. Four kilometres below Nᵃ Sĕbilit, coal is exposed in the right bank; here the section A. Map IXᵃ was taken. The coal is fissile and of inferior quality. In the boulder-banks below Nᵃ Sĕbilit I found sandstone, fine conglomerate, siliceous slate, amphibole-andesite, quartzite with muscovite and quartz.

At 1 p. m. we reached the Sĕbilit, an important left tributary of the Tĕbaoeng. I wished to go a few day's journey up this river. The water was very low and in many places so many tree-trunks had been thrown into the river when the ladangs were made, that the navigable areas were almost entirely stopped up, and the trailing of the boats over all these obstacles cost us much time and labour.

The banks consist of sand and clay with vegetable refuse, sometimes of gravel, and in a few places of sandstone. In the boulder-banks on the side of the river I noticed several large pieces of coal, a sure indication that at no great distance from there coal would be found as solid rock. And indeed shortly after, about one kilometre below Nˣ Sirah, in a cliff on the left shore a beautiful section with several layers of fairly good but somewhat fissile coal is exposed (see profile B, map IXᴀ).

The strata dip here 23° to N. 15° W.

Sept. 9   After spending the night at the Malay settlement Nˣ Sirah, I rowed up the Sĕbilit next morning, in a small open sampan which I had hired. Above Nˣ Sirah this stream continues its course from the south, but at the house Bĕtoeng, the river curves round and flows in a west to east direction along the northern slope of some hilly ranges with a W. 10° S.—E. 10° N. trend. The banks continue low, and only occasionally a white or yellowish white rough sandstone appears as solid rock. Slowly the river approaches the hilly district and just above the Chinese settlement Kĕbijau it curves to the south, the bed becomes narrower, the shores higher, even steep and rocky, and we reached the place where the Sĕbilit breaks through a threefold ridge of sandstone hills. Here the valley of the Sĕbilit is extremely picturesque, and at various places beautiful rock-groups of sandstone stand up on the shore or in the river-bed. There are some beautiful slender columns of sandstone in the middle of the stream about $3^1/_2$ kilometres above Kĕbijau, which, according to the tradition of the natives, are petrified men. The sandstone is deposited in thick beds, with a general north or north-westerly

BOEKIT AMPAM.

dip, and showing in several places vertical cleavage. It is of a yellowish or whitish colour and almost always contains coal, either as very thin veins, or as fine patches which give the rock a speckled appearance. However uniform the geological build of the shores in this portion of the river may be, there is enough variety in the boulder-banks. In one boulder-bank, three kilometres above Kĕbijau, I noticed the following rocks: tourmaline-rock with a yellowish variety of muscovite[1]), tourmaline-granite, quartz-diorite, tonalite, chlorite-chist, amphibolite, harzburgite, olivine-pyroxene-rock, serpentine, kersantite, diorite-porphyrite, amphibole-andesite, pitchstone, sandstone, siliceous-slate and quartz. Further up-stream I was therefore sure to find more variety in the geological structure of the ground. Continuing up-stream in a southerly direction through little elevated territory I found that sandstone forms the bottom-rock as far as the house Inggai, inhabited by a Chinaman. Here a north-easterly spur of Mt. Oendau extends right into the river which there consequently cuts its beds in hornblende-biotite-andesite with porphyritic character containing some quartz. Mt. Oendau is a cone-shaped mountain 743 metres high, grooved by erosion, in consequence of which here and there small groups of hills have become more or less detached from the main cone. All these secondary cones, like the principal one, have a rounded cupolar shape. Just below Inggai on the left shore of the Sébilit a footpath leads to Mt Oendau. Through low brushwood — old ladangs — this path brought me in a quarter of an hour to the foot of the mountain. On the northern and north-eastern incline there are a few scattered Dyak houses, surrounded by ladangs, which extend as far as the nearest tops of the mountain. Another quarter of an hour's walk brought me to one of these tops (325 metres) from whence I had an extensive and unobstructed view in all directions, except towards the south and south-west, where the view is intercepted by the thickly wooded

---

[1]) This rock corresponds entirely with a type found to the south of Mt. Kĕnĕpai in the contact-zone with augite-tonalite (see p. 39).

main top of Mt. Oendau itself. The panorama from Mt. Oendau gives a capital survey of the western part of the Müller-mountains, and may be considered the continuation of the panorama from the top of Mt. Sassak, as sketched on p. 72. From Mt. Oendau, as from Mt. Sassak one overlooks in a northern direction the far reaching Upper Kapoewas plain, and just as this plain, seen from Mt. Sassak, is bound towards the south by the volcanic tuff-ranges of the Mandai which are divided into a series of strips trending N.N.W.—S.S.E. by small streams running down to the Kapoewas, so in the present case the plain is bordered almost abruptly by a mountain-land, consisting of a large number of cone- or cup-shaped andesite-mountains divided into strips with a N.N.W.—S.S.E. trend by streams which flow towards the Kapoewas. Seen from Mt. Oendau these hills look like a series of undulating profile-lines rising one behind the other, in strong contrast with the sharp outlines of the table-shaped tuff-mountains of Mt. Soenan and Mt. Sassak. Mt. Oendau is entirely built up of quartz-bearing amphibole-biotite-andesite of porphyritic character, closely resembling the andesite of Mt. Ampan on the Embahoe-river. The slopes which are strewn with colossal blocks of andesite, are covered with a yellowish brown loam, the result of weathering of the rock, which seems to make a fairly good soil for the cultivation of rice.

Sept. 10. On September 10, I left Inggai and rowed further up the Sĕbilit which here describes a wide curve round Mt. Oendau. Above Inggai, rocks of Oendau porphyrite stand out in several places on the banks of the river and in its bed. This porphyrite, in some places entirely changes its habitus, on account of the appearance of large phenocrysts of amphibole. At this point and higher up, the river is very picturesque, with masses of rock, overarched by high trees; about $1\frac{1}{2}$ kilometres above Inggai, at the mouth of the Genting rivulet, the andesite gives place to cleaved gray sandstone with flakes of muscovite here and there mottled with particles of coal. These thick beds of sandstone dip about 8° to the N. or N.N.E. Just above the mouth of the Boenjau there was a rapid, caused by a dyke of amphibole-dacite of porphyritic character, cutting through saccharoidal sandstone. This is followed by other beds of sandstone with a slight northerly dip, occasionally rising out of the water in flat quadrangular though water-worn rocks, which offer no obstacles to navigation. After the little river Bakoel is past the disturbance of the water, and the appearance of other types of rocks forming narrow sharp ridges which run out into the watercourse, clearly reveal an area of a different geological build. Indeed a much older formation is here exposed. First of all we see on the left bank a much cleaved, beautiful blue glaucophane-amphibolite, and a little higher up, near the goeroeng Simpak Tandjan, the river-bed is narrowed by sharp rocks of amphibolite with vertical strata, trending N.E.—S.W. Further on, at goeroeng Kien, the amphibolite again contains glaucophane; the vertical strata here trend E. 30° S. About 800 metres above goeroeng Kien a ridge of andesite rocks with porphyritic character runs transversely through the river, it is probably a dyke crossing the amphibolite which higher up continuously forms the rock in which the river has cut its bed. The strike continually varies between N. 20° W. and N.W.; and the strata with a generally vertical position are much contorted. Serpentine makes its appearance on the left bank near the mouth of the Sabat. Here I left my sampan and took a forest path in a north-westerly direction to visit a place in the neighbourhood which some years ago had been prospected for gold. First the path crossed the Sabat several times and then led over a little hill, 20 metres high, the upper part of which was formed of almost horizontal, rather coarse, friable sandstone. Three quarters of an hour's walk brought me to the place where excavations had really been made with the view of searching for gold.

The river-bed is here full of large loose blocks of amphibole-andesite of the Oendau type, and on the banks a clay-stone formation is exposed, in which there is a layer of quartz, about 2 metres thick, which has led to the prospecting operations. It was proved however that the percentage of gold in this quartz was very low. The appearance of the quartz was quite like what prospectors sometimes call "hungry quartz". The strike of this entire complex of strata is W. 35° N., the dip 75° N.E. It is impossible, because of the absence of fossils, to determine the age of the clay-stone formation in which this layer of quartz makes its appearance, it is however reasonable to regard it as younger than the amphibolite and older than the sand-stone. In support of the latter it may be stated that the dip of the clay-stone is considerably steeper than that of any of the sand-stones on the Sébilit. Regarding the relation between the sand-stone and the Oendau andesite I consider the andesite to be more recent than the sandstone through which it has forced its way. This opinion is based upon the fact that the sandstone is prin-cipally composed of quartz-grains, flakes of muscovite and particles of clay and coal, with a total absence of any fragments which might proclaim an andesitic origin.

Returning by the forest-path I noticed several gĕtah-trees, and I learned from my guide that in the mountains of the Upper-Sébilit and of the Kĕrĕmoei, there are some excellent species of them, and that, although greatly reduced each year, they are as yet far from being eradicated. The return journey down-stream was so rapid that the distance between the mouth of the Sabat and Nᵃ Sirah was accomplished in three hours. I spent the night in my bidar, but did not get much sleep, for a heavy thunder-storm raged all night with floods of rain; and the water of the Sébilit rose so boisterously that the vessel was tossed to and fro continuously and threatened to be set adrift. That night the water rose fully 4 metres.

Sept. 11. We started again at 6.15, and the foaming stream carried our boats down with great rapidity; within half an hour we had reached the mouth of the Sébilit. I now perceived that when the water is high the Sébilit partly discharges into the Tĕbaoeng through a short cut, a pintas, which reaches the Tĕ-baoeng half a kilometre up-stream from Nᵃ Sébilit. As we rowed up the Tĕbaoeng I noticed that the water here was not nearly so much swollen as in the Sèbilit, and the banks were still for the greater part above water. I soon verified that the Tèbaoeng here flows through the same sandstone formation as the Sèbilit. The beds of sandstone dip here likewise at a small angle to the north, and here also layers of clay-stone and coal are intercalated in the sandstone. I obtained a good section of this formation on the right bank of the Tèbaoeng 7 kilometres above Nᵃ Sèbilit. It is given in section C, map IXᴬ. There are in this section three coal-beds, one above the other, respectively 4, 1.40, and 2 metres thick. They are separated (from one another) by thin layers of clayey sandstone and clay-stone with harder concretions containing impressions of plants, and the whole of this complex lies between beds of sandstone. The coal of the two lower beds is fissile, but the top bed appeared to be of good quality. It is a black, very lustrous pitch-coal. A little higher up is the large, wooded boulder-island, Poelau Karangan Boenga. At 10.15 we passed the kampong Sèmangoet, inhabited by Malay Dyaks ¹), and I took the son of the kampong-chief with me to act as guide up Mt. Loeboek. The footpath leading up this mountain, begins on the right shore of the Tèbaoeng at the mouth of the Tèngadak, a little stream which rises on Mt. Loeboek. Mt. Loeboek is a narrow cone-shaped mountain 487 metres high, and consists, like its neighbour the cup-shaped Pètoesi, of quartz bearing amphibole-biotite-andesite with porphyritic character, much resembling the rock of Mt. Oendau. The path leads through ladangs and through the valley of the Tèngadak rivulet up to the foot of Mt. Loeboek, which can there be scaled on its north flank. The ascent is not difficult but very steep, and therefore tiring; the general slope is about 40°. When on Mt. Oendau I had made up my mind to ascend Mt. Loeboek in any case, because I thought that its isolated position and comparatively great height promised to give me a first class view. I was not disappointed

1) Malay Dyaks = Dyaks converted to Islamism.

in my expectations, for the view from the top of Mt. Loeboek
proved to be in all directions extensive and unimpeded. Unfortunately the sky was cloudy so that the panorama did not show
to the best advantage. It teems with interesting contrasts, and is
one of the most richly variegated, and most instructive views
that I have seen in Borneo. To the north and north-west the
broad Kapoewas plain disappears in the hazy air of the horizon;
to the west and south-west, cone- and cup-shaped mountains,
have the predominance, they are the andesite-mountains of the
Raja, Dĕlapan [1]), Oendau and other ranges; to the east and
north-east the foreground is occupied by a few cone-shaped mountains, the Mĕnala, the Dapan etc., behind which, on the other
side of the Soeroek valley, the sharply outlined and grotesque
shapes of the volcanic tuff-mountains, are discernable; to the
south, in the foreground, stand a few cup-shaped hills and behind them elongated mountain ridges rising one behind the
other, some of them, as for instance Mt. Madi, extending from
west to east, as far as the eye reaches, at nearly the same
height. From Mt. Loeboek one can see distinctly that a strip of
relatively low land lies between these last named mountains,
which for the sake of briefness I will include under the name
of the Madi mountains, and the volcanic Müller mountains; this
strip is not confined to the western part of the Müller mountains,
where the andesite-cones predominate, but extends also to the
middle portion of the other side of the Soeroek-valley, therefore
to the south of the tuff-mountains of the Mandai. This lower
ground, as we shall see presently, is, as a matter of fact,
merely the gentle northern slope of the Madi mountains which
consist of a sandstone-plateau, remodelled by erosion and dislocations, and with a gentle northern incline. The volcanic Müller
mountains therefore form a strip of mountain-land at the northern

1) With regard to the Dĕlapan range it should be mentioned that it is divided by valleys
into portions, which have a steep incline towards the Sĕbilit valley, but on the opposite
side a much gentler slope. I presume that the Dĕlapan range is built up of porphyrite or
andesite, but its outward form does not point to this structure as undeniably as for instance
the shape of Mt. Oendau or the Loeboek mountains.

base of the Madi mountains, and they are in turn separated from the Upper Kapoewas plain by low hilly ground built up of sandstone. The width of the Müller mountains in the valley of the Tèbaoeng does not exceed 4—6 kilometres, but increases considerably in an eastern direction in the Mandai and Kèryau districts. The older formation, to which amongst others belongs the amphibolite of the Sèbilit, nowhere lends a distinctive character to the landscape; the later formations monopolize the features of the country, and the older formations only make their appearance in the valleys where the erosive force of the water has carried away the more recent rocks.

Most probably the andesite of Mt. Loebuek and Mt. Pétoesi has been forced through and has partly overflowed the sandstone which, at the foot of Mt. Pétoesi, about 15 metres above the water-level, is exposed in horizontal beds in the river, and is there partially overlain by a loose soil full of blocks of andesite. We went for some distance further up the river and spent the night on the boulder-island Poelau-Bètoeng.

The banks remain low till just below Sapoet, where on the Sept. 12. left shore a gray, hard quartzite becomes exposed, the strata of which dip 75°—85° to S. 30° W. Just above Sapoet some sharp rocks of amphibole-andesite rise out of the water, and still higher up the river is narrowed by projecting rocks of fine-grained sandstone with a dip of 16° to the north. These cliffs are of the same sandstone which, higher up-stream, makes the water rough and causes the riam Péredjoek [1]).

On the right bank, just below Poelau Noesa, andesite is exposed and the upper part of the island itself is composed of that rock, while the lower portion is thickly wooded and the ground covered with boulders.

Below this island, as indicated on Map IXᴬ, sandstone and andesitic porphyrite alternate continually along the banks, and

[1] On account of the strong current and impeded by my rather heavy bidar I could only examine the rocks below the riam; it is not impossible that the rocks higher up in the riam proper, which appeared to me of somewhat different shape, may consist of andesite.

the ridge of andesite which run diagonally or obliquely across the river-bed are generally the cause of the rapids which occur there. Near the mouth of the rivulet Kĕlibang bĕsar we met a sampan with some Dyaks whom I recognised at once by their tattoo-marks and their immense ear-discs as belonging to the tribe of

Fig. 57. Wangsa Patti.

the Oeloe-Ajers. I stopped them and learned that one of them was Wangsa Patti, the Kampong chief of Lĕ-matak, the highest settlement on the Tébaoeng. He was the very man I needed for my projected expedition into the Madi mountains. He told me that they suffered from a great dearth of rice, and that he was going down the river to buy some at Boenoet. After a good deal of talking and bargaining I at last succeeded in persuading him to accompany me, and from the moment he entered my bidar he proved the most faithful and trustworthy guide imaginable.

Wangsa Patti, as represented in fig. 57, is a capital specimen of a Dyak, and in many respects may serve as a type of the Oeloe-Ajers of the northern slope of the Madi mountains. He was most beautifully tattooed [1]; a large consecutive pattern reached from

1) Unfortunately nothing can be seen of the tattooing on fig. 57. The tracings are done in indigo colour and probably the portion of the skin operated upon by the coloring matter has made the same impression upon the photographic plate as the untouched part of the skin, so that nothing can be seen of the tattoo-marks although the smallest unevennesses and

Pl. XXXVII.

The Kǎǎnozi-river.

the neck, across the chest down to the hips, and gave the impression of being a finely woven vest. The tattooing of the arms was exceptionally beautiful and elaborate, and the tasteful pattern extended from the arms over the hands which looked as if they were covered by open-worked gloves. The pattern on the back was much simpler. The thighs were adorned with a few detached tracings, and on the calf of the leg, the round dark-blue disk, the distinctive mark of the Oeloe-Ajers, had been introduced. All the men of the tribe bear this sign, but elaborate and well-finished patterns, such as adorned the skin of WANGSA PATTI, are only granted to the principal men of the tribe.

The acquirement of these bodily adornments takes years, and requires great perseverance and disregard of pain on the part of the patient who has to undergo the operation.

I spent an anxious moment in the goeroeng Sémoenti where again a porphyrite dyke breaks through the sandstone. The rattan rope by which the bidar was being towed, broke, and the vessel was thrown with great force against the overhanging branch of a tree which tore away the awning. Fortunately the Dyaks all jumped into the stream and quickly mastered the boat, and none of my baggage was lost. Just above the goeroeng Sémoenti, on the high sandstone bank to the right of the river, is the deserted house Mijaboeng, on which are some sacrificial poles. The rocks along the banks are of sandstone with a northerly dip, as far as the goeroeng Bëlingin, where a dyke of quartz-porphyrite runs transversely across the river. A little higher again, near the goeroeng Sédoengan, navigation is impeded by sandstones dipping 24° north. Overlying the sandstone, on the right bank, is a beautiful section of a dark clay-stone containing coal, (D, map IXᴬ). The clay-stone has a gentle dip north, and has been disturbed by various

---

wrinkles in the skin are clearly marked. The use of isochromatic plates or glass-screens which are impervious to the violet rays, would no doubt have given a better result.

dislocations. Sandstones, clay-stones and porphyrites remain the predominating rocks as far as Loepak Djawit, the house of WANGSA PATTI, where I arrived at 1 p. m. It is a very long house consisting of 19 pintoes and reminds one strongly of the large house of the Oeloe-Ajers at N⁴ Raoen, on the Mandai (see page 52). I knew that a little way above this house the dangerous rapids and falls of the Tébaoeng began. As the water was low, and therefore favourable to the passage up-stream, I wished to avoid any unnecessary delay, and asked WANGSA PATTI to accompany me at once with a few guides who were thoroughly acquainted with the falls, so as to continue my journey. I intended to keep these people in my service as carriers up the Madi mountains. WANGSA PATTI, however, told me that his men were all dispersed over the ladangs and he would therefore have to collect them, which would take at least two days. He also doubted whether I should be able to pass the rapids with my bidars, so that in any case I should have to wait until his people had fetched some boengs, which happened just then to be away from Loepak Djawit. But as I had but little food with me and could not get any here, so that by waiting I should probably have had to give up my intended expedition to the Madi mountains, I made up my mind to attempt the passage of the falls without a guide, trusting to the very favourable condition of the water, to my good fortune, and to the experience I had already acquired in the passage of other rapids. WANGSA PATTI promised to follow me with his men by a shorter, overland route and to join me above the falls, in order to assist me with his guides and carriers up the Madi mountains.

So I left at 3.40, after having bought a few unusually, long rattan ropes to pull the bidars through the rapids. The little stream Lématak discharges into the Tébaoeng just above the house Loepak Djawit, and at that point there are two more Dyak houses, together known as the kampong Lématak. WANGSA PATTI is the acknowledged chief there also, and he calls himself chief

of Lĕmatak. Soon we were amongst the rapids; the riam Sélam-
boel lies in sight of Roemah Lĕmatak. This riam is caused by
two rocky ridges, one of amphibole-porphyrite, the other of
sandstone. A little further on, near the mouth of the Saroeng
rivulet, the river-bed is narrowed considerably by a large sandstone rock connected with the right bank. This fine-grained,
brownish sandstone has a dip of 10° N.N.E. Shortly beyond a
porphyrite dyke runs in a north-west to south-east direction.
Near the goeroeng Empakan our difficulties became really serious,
as for a distance of over 250 metres the river-bed is here
thickly covered with boulders of arkose, diabase, amygdaloidal-
diabase, and, at the lower end of the fall, also of sandstone, overlying the other rocks. The trend of the beds of sandstone is here
E. 20° S., the dip 16° N.N.E. The water rushed so wildly in
between the rocks, and so much time was spent in the passage,
that it began to get dark when, at last, we had brought all the

Fig. 58. THE PĔLAI RAPIDS.

boats safely through the riam. Fortunately, a little further up-stream,
just below the goeroeng Pélai, we found an excellent place for

our night quarters. As indicated on the plan (map IXᴬ), the river is at this point enclosed by groups of rocks which at low water contract its bed to a width of three metres. Above the fall the river widens into a pool and then abruptly curves to the south. The rocks on the right bank, on which I had taken up my night quarters, rose eight metres above the water-level, and I noticed on the ragged top a pile of tree-trunks, between which small drift-wood and some gravel had accumulated, a proof that at high water the entire mass would become submerged. As I stood on the top looking over the riam, I noticed that even at this low stage of the water, the waves were rushing through the narrow channel with so much noise and tumult that the human voice became perfectly inaudible; and then I also recognised the truth of what WANGSA PATTI had said, viz., that at high water the falls of the Tébaoeng are absolutely impassable. At different elevations and in several places the water has washed potholes 2 metres in depth out of the rock, which were still, for the greater part, full of water, and formed a series of miniature lakes. The shores were thickly wooded, and when the full moonlight lighted up the panorama I congratulated myself upon the exceptional beauty[1]) of the place which I had fixed upon for our bivouac that night. The night was far advanced before I could tear myself away from the glorious scene.

Sept. 13.   Not until the next morning did I fully realize how extremely difficult the passage of the riam was, more particularly as there was no possibility of unloading our goods at the bottom of the rapids, carrying them over the rocks and reloading them above the riam, because everywhere the rocks could only be scaled from one side. Consequently our boats had to be unloaded on a rock in the middle of the riam, then trailed over a rock-dam, and immediately reloaded. The route by which

---

1) The dozen photographic plates which I used at Boenoet and on the Tébaoeng, were entirely spoiled so that the takings gave no result.

boats have to pass is indicated on the plan. In the morning
WANGSA PATTI came to see me near the goeroeng Pélai; he
told me that he would join me the next day with a sufficient
number of carriers. Above the goeroeng Pélai the water was
much calmer, but the enormous quantity of foam on the surface
indicated the nearness of fresh obstacles, and so it proved, for no
sooner had we turned round the corner and veered southward,
than the goeroeng Kënsoelit, and close behind it the goeroeng
Nëkan, lay before us. At the goeroeng Kënsoelit the difficulties
were comparatively small, and with our united strength we managed
to pull the boats against the heavy current close up to the left
bank, but a little further on, near the goeroeng Nëkan, this
was impossible, for here the water is contracted to a width of
only nine metres, and it falls from a height of two metres over
sharp, pointed boulders. So I ordered all the boats to be unloaded; the bidars were partially taken to pieces, and were
then trailed over the rocks, four metres high, on the right
bank, until the calmer water above the fall was reached. On
the plan map IX^A I have marked the rocks displayed in this
portion of the Tébaoeng, they belong to a system of vertical
or almost vertical strata the strike varying from E.—W. to
E.S.E.—W.N.W. Going from north to south in the goeroeng
Pélai, we meet first with strata of diabase and diabase-porphyrite impregnated with pyrrhotine and an intercalated bed of
silicified diabase-tuff, this complex is succeeded, near goeroeng
Kënsoelit, by beds of sandstone, here and there passing into conglomerate, then comes hard, variegated clay-slate and whetstone,
and finally, near goeroeng Nëkan, once more sandstone and fine
conglomerate. A bed of this conglomerate here narrows the
river to about 6 metres and gives rise to the real fall, which is
two metres high. I believe these rocks to be older than the
sandstone with coal and the clay-stone exposed near Lématak
and further down-stream on the Tébaoeng and the Sébilit. The
territory above goeroeng Nëkan is composed of sandstone with
a few intercalated beds of conglomerate; this system of strata

is much contorted and the dip alternates between 50° northward or southward. The sandstone rocks projecting in the river-bed give rise to a few minor rapids, of which the goeroeng Hénap is the principal.

About one hundred metres above goeroeng Hénap, a bed of sandstone juts out from the left bank and extends to the middle of the river; it consists of soft sandstone with feeble N.N.E. dip. This I hold to be more recent than the steeply inclined, harder sandstones of the goeroeng Nékan and neighbourhood, and identical with the sandstones with feeble northern dip, which play so conspicuous a part further to the north, in the neighbourhood of Lématak and down-stream along the Tébaoeng. Quite suddenly the water now becomes perfectly calm, the bed of the river deep and entirely free from rocks. Opposite the mouth of the little stream Sémoerang bésar, lies the pangkalan Sémoerang or pangkalan Kapala riam, from whence a footpath leads through the wood and over the hills to Lématak, which can be reached in an hour and a half. Near this pangkalan we took up our night quarters.

Sept. 14. The next day, at 1.30 p. m., a messenger from WANGSA PATTI came to tell me that the chief with six carriers would join me near the house Tandjoeng Béloewan, a little further up-stream. I moved on at once and to my surprise I found that above the pangkalan Kapala riam, the water of the Tébaoeng remains deep and perfectly calm. The solid rock only appears very occasionally; the banks are low and sometimes boggy. The day before it had puzzled me to see, that the water of the Tébaoeng, although not actually muddy, retained a yellowish colour all the way up, and resembled exactly the water of the great lakes (see page 42), while in all the other rivers of Borneo, I had found that in their upper courses where the rapids begin, the water, except just after rain and when rising, is clear as crystal and perfectly colourless. For some little time I thought this phenomenon was owing to the low and apparently marshy region through which we passed, but I soon found out my mistake, and learned that the explanation of it (as will be shown later on) must be sought, not here, but at the sources of the Tébaoeng. The solid rock which now and then crops up along the bank in this low lying area, is invariably a fine sandstone dipping N. or N.N.E. at a small angle and occasionally alternating with layers of clay-stone. At 2.30 p. m. we reached the small house Tandjoeng Béloewan on the left bank This is a temporary settlement of Mëlawi-Dyaks, who are related to the Dyaks of the Tébaoeng. Here we were joined by WANGSA PATTI and his six followers, and my small boats were now laden to the full. The water remained calm and the river wound amongst hilly ridges trending W.N.W.—E.N.E. The rock *in situ* continues to be clay-stone and sandstone, the layers dipping feebly N., N.N.W., or N.N.E., hardly ever at a greater angle than 25°. Near the goeroeng Batoe Bélika the river breaks right through one of the above named low ridges. Beds of fine, hard, quartzose sandstone jut out in two places from the banks far into the water and cause a double waterfall, which, however, it was not very difficult to pass, because both above and below the fall the river is free from rocks. Above the riam Batoe Bélika the river takes at first an easterly direction, but soon after it curves S. and S.S.E. and continues in this direction for a distance of two kilometres. The condition of the ground remains unaltered; sometimes the navigable passage is narrowed by flat rocks of sandstone projecting from the banks, causing insignificant rapids, but on the whole the river is navigable and quite free of rocks. This is perfectly in accordance with the nature of the ground and wherever, as here, the substratum is composed of feebly inclined beds of sandstone, the same phenomenon may be expected. The inclined strata, occasionally cause step-like obstacles, in the shape of bars, running right across the river, but they offer such a uniform resistance to the erosive force of the water, that it is only very seldom that hard or tough parts are isolated to form separate masses, which stand up out of the water and make navigation dangerous. I spent the night on a sandbank opposite the mouth of the little river Tapang.

Sept. 15. Just above Nᵃ Tapang the sandstone which up to this place dips gently to the north or lies horizontal, begins to incline more to the south, and at the point where the river bends to the west, the sandstones stand vertical and the whole aspect of the water-course changes. It is once more full of elongated or square rocks with rounded edges, partially rising above the water. They cause a rapid, the riam Pandjang, which, however, with a little caution may be passed quite safely. Above the riam Pandjang the river curves to the south, and rapidly widens into a beautiful round pool, about 200 metres wide, into which, on the opposite side, the Tébaoeng discharges by a waterfall, the goeroeng Bénoewang. This fall is without exception the most beautiful I have seen in Borneo. Below the fall stretches the pool, its sloping banks lined with richly variegated trees, amongst which the wild sago-palm (ransah) and several kinds of Conifers and Nepenthes are conspicuous. In its fall the water rushes over hard quartzitic sandstone which dips 63° S.S.E. The sand-stone, however, is cleaved in thick beds dipping 25° N.N.W., thus showing a very marked false stratification, which makes it difficult to ascertain the true dip. We had here to deal with a serious difficulty and I considered myself very fortunate in having WANGSA PATTI with me, who, so to speak, knew every stone of the Tébaoeng. We had to drag the sampans over the rocks along the right bank, and wherever the distance between two points of rock was greater than the length of the boats, supports of tree-trunks had to be manufactured. Once, one of the boats, after having been dragged above the principal fall, slid again down the rocks, when the rope broke, and immediately the vessel was seized by the foaming current and with headlong speed was hurled back down the fall and into the pool below. For a wonder it did not sustain any serious damage. Next time we took better precautions, and at 10.45 I again entered my bidar above the fall, with the pleasant feeling that the greatest obstacles were

Pl. XXXVIII.

Batoe Ati, Sr Mélawi (Sandstone).

now past and that the land-route to the river-basin of the Mĕlawi lay clear before me. One hundred metres further on the Kéré-moei [1]) discharges into the Tébaoeng, and shortly after this river ceases to be navigable. We turned into the Kérémoei and reached the beginning of the land-route at 12.50 p. m. I immediately ordered the boats to be pulled up on dry land and the roof of my bidar to be repaired. The geological specimens which I had collected along the Tébaoeng and some minor articles of my baggage, I left behind in the boats for WANGSA PATTI and his men to take back with them on their return journey to Lĕmatak, and from there to take them on to Boenoet.

At 9.15 a. m. I commenced my overland expedition with two Sept. 16. pradjoerits, my own servant and my Chinaman, 13 coolies from Sintang, WANGSA PATTI and his 6 Dyaks from Lĕmatak, alto-gether 24 men. For the first few days I had some difficulty in distributing the baggage, of which there was a large quantity, although we only took absolute necessities and I had reduced my personal requirements to a minimum. I had often to inter-fere when the Malay coolies tried to increase the loads of the Dyaks with some of the burdens which had fallen to their share. WANGSA PATTI at once set a good example by taking for his share the heaviest article of all, viz., the water-tight box, entirely lined with sheet-iron, in which my photographic apparatus with the plates, the diaries, and the silver money were stored. I also noticed that over and over again he took part of the load of one of his men who seemed sickly and coughed a great deal, and strapped it on to the top of his box.

Our path led first southward and then south-eastward, with a moderate incline up to Mt. Béransa. On the slope of this mountain the rock *in situ* is a fairly coarse, white sandstone, with vertical strata trending E. 30° N. At 8.30 a. m. the west-

---

1) The full name of this river is Kĕrĕmoei Tĕbaoeng in order to distinguish it from another river Kĕrĕmoei which empties itself into the Mĕlawi-river. The latter is accordingly named Kĕrĕmoei Mĕlawi.

top, 677 metres high, was reached. Here the sandstones had a dip of 82° S.S.W. It threatened to rain, and the only view we had was towards the south-east, where the ranges of the Madi mountains could be seen. The natives distinguish these ranges, going from north to south, as Madi Soewah, Madi Bĕlimis, Madi tentoe or true Madi, and behind these Madi Gĕriga, which was now invisible. We now proceeded in an easterly direction by a path which led across the ridge of Mt. Bĕransa, where the sandstones dipped 85° S. 10° E., to the principal eastern top of the mountain which lies 714 metres above sea-level. The forest here consists for the greater part of Conifers and it is rich in Orchids. In pouring rain we descended the mountain first in an E.S.E. and afterwards in a southerly direction, and at 2.30 p. m. we crossed the little stream Tĕbaoeng Tĕnang, which at this point forms a beautiful fall, ten metres high, over banks of sandstone with a feeble northern incline. We spent the night a little further down at the mouth of the Tĕbaoeng Tĕnang Kĕtjil which flows merrily along, and tumbles, in pretty little waterfalls, over banks of sandstone with a feeble south-eastern incline, on its way to the Tĕbaoeng.

Sept. 17.   Below this point and till close to the mouth of the Kĕrĕmoei river numerous little waterfalls make the passage up the Tĕbaoeng absolutely impossible, then further up-stream for a distance of four kilometres the water is perfectly calm and smooth. The Dyaks of Lĕmatak and neighbouring districts who often visit the head-waters of the Tĕbaoeng in search of rattan and gĕtah in the forests, have built canoes, which they use exclusively on this calm portion of the river, and which each time after using they pull on land at certain appointed places. When returning home from their searches in the forest these boats are left behind near the mouth of the Tĕbaoeng Tĕnang. Unfortunately, when we arrived there was not a single boat left, for a party of Dyaks from Lĕmatak had shortly before left for an expedition to the forests of the Madi mountains. They had taken all the boats and left them four kilometres up-stream at the appointed landing-place. We were therefore compelled to proceed by a difficult, boggy, and indistinct path on the left bank of the Tébaoeng, and after losing ourselves for some time in the forest we reached the mouth of the Tébaoeng Gĕgarak at noon. The march had occupied 5 hours. The rock *in situ* is gray, soft sandstone, alternating with clay-stone; both contain particles of coal. From thence a very bad path, frequently necessitating our wading through deep marshes, brought us at 2.30, to drier ground at last, and led from there uphill, first in a southerly and afterwards in a south-easterly direction. We fixed our night quarters near a little stream, a branch of the Tébaoeng Gĕragak, at a height of about 400 metres. Here the rock *in situ* was white sandstone with a feeble northern dip.

This day I proposed to reach the Babas Hantoe one of the Sept. 18 chief heights of the Madi mountains; so we set out in the early morning before 7 a. m. and ascended in a southerly direction up a gently sloping incline. After we had climbed about a hundred metres this slope gradually widened and finally formed part of an extensive plateau with a feeble northern slope. The higher we ascended the more numerous became the Conifers in the wood; at an elevation of about 700 metres they reign paramount. The forest is thin, and when a gust of wind stirred the tree-tops, it created a peculiar moaning, grating noise which reminded me of the European fir-forests. Soon the character of a forest swamp began to prevail, the soil became soft and we had reached the borders of the moss-vegetation. The lower part of the tree-trunks is here covered over with a thick cone-shaped mossy coating, widening towards the bottom, on which the tree-trunks seem to stand as if on pedestals. The deeply-trodden narrow path winds between these wet spongy, cushions of moss. Accompanied only by WANGSA PATTI and my Chinese boy I continued the slow ascent. At an elevation of 1000 metres the ground was a genuine morass, and we could only proceed by jumping from the root of one tree to another. Then we gradually reached flatter ground, the wood opened out, and at 4 p. m. after a troublesome climb over a chaos of rock masses, the Babas Hantoe suddenly lay before me. The Babas Hantoe is an extensive, slightly undulating plain consisting of pure white friable, sandstones, having a scarcely perceptible dip to the north. The ground is still partially swampy and covered by shrubs one metre high, principally Myrtacea, very much like heather, and belonging mostly to the genus *Astartea*. The dryer places are recognised by low trees and higher shrubs. The naked white rock appears here and there in elongated strips. Very appropriate is the Dyak name Babas hantoe = ghost-ladang, or ghost-garden, for in truth it appears, especially in the moonlight like an extensive, beautiful, and tastefully laid out park with pure white paths winding in and out amongst the copses. In full daylight this landscape vividly reminded me of the marshy parts of the fens of Holland or perhaps still more of the Hornishgrinde in the Schwartzwald. The delightfully cool temperature, the heathery vegetation, and the fragrant smell of the pines, all combined to make this landscape so truly European that it needed the miniature Nepenthes' with cups not thicker than pipestems, the sulphur-yellow Orchids, which sparkle here and there amongst the bushes, and the webs, as large as cart-wheels, of the many coloured giant-spiders, to remind me that I was below the equator and not in my native land. From the highest point of the plateau, there is a really impressive view extending over a large portion of the mountain-land of West Borneo; and towards the north and the north-east it is especially fine. In the far distance arise the table-mountains [1]) of the Müller-mountains, amongst which Mt. Maroeng (1370 metres), as seen from here, is the most conspicuous. The nearer mountains, situated to the south of the Soeroek-valley, Mt. Hitam (1305 metres), Mt. Laboe (1120 metres), and also Mt. Liang Engijo (1091 metres), much remind one in their shape of types which I noticed elsewhere in the volcanic Müller mountains. The long ridge behind the Liang Engijo resembles the Madi plateau itself, but the southern

---

1) See panorama plate XIV on page 22 of the atlas.

continuation of this ridge, with Mt. Batoe Ensambang [1]) rising to a height of 1770 metres, appears to possess a more independent character. Along the southern edge of the Madi plateau one looks into the valley of the Upper Mélawi, which in the far distance, appears to be closed in by the Liang Koengkam plateau. I expect that future investigations will show that this marvellous plateau-land, which covers an area of nearly 17 kilometres, and is about 1300 metres high, is built up of almost horizontal strata of the Tertiary sandstone formation. A portion of this plateau is hidden from view by the beautiful double cones of the Bakijo mountains which rise high above the Madi-Gériga. To the south-east, south, and south-west, the horizon is limited by a long ridge trending from east to west, and almost uniform in height, the Madi tentoe, or true Madi mountains. East of the Babas Hantoe, the Madi mountains appear in their simplest form, as merely a stretch of plateau-land broken off abruptly towards the south. This is distinctly shown in the panorama. Near the edge, the average height of the ridge of the true Madi mountains is 1200 metres, and from there the plateau inclines northward at an angle of about 7°, until the valley of the Soeroek is reached.

To the west the shape is more complicated. The northern slope is here intersected by the Babas Hantoe where the plateau reaches about the same elevation as the southern ridge, and it only reassumes its gentle northern incline on the other side of the Babas Hantoe. Between the Babas Hantoe and the true crest of the Madi mountains lies a large shallow valley, draining to the eastward. This valley, as well as the greater part of the slope of the Madi mountains connected with it in an easterly direction, rises about 1000 metres above sea-level and this por-tion of the mountain-land between Kapoewas and Mélawi may therefore be properly called a high-plateau, the Madi high-plateau.

Next morning the sun rose with exceptional glory, and after Sept. 19. dispersing into vapour some small loose cloudlets which chased

---

1) In sketching this panorama on the spot, this mountain has come out decidedly a little too high, so that it appears to be at a relatively short distance off.

each other on and round the Babas Hantoe, the grand mountain-panorama unfolded itself with indescribable beauty in the fine pure morning air; and so entrancing was the sight both here and also over the entire dew-strewn Babas Hantoe, that after verifying the sketches which I had made the previous evening, I gave myself two full hours to enjoy the view to my heart's content. I followed all the natural sandstone paths of this fairy-garden and visited the different vantage grounds within my reach, and enjoyed myself like a school-boy on the first day of his holidays. At 8 a. m. we had to resume our journey, for WANGSA PATTI had warned me that the road to the ridge of the Madi mountains proper, although it looked so short and easy from here, was in reality rather troublesome. By natural giant-steps of pure white sandstone we descended without much trouble into the valley between the Madi tentoe and the Babas Hantoe. At first the soil is much the same as that of the Babas Hantoe, sometimes solid rock, then again, immediately adjoining, marshy forest and peat. As we neared the bottom of the valley the patches of peat became more connected, and I soon realized that the whole of the valley, and probably the whole of the Madi plateau, was covered with marshy forest, standing in a thick layer of peat, originating from the half decayed remains of all kinds of trees, shrubs and mosses, a true tropical peat-bog. In contrast to the fens of moderate zones, originating princi-pally from a limited variety of shrubs and mosses, the tropical fens are principally composed of the remains of various trees. This valley and also a considerable part of the Madi plateau eastward, drain into the Tébaoeng, in fact the principal sources of this river occur in this territory. Now I found where the yellow colour of the water of the Tébaoeng came from, it is derived from the yel-lowish-brown fen-water which flows down from the Madi plateau.

It was difficult to follow the path here, and matters did not improve when a little further south the ground became strewn with great masses of rock. At 1.30 p. m. WANGSA PATTI, whose lead for some little time had been rather uncertain, declared that he had lost the way; so we halted by the side of a little brook, and placed our goods all together on a fairly dry spot. Attempts were now made in different directions to recover the lost path, but without result, and we were compelled to spend the night in the marshy forest.

On this spongy soil, completely saturated with water, our night quarters were not the most comfortable, especially for the coolies, and to enhance our discomfort a steady rain, which held on most of the night, soaked through everything that was not already thoroughly wet. In the night the temperature fell to $17°.5$ Cels. and even at this position of the thermometer, the excessive dampness made us feel quite chilly, so that we all longed to be on the move again, and we set out at 5.30 a. m. Guided by my compass I decided to strike out a path to the south, because I knew that in this direction we must somehow reach the edge of the Madi mountains, where we should have a chance of taking our bearings.

The ground was composed partly of loose pieces, and partly of solid sandstone rock, much cleaved and consequently divided into enormous masses, separated from one another by deep, narrow cracks. The cracks between these masses are filled up with deep peat, and several of our party sank down in them up to the neck. Indeed it soon became evident that we should be obliged to keep to the solid rocks, and try to jump from one rock to another. This is certainly the most difficult piece of ground I traversed anywhere in Borneo, and as I think of it, I am still full of admiration for my carriers, who transported my goods through this indescribably chaotic wilderness. We certainly did not get on very fast, but at 8.45 a. m., before we expected, we found ourselves on the high ridge of the Madi tentoe, 1255 metres above sea-level, with the steep southerly incline of the mountain in front of us. The unobstructed view to the south enabled us to ascertain that we had gone too far west, and that we had reached the ridge of the Madi at a point, a little west of, and 175 metres higher than the

(margin note: Sept. 20.)

place where the path, which we ought to have followed, crosses the ridge.

Seen from the south, the upper edge of the Madi with its scanty vegetation, and its gigantic masses of sandstone, looked like a formidable wall, which, as far as the eye reaches, may be traced as a feebly undulating line. The sandstone of the Madi tentoe has a gentle northern dip. It is a pure white coarse quartz-sandstone, in every respect resembling that which forms the rock on the Babas Hantoe plateau. As we descended the steep southern slope of the Madi ridge I noticed that underlying the sandstone, and 25 metres lower down, clay-stone is exposed (Section E on map IX⁹) and that in several places tiny springs arise at the spot. Below this clay-stone, sandstone again crops out. Against the upper, steepest portion of the incline there is an opening in the forest and the ground is covered with a great variety of plants, amongst which several kinds of Orchids and Nepenthes are conspicuous by their rich colouring and exquisite shape. Soon our small caravan once more disappeared in the forest, and now came a very hard piece of our route, down the slope of the mountain which was strewn with huge pieces of rock and covered with a tangled wilderness of vegetation, until at 12.45 p. m., we reached the bed of the Midih-river, at a point about 734 metres below the ridge of the Madi. Both the solid rock and the boulders of the Midih are exclusively sandstone and clay-stone, with here and there particles of coal; in fact along the whole slope of the Madi I found no other kinds of rock. The Midih at this point is but a small mountain torrent, and we followed its bed down-stream in a south to east direction. Here we caught an unusually large tortoise, and when I expressed my desire to take it with me, alive, WANGSA PATTI tied it on the heavy box he already carried, thus increasing his burden by at least 25 kilograms. I mention this as a proof of the great willingness and also of the extraordinary bodily strength and toughness of this Dyak chief. On this day the only kinds of rock I saw *in situ* in the Midih were clay-stones alternating with beds of sand-

Pl. XXXIX.

TOBAS AND TENADORS OF ROEMAH MÉRAKAU.

stone. The strata are mostly disturbed and here and there folded. The predominating strike is E. 15° S. the dip is generally steep towards the south. At 4.30 we took up our night quarters at a suitable place on the Midih. The Dyaks cleverly caught some small fishes which were of excellent flavour. We were now at an elevation of 440 metres, therefore 815 metres below the ridge of the Madi mountains. A steady rain which fell during the night made the stream to swell half a meter.

At 6.30 we started again and followed the winding Midih Sept. 21. down-stream, occasionally cutting off a curve by going through the forest. As a rule the banks were low, and the solid rock, always sandstone and clay-stone, only appeared at rare intervals. At 8 a. m. we reached the point where the Midih Běhansap, a little stream about as broad as the Midih, joins this latter, and here we recovered our lost path. The Midih is fed by a number of little rivulets flowing down from the southern slope of the Madi tentoe, and it discharges into the river Měrya, by which we should have been able to reach the Kěrěmoei and also the Mělawi. Generally however, the Dyaks make use of a path, which from the Midih leads in a south-westerly direction along a northern spur of Mt. Toejoen (1172 metres), to the river Panai, which empties into the Kěrěmoei. WANGSA PATTI recognised the place where this path turns away from the Midih, and at 9.30, to my great joy, we came to the end of the "djalan ajer" through the bed of the Midih and began to mount the steep incline. The highest point was reached at 1 p. m. (681 metres), and from there the path runs for some considerable distance at a uniform height on the mountain ridge. In more than one place the path widens into a kind of plain, quite clear of vegetation, which forms the play-ground of the Argus pheasants.

Several times we crossed deep-trodden rhinoceros-paths with fresh footprints. Judging from the absence of comparatively fresh cut twigs and stems, our path had evidently not been recently used, and the movements of my guide again became uncertain. At 2.30 we discovered that we had been misled by the traces of old tracks of gĕtah-seekers, and that once more we had lost our way. The provoking shrieks of the laughing-birds (with their gok, gok, gok, háh, háh, hah, háh), which congregated here in large numbers, sounded like mockery in my ears. It seemed plain, however, that by going in a south-westerly direction we should ultimately arrive either at the Bĕtoeng or at the Pĕnai, and I therefore ordered a path to be hewn in that direction through the thick forest. But the falling darkness compelled us to make a halt by the side of a little stream. All the day we had past through a sandstone region, and the rocks in the bed of the little streamlet were also composed of gray sandstone. During the night it rained heavily, and it was unpleasantly damp, with a temperature of 20° Celsius.

Sept. 22.   At 6 a. m. I sent Wangsa Patti to explore the way, with my servant Aboe, who had proved very dextrous in getting through the forest. They returned at 9.30 with the news that they had found a little stream, which, according to Wangsa Patti, was the Pĕnai. We broke up at once, and at 10 a. m. we reached the little river, which afterwards prove to be indeed the Pĕnai. Its bed is broad and thickly covered with boulders through which the water runs in various gullies.

There is a striking contrast between this river and the Tébaoeng; the Tébaoeng rises in the high peaty area on the gentle northerly slope of the Madi plateau, its water is of a yellowish-brown colour and its bed is remarkably devoid of boulders; the Midih and the Pénai on the other hand, rise on the steep southerly slope of the Madi mountains, their waters are clear as crystal and their beds are covered with an exceptionally thick layer of boulders.

At 11 a. m. we reached the point where the Bĕtoeng and the Pĕnai flow together, and where, if we had not lost our way, we ought to have reached the river. All that day we followed the bed of the stream, sometimes wading through the deep water, sometimes jumping from rock to rock. To me personally, this mode of travelling was a real trial, but the natives differed greatly from me in their appreciation of the circumstances; a dip in the water never comes amiss to them. We spent the night at a pretty spot, where the deep water offered a splendid opportunity for a swim, and where there was also abundance of fish. The solid rock in which the Pénai has cut its bed, I found to be sandstone, in a few places passing into biotite-arkose, and clay-stone, the layers of which are generally bent and displaced; the strike and dip are irregular, only in the upper course of the Pénai an easterly or north-easterly dip seems to prevail. At one point, two kilometres below the mouth of the Bétoeng rivulet (see Map IXᴬ), we discovered a dyke of quartz-amphibole-porphyrite[1]. The boulders in the river-bed consist almost exclusively of sandstone and clay-stone, but in a few specimens we found andesitic amphibole-porphyrite, quartz-porphyrite, kersantite and amphibole-andesite-pitchstone. The obstructions, rapids, and falls, are caused, without exception, by great masses or beds of sandstone.

At 7 a. m. we continued our journey along the bed of the Sept. 23. Pénai. Sandstone and clay-stone still form the solid rock, but the strata are less disturbed, and as a rule, they dip at a greater or lesser angle to the south. Both clay-stones and sandstones often contain a fair amount of biotite or feldspar and then pass into flagstone and biotite-arkose. This latter variety is well represented at Noesa Nĕkoe, where it occurs in beautiful, sharply defined strata, dipping southward at an angle of 65°. A fine view is given below, fig. 59, where the Pénai river forms a rapid between two high walls of sandstone. This view was taken with the light falling vertically. The layers of sandstone dip 30° S.S.E., but the rock shows a well-marked, almost vertical cleavage, so that the stratification is not very conspicuous. In the foreground of the figure is the Dyak Tooi, from Lĕmatak. Tooi is married both on the Lĕmatak and on the Mélawi, and he was now on his way to fetch one of his children from his wife on the Mélawi, to show to his wife on the Lĕmatak.

---

1) This rock has an andesitic character and might perhaps with equal right be called amphibole-dacite.

The passage through the river this day was even more difficult than that of the day before, because the water was deeper, and

Fig. 59. GOEROENG TOOI IN THE PĔNAI RIVER.

there were more obstructions in the shape of rounded and generally very slippery sandstone banks.

Shortly before 3 p. m. we heard a shout in the distance, and soon after we saw before us the Kĕrĕmoei, in which the Pĕnai discharges, and here we found the men whom I had sent from Sintang to this spot with the boats and provisions. They had only just arrived; having been 17 days on the way, and evidently they had journeyed leisurely. With them was a Malay, BOENDJANG by name, who had acted as their guide in the passage up the Kĕrĕmoei.

Their arrival was most welcome, for our store of rice would have only served for two or three more meals.

Sept. 24.   One of the Dyaks from Lĕmatak had slipped from a rock in the Pĕnai the day before, and as he had hurt himself he had stopped behind. As he did not join us again WANGSA PATTI went

back to look for him, and to take him some rice. In the afternoon he returned to N° Pénai with the load of his tribesman,
and told me that the man preferred to remain where he was,
and intended after a few days rest, to return with the others
straight to the Tébaoeng. I used this delay for repacking and
rearranging my collection which had got wet. I also rowed a
little way up the Kérémoei, and visited some boulder-islands and
boulder-beds, where I found nothing but sandstone and quartzite,
from which it appears that no igneous rocks have been transported
by the Kérémoei above the mouth of the Pénai. In fact the Kérémoei receives its water supply almost exclusively from the Madi
plateau, which we already know to be of sandstone, and it rises
in the immediate vicinity of the river of the same name which discharges into the Tébaoeng. As a rule, the natives distinguish these
two rivers as the Kérémoei Tébaoeng and the Kérémoei Mélawi.

The Kérémoei is a brisk, rapid stream from 40 to 50 metres
wide. Between N° Pénai and the Mélawi it has a considerable
fall (1 : 410), and several strong rapids. These are only very
seldom caused by rocks in the river-bed, but are mostly due
to high karangans.

B. The Mélawi valley, the Lékawai, and the Raja mountains.

At 6.35 a. m. I left N° Pénai where I took leave of my Dyak
carriers and their able chief WANGSA PATTI, who intended to
return from here to Lématak. The photograph I took of them is
given in fig. 60; the individual on the left of the picture is the
tallest Dyak I have met with in Borneo. His height was 1.72
metres while the average height of the Dyak is only 1.60. This
giant was sickly, he coughed a great deal and seemed to suffer
from pulmonary disease.

The passage down the Kérémoei was most enjoyable and sometimes quite exciting, because of the dangers involved in shooting the rapids, which we did at a great rate. As shown on
Plate XXXVII, this river, although fairly wide, is in most parts

overshadowed by a vault of green, and in many places it might serve as the type of a beautiful river in Central Borneo. On map IXᵇ further particulars are given concerning the condition of . the river and its banks. The highest banks of the Kĕrĕ-moei and the most awkward rapids are in the neighbourhood of Noesa Ménatau, where the river makes a great loop-shaped curve round a spur of Mt. Doeryan. The component rocks in

Fig. 60. OELOE AJER DYAKS (MENTEBAH-DYAKS) FROM LĔMATAK, WITH THEIR CHIEF, WANGSA PATTI.

Taken at the pondok at the mouth of the Pĕnai, on the Kĕrĕmoei Mĕlawi river.

the Kĕrĕmoei valley are exclusively sandstone and clay-stone, near Nᵃ Pĕnai the strata dip feebly northward; a little further on they lie horizontal, and then as far as the Mĕlawi they dip southward at varying angles. In most places the sandstone contains particles of coal.

At 10 a. m. we reached the Mĕlawi at the place where the

stream has already a width of 50—60 metres. Sandstone continues the predominating rock both on the banks and in the bed of the stream. It is a fairly rapid river, but quite safe for navigation; at any rate, when going down-stream. The banks are thickly wooded and the solid sandstone rock, is but very seldom exposed.

The house of BOEDJANG, near the mouth of the Gilang river, a dirty old place, was reached at 11 a. m. Here I stopped to divide the provisions, and to dismiss six of my coolies with instructions to follow WANGSA PATTI, and join his party on their way home. They were to call for the luggage which I had left behind at the pangkalan Kérémoei, and take it to Sintang by way of Boenoet, which order they faithfully executed.

We resumed our voyage at 1 p. m. The character of the Upper Mélawi is quite different from that of the Upper Kapoewas. While the latter abounds in boulder-banks and rocks protruding in the river-bed, and consequently in rapids and cataracts, the Upper Mélawi is especially free from boulders, and consequently there are no rapids of any size to speak of. The solid rock is rarely in detached blocks but generally in nearly horizontal slabs, locally narrowing the stream, and at such points forming small cataracts.

The first type, that of the Upper Kapoewas, we might call typical of a river in an area of disturbed and highly tilted strata, the second type, that of the Mélawi, is typical of a river in an area of horizontal or feebly inclined strata.

The banks of the Mélawi below Nᵃ Gilang, are, as already mentioned, wooded down to the water's edge, only occasionally when the river breaks through one of the ranges of hills, trending E.—W., are sandstone escarpments seen.

One of these points is figured on plate XXXVIII. The enormous rock-mass known as Batoe Ahi, lies in mid-stream, and the illustration gives a very adequate idea of the scenery on the Mélawi. Batoe Ahi is not a separate mass, broken off and hurled into the stream from one of the cliffs above, but it is remainders of a bar of sandstone which at this point crosses the stream, now mostly worn away by erosion. The almost horizontal stratification, which is clearly discernible both in the Batoe Ahi and in the river-banks, proves indubitably their original connection. About 2 kilometres below Batoe Ahi, and on the right side of the stream, the Sinoet flows into the Mělawi over horizontal beds of sandstone. The peculiar features of this tributary are illustrated in section F. map. IX². The sandstone is in thick banks with distinctly marked vertical cleavage. The rocks in the bed of the stream and on the banks are full of pot-holes being the effect of erosion assisted by the constant friction of the boulders. With the gradual sinking of the river-bed to a lower niveau, the water found an outlet through a joint-plane at the points and flowing between the banks *k* and *c* it spouts out 10 metres further down-stream from the pot-holes at *d* which must originally have been ground out from above, down towards the plane of juncture between the strata *c* and *k*.

The entire course of the streamlet runs through a deep gorge, cut in sandstone, and here and there one can see in the sheer precipitous heights which line the river, the distinct impressions of the old pot-holes, ranged in rows or one above the other, thus marking the former levels of the stream. These holes occurring at regular intervals, give the rock-face in places the appearance of the piped frontage of some colossal organ.

Some of the houses inhabited by Mělawi-Dyaks of the great tribe of the Ot-Danoms, differ slightly from the ordinary plan of those, found on the Mandai and Tébaoeng, as given in figs. 69 and 70. The larger houses present the common type of architecture, but the smaller ones markedly differ in the peculiar shape of the roof. This is boat-shaped, the centre being lower than the sides, which moreover jut out all round, thus making the roof considerably larger than the base of the building. Next to the houses stand sheds built on the same plan, in which the rice is stored. Another difference is, that the part of the house used in common, which adjoins the verandah, has

Pl. XL.

TĔMADOKS OF ROᴇMAH, MÉRAKAU.

often no side wall and is entirely exposed; where this is the case the verandah only extends along part of the building. Close to the houses, there are generally to be found a greater or smaller number of memorial pillars (tĕmadoks) representing human figures carved in wood. Sometimes, as was the case at Nᵃ Biai these poles are protected by an awning being an exact miniature reproduction of the roof of the house just described.

The area through which the Mĕlawi flows is exclusively composed of sandstone in thick horizontal strata, or in beds slightly inclining southward. The current is here in a southern direction, and the river cuts in succession through several ranges of sandstone hills trending east and west. At each of these obstructions the current is impeded by rocks which protude from the banks and form more or less connected bars across the river. In this way at Batoe Mĕngkoedoe, 800 metres above Nᵃ Embalau the bed of the stream is narrowed to 25 metres, and the water rushes with tremendous velocity through the one remaining gap, forming a furious whirlpool, where many an ill-guided boat comes to grief.

At 4 p. m. we passed Nᵃ Embalau, an astronomical station of the topographical survey, 86 metres above sea-level, and an hour later I reached the kampong Kĕmangai the most important settlement on the Upper Mĕlawi, inhabited principally by Malays. The Malay kampong chief MASRI WANGSA is here the representative of the Dutch Government. Although most of the Dyaks in this district have been converted to Islam many of the memorial pillars, remnants of olden days, have been left standing close to their houses. Conspicuous amongst these is a stone column, 2.10 metres high, 0.35 metre wide and 0.17 metre thick. This is an interesting monument because it is the only stone pillar I have seen in Borneo. According to the natives this column is very ancient and was in existence when the present population settled here.

I stayed in the half-finished house of the kampong chief. A Sept. 26. heavy thunderstorm raged all night and at 7 next morning I started for the Lĕkawai river. Close to Kĕmengai I met RASSAH, a converted Dyak from N^a Lĕkawai, who had been recommended to me some time previously by the controller BARTH, as the best guide for the Upper Mĕlawi district. I secured his services for my proposed journey to the head-waters of the Lĕkawai and the neighbourhood of Mt. Raja, for a remuneration of one ringgit per day. Continuing our journey down-stream through an area composed entirely of sandstone we arrived, at 8.30, at N^a Lĕkawai where I remained till 12 to buy the necessary tools, especially hatchets, in case we needed to make boats to go down the Samba river. RASSAH, meanwhile, gave me some valuable information concerning the head-waters of the Mĕlawi and the passes which lead from there to the Siang-Moeroeng district. He told me that a journey from N^a Lĕkawai to Moeara Teweh on the Barito could be accomplished in 23 days. It would take 12 days to reach the farthest point on the headwaters of the Mĕlawi where the river is still navigable, from there it is two days by overland route to reach the Djoentoep, a small right tributary of the Moehoed, which latter river can then be reached in another two days' journey. The Moehoed is a right branch of the Djoloi, an important affluent of the Batang Moeroeng or Barito river. From the mouth of the Moehoed to Moeara Teweh takes 7 days. No European had ever before accomplished this journey, nor until quite lately was it considered safe for anyone to travel in those parts, because a pretendent Sultan of Banjermassin, GOESTI MAT SĔMAN by name, had established himself on the Batang Moeroeng. When the Dutch Government refused to recognize his claims, he fled to these regions with a large body of followers, and organized a strong party hostile to the Dutch Government. RASSAH went on to relate that this Sultan being now dead [1]) the hostile disposition of the Malay populace was greatly abated, and he guaranteed to escort me safely to Moeara Teweh, as he himself had lately made the journey there and back, without meeting with any difficulties.

The geographical features of this part of Central Borneo, as

1) This report about the death of GOESTI MAT SĔMAN has not been confirmed later.

far as the river-basin of the Mélawi is concerned, had been
explored during the topographical survey, but of the river-basin
of the Batang Moeroeng above Moeara Teweh we know nothing
beyond the mouth of the Djoloi, the highest point reached by
Colonel HENRICI in 1833. Since that time no European has penetrated so far into the interior. We are altogether ignorant of the
character of the vast area which lies between the mouth of the
Djoloi and the water-parting separating the Mélawi from the
Batang Moeroeng; and we have nothing to depend on but the
vague scraps of information gathered from the natives. Therefore,
geologically speaking, all this region is a "terra incognita" and
I was very much in doubt, whether it would be better to keep
to my original plan and cross the country in a southerly direction
or whether I should undertake this expedition which RASSAH
proposed. But, as mentioned before, I had every reason to suppose that the journey east would lead over vast areas of the
same geological structure, and this opinion was strengthened by
the fact, that up to the point to which I had now advanced on
the Mélawi, nothing but clay and sandstone had appeared, either
as solid rock or as loose boulders. Moreover I was aware that
HENRICI[1]) only makes mention of sandstone as the rock occurring in the banks of the Batang Moeroeng, both at the mouth of
the Djoloi and further down-stream, so that I thought, and still
think, that the proposed line of route through the Mélawi valley
and as far as Moeara Teweh, would probably lead entirely
through a sandstone region. But by adhering to my first plan
and continuing my journey southward, I had every reason to
expect that I should pass through a much more varied territory,
and be enabled to contribute in a far greater measure to the scientific knowledge of the geological structure of Borneo. The journey
as proposed by RASSAH, first to Moeara Teweh and from there to
Samarinda, was very tempting, as it would afford me the gratifying
opportunity of becoming the first European who had actually
crossed Borneo from west to east. But in the end scientific con-

1) S. MÜLLER *42*, p. 234.

siderations prevailed, and I decided to carry out my original plan.

At 12.30 p. m. we commenced our voyage up the Lĕkawai. This river is on an average 30 metres wide, and the water being low we made fairly rapid progress. The solid rock was rarely visible but continued to be sandstone in horizontal or feebly inclined beds. At 3 p. m. we passed the Dyak house Nanga Riang. The space underneath the house was on all sides closed in by palisades. In the centre was the staircase, which can only be approached by a door in the palisades. I noticed this same type of building higher up on the Lĕkawai and amongst the

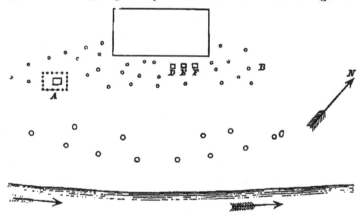

Fig. 61. PLAN OF THE DYAK HOUSE MÉRAKAU ON THE LEFT BANK OF THE LĔKAWAI RIVER.

A. Shed for skulls. Within the fence around this shed are placed several small flat wooden posts, on which human faces are carved.

B. Toras placed without order along the front of the house.

C. Tĕmadoks, hampatongs or ampatans erected in a double arch facing the river.

D. Sandong toelang.   *E*. Tĕmadok.   *F*. Singaran.

Ot-Danoms on the Samba river in South Borneo. We spent the night on a bank facing the mouth of the Pamai, about 10 kilometres beyond Nᵃ Lĕkawai, where I found indistinct shell-impressions in the sandstone on the opposite bank.

Sept. 27.   We left again next morning at 6 in glorious weather. The panorama remained the same as on the preceding day; forest on both sides, principally a secondary growth on the old ladangs, and low banks, sometimes interrupted by a sandstone cliff. The sandstone contains particles of coal in several places, and as far as the mouth of the Séboengoei the strata are horizontal or have a very gentle dip southward. A little way up-stream from this tributary of the Lĕkawai the dip changes gradually to about 5° to the north.

At 8.30 we arrived at Roemah Mérakau, an old Dyak house on the left shore of the Lĕkawai. Mĕrakau being such an excellent specimen of an Ot-Danom-house and its surroundings, a short description (compare Pls. XXXIX and XL) may not be out of place here. Fig. 61 shows its situation. The simple unpretentious looking house is surrounded by toras or pantars, wooden pillars from 4—10 metres high, tapering towards the top and crowned with a thick knob. They surround the house like a forest, they are sometimes crooked and put in the ground without any order. Whenever an inmate dies one of these toras is erected. At some little distance from these and in a semicircle facing away from the house is a row of tĕmadoks or hampatongs, i. e. poles put in memorial of sacrifices made at those spots; a second row nearer the house is not yet completed. These tĕmadoks are columns of very hard wood, upon which male or female figures have been carved, sometimes in quite an artistic style. They are carved with hatchets, the whole column including the figures being always made out of a single block of wood. The men are represented standing and in their most gorgeous apparel, with a parang at their side, sometimes wearing a Dutch cap, sometimes, though very seldom, carrying a musket over their shoulder. The women are in a sitting attitude their hands in their laps holding a sirih-box. On the pedestal of some of these pillars a face has been carved, with an enormously long tongue hanging out of its mouth, reminding one of the old sign-boards of Dutch apothecary shops, but in this case they are intended to ward off evil spirits. In the olden days

on solemn occasions men were slaughtered, nowadays a bull
is sacrificed at the offering-place and at the very spot a tĕmadok is
erected. Immediately in front of the house are three poles, represented separately on the first page of the atlas. The centre one
is a small but exceedingly well executed tĕmadok, erected after
the death of one of the kampong chiefs. To the right (F on
the plan) is a singaran about 10 metres high. To the left
(D on the plan) is a sandong, an exquisitely carved, solid wooden
column, on the top of which, in a little shrine with prettily deco-
rated roof, is placed a tĕmpajan[1]) containing the ashes of another
kampong chief. More in the foreground is a small separate building, 1.80 metre high, in which under an awning are ranged
on laths eight human skulls (the lower jaws are missing), trophies
of former head hunting raids. Round this building is a fence of
posts 1 metre high; the fence itself is 1.90 metre long and 1.20
metre wide. Inside this little enclosure stand several flat posts of
small size upon which, in most cases, human faces have been carved.

About 1½ kilometre beyond Mérakau the Toendoek or Tondok
flows into the Lĕkawai. This tributary is scarcely less important
than the principal river. Both arise on the boundary range between West and South Borneo, and in order to reach the pass
across Mt. Boenjau into the river-basin of the Samba, it is im-
material whether one follows the Tondok or the Lĕkawai. I was
told that the Tondok was the shorter route but offered more
difficulties to navigation. The choice, however, was soon made in my
case, for I had Mt. Raja on my programme, and in order to reach
that mountain I must needs follow the Lĕkawai. Above the mouth
of the Tondok, the stream became narrower, although the general
condition of the current continued unaltered, and there were no
obstacles of any importance. Near Batoe Njépit the bed is considerably obstructed by sandstone rocks, and one has to pole
with all one's might to keep control over the boat in the furious

1) Tĕmpajans (Dyak) = Vases or urns of pottery in which the ashes of the bodies of
dead chiefs are deposited, by certain tribes. Those of ancient make are considered of
great value.

current. Another day went by during which I saw only clay and sandstone with the usual horizontal or slightly inclined stratification. We spent the night on a boulder-bank, opposite the mouth of the Riang having accomplished that day 20 kilometres. Continuous fine weather prospered my journey, and the charm of the moonlit evenings and nights was not marred by mosquitoes. These, or at any rate the kinds which fly by night, are unknown in the mountainous districts of Borneo.

At the mouth of the Riang and for some distance beyond, the Sept. 28. Lékawai passes noiselessly and undisturbed through the most lovely scenery. To right and left are hills covered with ladangs, and on the shores one sees houses occasionally, amongst which the new, partially completed, house N* Térakoei with 13 lawangs[1]) commands attention. The banks are sandy sometimes alternating with low sandstone rocks. At 8 a. m. the scene suddenly changed; high, thickly wooded, hills come close to the water's edge, the bed of the river is narrowed by enormous masses of rock, the current becomes impetuous and riam Batoe Tossan lies before us[2]). The most difficult part of the riam is represented on plate XLI, a bar of rocks checks the progress of the stream and forms a slight fall. This dam consists of rocks of diabase-porphyrite and olivine-diabase. In the precipitous slopes on the left one can distinctly see that the undulating surface of igneous rock has been overlain by claystone and subsequently by sandstone. The clay contains indistinct shell-impressions. Above riam Batoe Tossan the current subsides again, sandstone with a feeble northerly incline prevails, and sandy clay-stone with shells becomes exposed in the river-bed. Half a kilometre further up-stream the channel is again narrowed considerably, its width ranging from $2^1/_2$ to 7 metres, and for a distance of $^3/_4$ kilometre the deep waters

1) Lawang = pintoe, family residence.
2) The difference between a goeroeng and a riam which is strictly observed on the Upper Kapoewas and its tributaries, is not regarded here, they only use the word riam, sometimes kiham, which has the same meaning in the dialect of the Ot-Danoms. In the river-basin of the Mahakkam and in most places, it seems, in East and South Borneo the term Kiham is used.

rush with tremendous velocity between the bare, high cliffs. These
are composed of dolerite or olivine-diabase, in which funnel-shaped
holes, sometimes 4 metres deep and wide enough for three persons to stand in, have been worn. Owing to the water being
extraordinarily low these holes were visible and I was thus
able to pass this riam in a comparatively tranquil manner.
Nevertheless we had our difficulties, one of which might have
been serious. The current was very strong, it was impossible to
scale the rocks and we had no means of unloading the boats;
it was therefore by no means an easy task to steer them clear of
the rocks in the narrow channel. There I had a narrow escape.
When nearly at the top of the riam the two rattan ropes by
which my boat was being hoisted up, suddenly broke, the un-
governed boat was seized by the furious stream, but at the very

Fig. 62. Riam Pandjang in the Lekawai river.

moment when I expected it to be dashed to pieces against the
protruding rocks, it suddenly swayed in another direction. Some
of my people following close behind me and realizing the danger

Pl XLI.

THE BATOZ TOSSAN-RAPIDS.

of my position, had jumped into the stream and succeeded in getting hold of the broken ropes. With a violent effort they turned the boat into an open space between two rocks, out of the reach of the current. Above the riam Pandjang the banks became lower, and I managed to land and to take a photograph, which is reproduced in fig. 62.

This riam is the only really serious obstacle in the Lĕkawai. At high water it is simply impassable, and during the months of November, December, and January, it often happens that boats have to wait two or three weeks before they can pass this point. Going down-stream the riam is dangerous at all times, because as the current extends over a considerable distance, the boats acquire tremendous speed, and in the middle of the riam the stream suddenly turns at a right angle to the north. The place is therefore rightly called Tandjoeng Kiham [1]).

At noon we had got clear of the riam and were once more in quiet waters. Diabase rocks stand out here and there above the surface, but after a while this rock gives way to sandstone, with horizontal or feebly inclined strata dipping in different directions. We bivouacked on a boulder-island half a kilometre beyond the house Mĕnjahai. The boulder-banks of the Lĕkawai, show a great variety of rocks. Sandstone predominates, but besides I noticed granite, quartz-diorite, amphibole-porphyrite, amygdaloid diabase, diabase-porphyrite, amphibole-andesite, pitchstone, quartzite, banded chert, and sandy clay-stone with white shells.

Although the low level of the water had stood us in good stead on the previous day, we now began to long for rain, for the river was so shallow, that the boats had constantly to be dragged over the pebbles at the bottom, a very fatiguing business and most damaging to the boats. Near the Dyak house N⁰ Bawai I met some Dyaks from the Samba river in South Borneo, and I tried hard to persuade them to enter my service, but they could not, as they were bound to assist in the making of the ladangs I think they must have been hostages. The house

1) Tandjoeng Kiham (Dyak) = Bend in the cataract.

N^a Bawai was not only enclosed on all sides by palisades but these were moreover furnished with two rows of sharp spikes

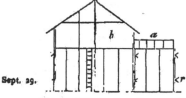

made of split rattan, the sharp points turned outside. This obviously was to prevent any ill intentioned individual from gaining access to the house (see fig. 63).

Sept. 29.

Fig. 63. SECTION ACROSS THE HOUSE N^a BAWAI ON THE LĔKAWAI RIVER.

r. Spikes of split rattan.

A little rain fell during the night and the water rose slightly, but the sampans had still to be towed in most places, and our progress was very slow. Near the mouth of the Boeran rivulet on the right shore clay-stone with shells of the same kind as those of which I had found loose pieces in the boulder-banks lower down-stream, crops out at several places. Section G. on map IX^B explains the position of the shell-beds. They dip slightly N.N.W. In the sandstone resting on the clay is a bed of brown coal 0.38 centimetre thick. This same formation of clay with shells is much better exposed higher up-stream in the Liang Boehies, a steep cliff about eleven metres high somewhat undermined by erosion. Strata of dark-grey crumbling clay-stone (see section H) dip feebly north-north-west. In the clay-stone three distinct shell-beds may be seen in this section, each about 20 metres thick. They are composed almost entirely of large shells, closely packed together, the majority of which belong to one species, *Corbula dajacencis*, KRAUSE[1]). Dr. KRAUSE, who has examined these fossils,

---

1) KRAUSE at first considered these deposits, as he communicated to me in a letter, as late-Cretaceous or possibly old-Tertiary. According to this opinion I separately distinguished these deposits on the Mĕlawai and the Lĕkawai, on the maps and sections, from the Tertiary deposits elsewhere in West Borneo. Later on, when the maps and sections belonging to this book were already printed, he found out that the deposits on the Lĕkawai must be considered to be of Tertiary age. Later examinations by MARTIN[1]) have confirmed this opinion. Hence there is no necessity to separate the deposits on the Lĕkawai from the sandstones of the Madi plateau and the Schwaner mountains: and the separation indicated on the maps and sections ought to be looked upon as non-existing.

1) K. MARTIN *38*, p. 302.

thinks that they are a brackish-water deposit of Tertiary, probably Eogene age, formed most likely in an estuary or bay close to the mainland. One might conclude from the position of the strata as shown on map IX², that the shell-beds at Liang Boehies represent a lower horizon than those at N⁴ Boeran. This however may be only in appearance, for it is quite possible that dislocations of the same order as those which may be seen in the section at Liang Boehies itself, have brought the shell-beds of N⁴ Boeran into a relatively, higher level.

We remained at Liang Boehies from 12 till 2 p. m., and about 3.30 we disembarked at the pangkalan Měroeboei from whence I intended to make several trips. This house is situated on a strip of land between the Lěkawai and the Měroeboei, which unite just below. In front of the old and rather dirty house stood a few old těmadoks. The interior was slightly different to the ordinary type of house. It is illustrated in fig. 64. My first care was to interview the kampong chief, Raden MATJAN SANGIN, and to ascertain something about Mt. Raja, which rose up in front of us shrouded in clouds. We did not however, get on very well, and when I men-

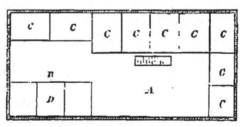

Fig. 64. HOUSE MĚROEBOEI OF OT-DANOM-DYAKS ON THE LĚKAWAI.

A. Large common room.
B. Small   „   „
C. Lawangs, private family rooms.
D. Room of Raden MATJAN SANGIN.

tioned my great desire to visit Mt. Raja, Raden MATJAN SANGIN became positively uneasy, and emphatically declared that none of his people knew the way, that nobody had ever been there, and, in fact, that it could not be done. Now I knew that the Ot-Danoms, both on the Mělawi and in South Borneo[1]), hold that Mt. Raja

1) The Ot-Danoms (Oeloe-Ajers) who have removed from the Mělawi valley and from

is the abode of departed souls [1] so I was prepared to meet with difficulties, and as I found that these people were in fear lest some great evil should overtake them if I persisted in the attempt, I allowed the matter to rest for the present. But in spite of this courtesy on my part the chief continued very disobliging, and when I asked him for guides and carriers across the water-parting and on to the river-basin of the Samba, he made all sorts of excuses. Then I made him clear that my work was considered urgent by the Government and advised him to

think it over, and in the mean time I ordered my men to put up two strong pondoks for me and for my coolies on the space before the house, where I also noticed some very pretty rice-granaries.

The night was clear and cool with a minimum temperature, at 5 a. m., of 18.5° Celsius, but the sun soon made its power felt, and as the sky was perfectly cloud-less all day, it was what one might call really hot, and I had the benefit of it on my journey to Mt. Sédaroeng. I had left at 7 a. m. rowing down

Fig. 65. RICE-FIELDS NEAR MT. SÉDAROENG.

the Madi mountains to the Kapoewas plain (Mandai, Soeroek, Mentébah, Tĕbaoeng, etc.), hold Mt. Tiloeng on the Mandai river to be the home of the souls of their departed.

[1] SCHWANER 55, p. 151 tells us concerning this: the abode of the departed is, according to the Ot-Danoms, on Mt. Raja, the highest mountain of the region round about, also on Kait Boeran and Mt. Njait, two mountains connected with the former. Mt. Raja was much higher once, it reached up to heaven, the seat of good and evil spirits. Botjong the spirit of evil descended by it to the earth, to destroy the human race, but Boeroeng Madeira touched it with his wings, and the mighty mountain crumbled down and part of it fell in the abyss below, whence it arose as Mt. Njait. Mt. Raja was left standing in its present shape. In a golden barge, Sansang Tandoko conveys the souls of the departed to Mt. Raja, followed by the prayers and exorcisms of the Bilians On their journey to heaven, which rests on Mt. Loemboet, they, like the souls of the Njadjoes, have many dangers and difficulties to encounter.

the Lĕkawai as far as N⁰ Mĕnsioeng, where I went ashore. A path led in a westerly direction, over an undulating sandstone plain, covered with rice-fields, as far as Mt. Sĕdaroeng. At about 8.30 I stood on the top of a hillock about 200 metres high and had a extensive view over the ladangs dotted about with graceful little ricebarns and over Mt. Sĕdaroeng, (Plate XLII and fig. 65), behind which to the south arise the high mountains of the water-parting Mt. Kĕpĕnjahoe, Mt. Kait-Boeran and Mt. Raja, the latter almost entirely enveloped in clouds. At 10.30 we came to the foot of Mt. Sĕdaroeng picking our way between large blocks of porphyrite. In the forest on the slopes of the Sĕdaroeng I saw several game-traps. Right across the valleys through which the game has to pass are placed long fences with occasional small openings, above each of which a heavy tree-trunk is suspended, one side resting on the ground, and the other slightly supported by a piece of wood. This piece of wood is placed in such a manner that any game passing through the opening must of necessity touch it, whereupon the tree immediately falls, and crushes the animal to death. The Sĕdaroeng is entirely composed of porphyrite, sometimes amphibole-porphyrite, sometimes plagioclase-porphyrite, with occasional phenocrysts of amphibole, whilst quartz, biotite and augite are found in the ground-mass. This porphyrite has an andesitic character, particularly conspicuous under the microscope. At the top (930 metres high), a clearing had been made on the south side on the occasion of the topographical survey. I had therefore a beautifully free and open view over the high mountain plateau which here forms the water-parting, between the Mĕlawi and the Katingan. The highest peaks however, Mt. Kait Boeran and Mt. Raja, continued for the greater part wrapped in clouds, and I had to wait several hours before I managed to get a fairly complete sketch in outline of the mountains as given on plate XIV of the atlas. Looking at it from that point of view, it is evident that the divide is formed by mountains having a one-sided structure, sloping from the highest ridge towards the north (on the sketch

towards the beholder), but on the other side passing with a very
steep incline in bold steps into the hilly district of South Borneo.

Several affluents of the Mélawi, flowing northward, have divided
this feebly sloping plateau into several strips, each in itself forming
a range of hills, the highest crests of which must be looked for
towards the south. Thus looking from east to west, we recognise
the Saroen range with Mt. Soeha (1118 metres), the Képénjahoe
range with Mt. Pintoe Bénoewang (1792 metres), the Kait-Boran
range with Mt. Raja (2278 metres), and the Balai Kéméloe
range with Mt. Pandjang (1613 metres). The highest peaks of
Mt. Raja are somewhat dome-shaped, and have not the same
clearly defined monoclinal structure which strikes one even at this
distance in the secondary peaks of the Raja range and the
Kait-Boran.

That this high mountain district, to which collectively I will
give the name of the Raja mountains, produces a really overwhelming impression can be inferred of from the fact that the
actual distance between the top of Mt. Sedaroeng (930 metres)
and Mt. Raja (2278 metres) is only 14 kilometres. Looking
W.N.W. from the Sédaroeng, one gets a splendid view of Mt.
Moeloe, an isolated mountain with almost bare vertical, rock-
faces to south-west and west. I supposed this mountain, like
the Sédaroeng, to be of an igneous nature, and I was confirmed
in this view next day, when the Dyaks whom I had sent to
break some pieces from it, returned with several specimens, every
one of which was of amphibole-porphyrite with andesitic character,
containing a great deal of quartz in the ground-mass. It took
only an hour to descend Mt. Sédaroeng and at 3.30 I was
plodding over the treeless ladangs in a temperature of 33° Celsius.
I was not returning to Ménsioeng but to Ménari, a small Dyak
house where one of my sampans was to wait for me.

We arrived at Ménari at 5 p.m. and I fortunately succeeded
in persuading the kampong chief Njaroh to promise to show me
the way up Mt. Raja as far as he knew it himself, and to
accompany me there with nine of his men. Our sampan was

there, waiting for us, but unfortunately so battered by the continuous scraping over the rocks, that the bottom was split right across, the gap became wider and wider, and soon the boat sank before our eyes. I was now obliged to get through the river on foot as best I could, and in the dark, it is no wonder that I was tired out when we reached Mëroeboei at 8.30.

The following day I rested, wrote letters and made my pre- October 1.
parations for the next excursion on the programme.

On the 2nd of Oct., Raden MATJAN SANGIN, who in the mean- October 2.
time had made up his mind that his best policy was to keep on good terms with me, five of his men and six of my coolies were sent overland to the pangkalan Oeloe Tëmangooi [1]) in the river-basin of the Samba in South-Borneo. Four of them were to go on to the house Toembang Habangooi on the Samba, buy two sampans, and return with them immediately to the pangkalan. The remainder of the party, waiting at the pangkalan, were instructed to cut down a suitable tree and make a third sampan, for which purpose I lent them the hatchets bought at Lëkawai. On the same morning eight of my regular coolies left in the opposite direction for Sintang. They had charge of one of the boats containing part of my collection and some letters.

I myself left for Mt. Raja at 11 a. m. with the guide NJAKOII, nine Dyaks from Mënari, ten coolies from Sintang, RASSAH and my two servants. Just before starting we had a little reverse, which annoyed me very much. One of my men stupidly let my last roll of tobacco fall into the water, and it was utterly spoiled. No one, except those who have travelled in Borneo, can conceive the seriousness of this loss. Tobacco is the magic power which enables the Dyak cheerfully to bear any amount of fatigue. Give a Dyak the choice between lack of rice and lack of tobacco, and without any hesitation he will choose the former. With a quid of tobacco in his mouth he will face almost

---

1) The natives sometimes call this river Kowin, the name of a Dyak kampong chief of the Mëlawi district, who about ten years ago had been murdered there..

any hardship, even if food be scarce, but deprive him of his tobacco, and there is no power of endurance left in him. I therefore was much troubled at the loss of this precious tobacco, but fortunately I had a little box in reserve, and by very careful management, I contrived to keep my Dyak carriers fairly well supplied.

A sampan took me just beyond Ménari. The rocks along the banks are clay- and sandstone dipping 20° to 45° N. or N.N.W. At Ménari a thin layer of brown-coal appears in the sandstone. A little past that point, some large blocks of porphyrite, in every way resembling the porphyrite of Mt. Sĕdaroeng, lie scattered in the stream, the rock *in situ* being dark grey clay-slate. There the Lĕkawai ceases to be navigable, and we proceeded by land, in a southern direction, through old ladangs and marshy jungle, between two ranges of hills, until we had to ford the river again at a point where coarse sandstone resting on coarse conglomerate crops out. The pebbles in this conglomerate consist of quartzite and silicified clay-stone. We followed the path on the left bank for some little distance; it led up the steep slope of a hill of quartz-amphibole-porphyrite and back again to the river. The porphyrite of this hill is in all probability akin to that of Mt. Sĕdaroeng. Along the river after passing the porphyritic hill, the rock *in situ* is at first coarse conglomerate, and afterwards thinly stratified, dark grey, clay-slate. Two rather important tributaries here join the Lĕkawai river, first the Sĕpan, rising on Mt. Kĕpĕnjahoe, and then the Bassi, rising between Mt. Kait-Boran and Mt. Pintas Bĕnoewang. The Bassi is almost as rapid a stream as the Lĕkawai itself. At the confluence lies a large boulder-bank, affording a rich assortment of samples of beautiful crystalline rocks. I noticed biotite-granite, sometimes gneissic, amphibole-biotite-granite, quartz-diorite, plagioclase-quartz-rock with titanite, presumably fragments of a quartz-diorite being very poor in dark constituents, tonalite, quartz-porphyry, quartz-porphyrite, amphibole-porphyrite, often containing a fair amount of quartz, pitchstone, amphibolite, chert, hornfels, arkose and conglomerate.

Pl. XLII.

LADANG NEAR BT. SEGAROENG.

We bivouacked at Nª Bassi. A heavy thunder-storm raged October 3. during the night, and the violent rains greatly swelled the stream. We followed the river for some little distance further west and forded it several times, no easy matter in its swollen condition. At 8.40, leaving the Lékawai to our left, we began the ascent of the Matan Poroet, a narrow ridge which separates the head-waters of the Lékawai from those of the Sérawai. A gradual ascent in a southerly direction brought us at 9.30 to the crest of this ridge, which we continued to follow southward and upward. At noon we had reached a height of 728 metres, to the south of us was a precipice affording a fairly good view over the valley of the Lékawai river and Mt. Pintoe Bénoewang. The summit upon which we stood is called Goedjang Piloeng, and it is composed of coarse sandstone allied to greywacke because of the many particles of clay-stone which it contains. The beds of sandstone dip feebly N.N.W., and overlie granite which crops out about 25 metres lower down. From the point where I left the Lékawai up to here, the solid rock is generally sand-stone, and only occasionally clay-stone. The precipitous southern slope of Goedjang Piloeng is about 30 metres high and it is joined at its base to a flat narrow ridge, not more than 2 metres across, with deep valleys on both sides, the water-parting between the Lékawai and the Sérawai. This ridge soon gets somewhat wider and steadily rises towards the south, the rock *in situ* being biotite-granite, succeeded higher up by sandstone. Ascending continually we came to a swamp where the forest was less dense but full of rattan. The ground was covered with a great variety of flowers and plants with beautifully tinted leaves. Little brooklets intersect this region, all flowing west towards the Sérawai. I bivouacked by the side of one of these, at an altitude of 851 metres. The rock, where exposed, is friable white sandstone, in almost horizontal strata. The night was cool and damp, the minimum thermometer registering 17.9° Celsius.

Next day there was no change in the scenery, and we con- October 4. tinued our ascent in a south or south-eastern direction. The rock *in situ* was at first the same coarse white sandstone, but after a while I noticed some loose blocks of porphyrite, and when I had reached an altitude of 1392 metres the country rock had changed to quartz-porphyrite. Ever steeper became the ascent, the ground was strewn with masses of rock, the forest was less marshy and conifers and tree-ferns became its most conspicuous features. Taken as a whole, the vegetation here is richer and more varied than in any other part of Borneo. At 12 o'clock I called a halt close to a wild mountain-stream, one of the feeders of the Sérawai; the ground was there covered with the most exquisite flowering Aroidea. Shortly after noon it suddenly became pitch-dark, and the cicadas began to shriek as they are wont to do at nightfall. Thick heavy clouds gathered over our heads, and then came a tremendous downpour which lasted for several hours.

I decided not to go any further that day, particularly as the place was very well adapted for bivouacking, and our next venture would be the scaling of an exceedingly steep height. My guide NJAROII declared that there was no other way of access to the top, but he also candidly confessed that neither he nor his men had ever ventured beyond the spot where we now were, this being considered the beginning of Mt. Raja proper. To penetrate any further would disturb the spirits of the departed which dwell there, and bring misfortune upon themselves and their whole tribe. 'Nevertheless', he said, 'we know that no spirit can compete in power with the white man'. In my company therefore he should feel safe and was willing to proceed. He had taken the precaution, however, of bringing a couple of fowls and a great many trinkets which were to serve as peace-offerings to the spirits. The fowls he considered most essential for counteracting any evil effects of this undertaking, and RASSAH told me that the Dyaks would assuredly turn back if they were to die before we reached the top. It was a precarious business, as the chances of life for fowls tied on the top of a basket full of rice or on some other package, and carried in this uncomfortable position over immense tracts of difficult and thickly

wooded ground were not of the best. Of course I did my share
in looking after them, and took a great interest in their well-being,
and further saw that at every halt they had plenty to eat.

The enormous rocks in the stream near our bivouac consisted
of quartzose diabase-porphyrite, similar to the porphyrite of which
the highest tops of Mt. Raja are composed, so far as I scaled
them, and which rock I have therefore provisionally called Raja
porphyrite; sometimes it is like diabase-porphyrite, sometimes
like andesite, and many varieties occur, one of which contains
so much quartz that it ought to be called quartz-porphyrite. The
temperature of the water was 15°.5 Celsius. In the course of
the afternoon the rain abated, and accompanied by NJAROH and
my Chinaman, I set out to explore the slope of the mountain
in order to find the best place of ascent. Not far from our
bivouac we came upon fresh spoor of the rhinoceros (badak)
which frequents these mountains. The minimum temperature in
the night was 15°.1 Celsius.

At 7 a. m. next morning, accompanied only by my Chinaman October 5.
and one coolie, I started on my expedition, the other men, who
were somewhat unfit for their work through the cold, were to follow the track which we should cut through the jungle. It was a
very steep climb, and at 7.30 we found ourselves in front of an
almost perpendicular wall of diabase-porphyrite, totally devoid of
vegetation and unscalable, so we had to turn off in a south-south-
western direction over very rugged ground. We had to cross
several narrow valleys and gorges, all trending west or north-
west towards the valley of the Sérawai. Those deep, narrow chasms
were a great nuisance, but at last, keeping on now in a south-
easterly direction and steadily mounting we came to the brow
of the ridge which joins Mt. Kait Boran and Mt. Raja, and forms
part of the divide separating the basin of the Mĕlawi from that
of the Katingan. This water-parting also forms the boundary line
between West and South Borneo. It was 9 o'clock when we
reached the brow, a little South of Mt. Kait Boran, and ± 1900
metres high. The rock *in situ* is decomposed Raja porphyrite. We

continued southward over very uneven ground, the deep clefts
being partially filled up with a tangled mass of trees and shrubs
so thick and solid that often we never touched the ground for a
good distance.

We climbed several peaks, but always to find that there were
others still higher just behind. The natives may well call Mt.
Raja the "many-peaked" mountain. At noon the wind rose, and
rain began to fall in torrents. The cold was so intense that I was
obliged to send my two companions back to hurry forward the
coolies with my things. Meanwhile I found a look-out in a tree
on the nearest summit; the clouds however allowed me but a
very limited view of the country round. As far as I could see,
all the other peaks were below me, and I congratulated myself
that the highest had been reached. Coming down from my
vantage ground, I saw no sign of my men, and went myself a
little way back to meet them, when to my surprise, I came upon,
my two faithful satellites numbed with cold and cowering over
a poor little fire which they had somehow contrived to light. I
encouraged them as best I could and they went on. At 4 p. m.
the coolies arrived, and I was then able to protect myself somewhat against the extreme cold and the violence of the storm.
The rain continued all day, and the mountain never shook off
its cloud-wrap, so that there was no chance of a view. The peaks
of Mt. Raja, in the neighbourhood of our bivouac, were all covered
with a low vegetation of brushwood and dwarfish trees, consisting
for the greater part of conifers and tree-ferns. The general
aspect is very curious. All the trunks and branches are covered
with tufts of moss, often as much as two feet thick. Even the
leaves are covered with many varieties of mosses, and long
bunches of lichens with their manifold ramifications hang down
from all the cross-branches. Near the base of the stem the mossy
covering assumes the shape of an obtuse cone, upon which the
tree appears to stand as on a pedestal. Amongst the conifers,
one variety closely related to the well-known *Gingko*, attracts
special attention. Many of the trees are overgrown with orchids,

of which there is here an enormous variety, greater than anywhere else in Borneo. Some of them were in full blossom and
their fragrance is beyond compare. Here again I found proof of
the fact, already noticed elsewhere, that not in hot low-lying
regions, but in a cool, damp, tropical, mountain-climate, orchids
attain their most luxurious development. Nepenthes' also abound,
amongst them are some specimens with truly gigantic cups. I
made a large collection of plants for the botanical gardens at
Buitenzorg, but unfortunately only a small portion reached their
destination in anything like good condition.

At 6 a. m. I commenced a series of barometrical observations
at hourly intervals, in order to ascertain the daily movement
of the barometer at this height, and I continued my observations regularly till 8 a. m. next day. The maximum temperature
in the daytime was 15°.2 Celsius, the minimum at night 11°.1
Celsius.

My coolies were quite demoralized with the cold, and could
hardly do any work. Some of the Dyaks from Mĕnari and my
Chinaman kept up better than any of them.

A little before sunrise I ascended the peak nearest to our October 7.
bivouac which was 2120 metres high, and enjoyed from there
the indescribably beautiful and imposing panorama which more
than made up for all the discomforts and privations of this excursion. The greater part of Dutch Borneo lay at my feet. Far
below me the clouds filled all the valleys as with a sea of foam,
out of which all the mountains more than 500 metres in height
rose clearly and sharply defined against the horizon. To give some
idea of the extent of the area included in this panorama, where
no high mountains intercept the view, I will just mention that
I could distinctly see the following: Mt. Saran in the Oeloe
Tĕmpoenak (1758), Mt. Koedjau (1322) and Mt. Kélam (936)
near Sintang, Mt. Kénĕpai (1156) in the Lake district, the Madi
mountains (1200), the mountains of the Oeloe Mandai, and those
of the Oeloe Melawi and Siang-Moeroeng territory. To the south-
east and south, my vision roamed unobstructed over the endless

hills and plains of South Borneo, reaching, I could almost believe, as far as the Java Sea. I was disappointed in one thing however, I found that I was not standing on the very highest peak of Mt. Raja. Separated by a valley from the place where I stood, to the west-south-west, there rose another summit about 100 metres higher than mine. This must have been the peak which, by the topographical survey had been estimated at 2278 metres. In vain the Dyaks of Mènari insisted that the top upon which we stood was really the highest point of Mt. Raja, and that the other one belonged to another mountain, Mt. Mèlaban Boeli, my own observations convinced me that the position I occupied was a little to the east of that marked on the topographical map of Mt. Raja as the highest. Scarcity of rice prevented me from adding a few more days [1]) to the trip, and so this highest peak remained unexplored, and I had to comfort myself with the fact of having been the first to set foot on one of the two highest peaks of this mountain [2]).

The highest peak of Mt. Raja, as I saw it, is somewhat dome-shaped and slightly arched, which in all probability indicates in a plain of some size at the top. With my fieldglass I could see quite distinctly that it is covered with low brushwood, whereas the sides are, as a rule, perfectly bare.

SCHWANER [3]) tells us, on native authority: «The top of Mt.

---

1) It must be borne in mind that the difficulties of the ground are such that one cannot reasonably expect to accomplish more than 3 or 4 kilometres per day, so that I doubt whether the comparatively short distance which separates the two highest peaks of Mt. Raja could have been got over in one day, as not unfrequently long detours have to be made to avoid the many deep chasms which intersect the brow of the connecting ridge.

2) In 1883, TEUSCHER made a botanical excursion on the Mèlawi and Sèrawai, and mentions a partial ascent of the Raja, starting from Nanga Sèpan on the Tjèloendoeng, a tributary of the Sèrawai. From there he followed for one day a path leading upward, and after having passed the night in the forest, he returned the next day to Nanga Sèpan. Now as the distance from Nanga Sèpan to Mt. Raja is 22 kilometres, not counting the ascent, and as TEUSCHER, according to his own account, was not at all a good pedestrian, he could not possibly have reached as far as the foot of the mountain in that time. From personal experience I have come to the conclusion, that any one in perfect physical health and in every way fitted for such an undertaking, could not possibly travel from Nanga Sèpan to the top of Mt. Raja in less than 4 days. See 56 p. 146 et seq.

3) C. A. L. M. SCHWANER 55, II, p. 154. Amsterdam 1854.

Raja is a plateau of ± 200 square feet, covered with low brush-wood. The trees are slender and thickly clothed with moss, while the roots are disproportionately thick and meander over the stony ground like so many serpents. The atmosphere is unpleasantly cold and there are very few animals. On one of the slopes is a small lake, only a few fathoms across, but extremely deep, with precipices all round it." This description must be fairly correct and applies equally well to the peak upon which I stood. I could not learn any thing very decided about the lake which SCHWANER mentions, but I was told that there is a hot spring somewhere on the south side. SCHWANER estimates the height of Mt. Raja at 8500 feet which cannot be far off the mark. He also tells us that the rock of which Mt. Raja is composed, is light coloured. I should not be surprised if the highest summit should ultimately be found to consist of porphyrite or considering its dome-like shape, it might possibly be granite, perhaps the same amphibole-biotite-granite, which is such a conspicuous feature amongst the boulders brought down by the Karang, a tributary of the Samba which rises on Mt. Raja. The peak which I ascended consists of decomposed Raja porphyrite.

From a geological point of view the panorama seen from Mt. Raja is also most instructive. The boundary mountains be-tween West and South Borneo evidently form a connected high, narrow, belt, which through erosion is now divided up into many pieces. The trend of the hills, at least over a considerable distance, is W. 7° S.—E. 7° N. The boundary mountains have a decidedly monoclinal structure, a steep sheer incline to the south and a generally feeble incline to the north. The geological formation explains this phenomenon. Granite forms the basis [1]) of all the mountains within the Raja district. Porphyrite has forced its way through the granite, and is the component rock of the higher summit such as Mt. Kait Boran and Mt. Raja. Resting upon the

---

1) The fact that pebbles of the 'hornfels' group occur amongst the boulders of the Bass and the Karang leads me to suppose that this contact-rock must also play some part in the granite area of the Raja mountains.

granite basis and with a feeble northern incline, there are beds of sandstone alternating with clay-stone. They are sharply broken off to the south, probably faulted down. This cliff of displacement explains the precipitous slope to the south, whereas the gentle northern incline is nothing but the natural position of the sand-stone itself. Further east, near Mt. Boenjau, where I eventually crossed the water-parting, the boundary mountains may be seen in their simplest form, showing the pure one-sided type. The granite basis if indeed it exists here, never comes to the surface, nor is there any porphyrite to be seen, so the mountains separating the head-waters of the Mĕlawi from those of the Katingan present simply a complex of beds of sandstone and clay-stone with a gentle northern incline, and broken off abruptly towards the south. Mt. Raja shows a rather more complicated development, because there the boss of porphyrite far surpasses the sandstone-plateau in height. Further westward again, and on the other side of Mt. Damar, in the river-basins of the Mĕntatai and the Ella, more complications arise. There sandstone seems to occupy but a very limited place, leaving full play to the granite and different contact rocks of the hornfels group.

The view which has now been obtained of the Raja mountain land leads me to believe that if there be still unexplored or hidden treasures of zoological or botanical interest in Dutch Borneo they certainly must be found here. The explorations of BÜTTIKOFER in Dutch Borneo, and of HOSE [1]) in Sarawak, have proved decisively that the low country and the hilly districts possess comparatively few unknown specimens, but that the highlands of Borneo, judging by the few places which have already been explored, form a most promising hunting ground for the zoologist. To ensure success, one should begin operations at an altitude of at least 1500 metres, and the area to be explored at this height should not be of too limited an extent. The Raja territory is the only one which answers all these demands, at any rate as far as West

---

1) CH. HOSE 24, p. 193.

Pl. XLIII.

Sⁿ Rassahooi near Toembang Karang.

Borneo is concerned. As already observed, the Raja and the neigh-bouring summits have all a steep incline southward, from Mt. Raja therefore one looks upon the almost perpendicular bastions which form the southern extremity of Mt. Kait Boran and Mt. Pintoe Bĕnoewang. To the north, however, all these mountains slope very regularly, whilst their summits are connected by broad shallow valleys, forming vast tracts of country partly of a slightly sloping, and partly of a very rugged character, situated at altitudes between 1400 and 2000 Metres above sea-level. Moreover there is sufficient variation in the composition of the soil, from the oc-currence of various rocks such as sandstone, granite and porphyrite, therefore a rich and extensive flora may reasonably be expected. Finally Mt. Raja can easily be reached by a well equipped expe-dition, so that in fact every thing tends to make this region a most promising centre for zoological and botanical research in Central Borneo.

Amongst the numerous characteristic mountain-peaks visible in the vast panorama from the summit of Mt. Raja, Mt. Pĕn-joekoei, although situated at a great distance, especially invites attention by its very curious, almost needle-shaped, outline. This mountain probably lies in the head-waters of the Mĕngiri a left branch of the Kahajan, and must be, as I was told, comparatively easy of access from the Upper Embalau river.

The Dyaks from Mĕnari, who accompanied me, were much excited in spite of the intense cold. They had managed to keep alive the fowls, which were now sacrificed to the hantoes, and many personal adornments especially bracelets, also prepared quids of sirih, and the wherewithal to make more, were left behind on the top as presents and peace-offerings to the hantoes. The Dyaks adorned themselves with bunches of a kind of lichen which here grows on the thin stems of the trees, in tufts about 0.60 centimetres thick. They also took with them large quantities of the exquisitely bronze-coloured scaly leaves of the daun-tali (a species of the genus *Elaeagnus*), the possession of which in their idea should ensure a good harvest.

SCHWANER must have known of this usage, for he says[1]): „The natives believe that Mt. Raja is the abode of good and evil spirits, and its vegetation is supposed to possess great virtue. No one ever goes up this mountain except for the purpose of fetching from there some special herb or root, which they carry about them as an amulet or talisman."

At 8 o'clock we were completely enveloped in the vapours which we had watched slowly rising from the valleys, and the panorama was of course entirely hidden from our view. The only time to insure a prospect from Mt. Raja is directly after sunrise; the deep valleys of course are then invisible by reason of the vapours below, but on fine days the tops stand out clear and entirely free from clouds at that early hour. In a comparatively short time the vapours rise from the valleys and surround the mountain-tops, which a few hours after sunrise are completely enshrouded. Still higher rise the clouds until sometimes on very clear days, Mt. Raja once more late in the afternoon, makes its appearance from under their canopy. Then only is the entire landscape visible from the top, the mist below having been dispersed by evaporation, and moreover, this is the only time that the people down in the valley can get a glimpse of the mighty mountain. But it is a rare occurrence, which hardly takes place more than three or four times in the year. I never saw Mt. Raja and Mt. Kait-Boeran thus entirely free from clouds, either when travelling in the lowlands of West Borneo or afterwards in the southern division. Under unfavorable circumstances the top remains enveloped in clouds even at sunrise, as was the case on the first day that I was there.

At 9 a. m., we left the mountain, taking the same path by which we had ascended. About 1.40 we took a short rest at our bivouac of the 4th, and an hour before sunset we were back again at the place where we had passed the night of the 3rd to 4th October. The higher temperature (19°.2 Celsius) seemed to in-

---

1) C. A. L. M. SCHWANER 55, II. page 123.

vigorate my men, who spent the greater part of the night chatting and laughing over their camp fires.

At 7 a. m. we resumed our journey and at 10.50 passed under October 8. the sheer precipices of Goedjang Piloeng. We walked on all day and spent the night at Kwala Bassi.

We struck camp early in the morning and at noon reached October 9. my headquarters at Roemah Měroeboei. Here, as I reviewed the situation, it became more and more doubtful whether I should be able to carry out my programme. My undertaking seemed doomed to failure. In the first place provisions began to run short, and I knew the utter impossibility of procuring rice in anything like sufficient quantity, even if I were prepared to pay a high price. There was certainly no time for delay, and if I meant to go on at all, I must do so at once. But then there was the second difficulty, I had not enough carriers to take my luggage across the water-parting. My own 10 men were not nearly sufficient, and Njaroh and his party maintained that they must return at once to Měnari to look after the ladangs. They had told me so from the first, so I could not blame them. As I sat cogitating over these matters a happy chance helped me out of the difficulty. At some little distance I happened to see a party of Dyaks, wending their way from Měroeboei to the mountains. I accosted them to ask the object of their journey. They were Dyaks from Boentoet Riam on the Tondok river, en route for the forests at the head-waters of the Kahajan to fetch rattan. I tried to persuade them to go with me first to the pangkalan on the other side of the water-parting, and I succeeded so far, that they consented to make a halt to think over my proposal. I now contrived that they should witness the payment of Njaroh and his men. After paying their wages, I added a little present in money for each man, and the marked gleam of satisfaction on Njaroh's face went a long way and had a most benificial influence upon the subsequent consultations of the Dyaks from Boentoet Riam. Within a very short time the eldest of them came to me and said that he and his companions would go

with me as far as the pangkalan Kowin. I fixed our departure
for the next morning, and spent the greater part of the night
in making my preparations.

C. ACROSS THE SCHWANER MOUNTAINS TO SOUTH BORNEO AND ALONG THE SAMBA AND THE KATINGAN TO THE JAVA SEA [1]).

October 10. I left one of my men in charge of the specimens of rocks,
dried plants, and other luggage which it was unnecessary to take
with us, and started at 9 a. m. for the boundary mountains, accompanied by my nine permanent coolies from Sintang, nine
Dyaks from Boentoet Riam, and my two servants, 20 men in all.

For about 1½ kilometres we walked through old ladangs in a
S.E. direction, then verged a little more to the east across the
steep northern slope of a sandstone hill and straight down
into the valley of the Sebangkang, a small tributary of the
Měroeboei, where I rested in a deserted house, waiting for
the carriers. The rock in this stream is grey sandy clay-stone.
The boulders were composed of sandstone and a small proportion of dark chert. At 2 p. m. we left again, travelling east,
over very awkward ground, old ladangs covered with a low
thick growth of brushwood. The prevailing rock was coarse white
sandstone. After a while the path began to rise again, winding
round a steep hill, first in a north-eastern then in an eastern, and
finally in a south-eastern direction. We left the highest summit
of this hill to the right, i. e. to the south of us; it is entirely
composed of sandstone and conglomerate in beds, dipping from
15° to 20° north and north-east. The ground is covered with
pure white or semi-transparent quartz-pebbles. At 3.15 we
forded a little affluent of the Měroeboei, and in another quarter
of an hour reached the river itself. Its bed is composed of
horizontal or slightly tilted beds of sandstone rich in kaolin;
the quartz grains of this sandstone have the same appearance

1) For this chapter consult Maps IXᶜ, Xᵃ, Xᵇ and Xᶜ of atlas.

as the slightly rounded grains found in granite. On the right bank
I found a good place for bivouacking, and as the coolies were
still far behind us, I decided to go no further that day.

Shortly after sunrise we set out again over slightly undulating October 11.
wooded ground until, at 7 a. m., we came to the little streamlet
Mĕroeboei Kĕtjil, which we forded, and climbed up the brow of
a sandstone hill 334 metres high, keeping to the south-east
and east. On the other side of this hill lies the valley of the
Kĕpingoei, a left tributary of the Toendoek or Tondok, which
river, as already stated empties into the Lĕkawai. Going down-
hill to the north-east, we came, at 8 a. m., upon the Kĕpingoei,
close to the place where it joins the Pĕlatti. The solid rock in
the bed of the stream was fine grey clayey sandstone upon which
lay coarser crumbling sandstone with indistinct shell-impressions.
The stratification is almost horizontal with a scarcely perceptible
northerly dip. On the other side of the Kĕpingoei, a gradual
ascent led us up to a height of 504 metres, and here we came
upon a succession of tiny feeders of the Kĕpingoei which all had
to be forded. The country rock continued to be sandstone, but
scattered on the ground were loose blocks of amphibole-porphyrite
with biotite and augite, originating from Mt. Lĕmoekoet. We
skirted the steep southern slope of this mountain, at an altitude
of 770 metres, leaving the top to the left, i. e. north of us.
Soon after we went downhill in a north-eastern direction. By noon
we could discern the valley of the Tondok river, deep down at the
bottom of the sandstone slope, and by 12.45 we had reached it[1]).

The bed of the river was almost dry, but as we forded it
we could distinctly hear the rush of water underground. I discovered afterwards that about 100 metres up-stream the water
is lost in a fissure which crosses the river-bed, and reappears
about 140 metres lower down. This „perte du Tondok" is
caused in this way: the bottom of the stream is sandstone

---

1) My companions said that it was a right branch of the Mĕnkoetoei and they called
it Mĕnkoetoei kanan, but I have followed the iudications on the topographical map which
gives this river as the beginning of the Tondok, which lower down-stream is called Toendoek.

lying in thick beds dipping feebly N. 5° W., i. e., in the same
direction as the current of the stream. The sandstone is much
cleaved and in one of the cleavage-planes the water disappears
and has cut itself a fresh passage between the beds of sandstone as represented in Sect. J. on Map IXc. When at last
the river reappears at the surface, the small stream, which had
continued its course over the original bed, joins the principal
stream, forming a waterfall 3 metres high. When the river is
much swollen the subterranean channel is wholly inadequate to
contain the enormous mass of water, and the surface water then
increases to a considerable volume. At such times of course,
the phenomenon of the disappearance of the river, cannot be
observed.

Leaving the river-basin of the Lĕkawai we went south, and
ascended the mountain ridge which separated us from the Mĕnkoetoei valley belonging to the river-basin of the Ambalau. At a
height of 500 metres we passed two large pondoks, temporary
residences occupied by Tondok-Dyaks when seeking for gétah.
These pondoks were fenced in by a double strong enclosure, in which
finely tapered rattan spikes were placed in different directions, thus
making it practically impossible to break through the fence. These
precautions were taken against the Poenans, who are supposed
to haunt the mountains. We had now a difficult piece of ground
to traverse. It was full of deep fissures through which a number
of small brooklets found an outlet to the Mĕnkoetoei, and high
masses of rock also obstructed our way. At last, about 3 o'clock, we
could faintly distinguish the Mĕroeboei valley through the thick
foliage, and half an hour later we reached the flat summit of a
sandstone rock standing up far above the highest tree-tops, from
whence we had an extensive view over the valley of the Mĕnkoetoei and the boundary mountains between North and South
Borneo. This rock is called Liang-Kĕnangka, on account of its
pure white, naked, vertical face, which is turned towards the
Mĕnkoetoei valley, and is visible at a very great distance, as
a patch of white amid the all-pervading green. The mountains

of the water-parting, which in actual distance are only about
4 kilometres away, appear from Liang-Kěnangka like a horizontal
ridge of no great height with a few excrescences, such as Mt.
Lobang Harimau (595 metres), Mt. Koeroeng Doehoeng (810
metres), and Mt. Boenjau (584 metres). From the place where
we stood we could see distinctly that the general slope of this area
towards the north is very slight, except of course in the deeply cut
river-beds. It answers in every respect to the description of
a plateau-land, inclining feebly northward, but abruptly cut off
to the south. From Liang-Kěnangka we proceeded south-east
and down hill through another area of horizontal sandstone beds
sometimes alternating with clay-stone. At 4 p. m. we reached the
Měnkoetoei river and followed its course southward. Seven times
we had to ford this torrential stream until, at 5.30, we found near
Noesa Marienpoeng a suitable place for our night-quarters. This
day, though sometimes under heavy rain and in a difficult and
very rugged country we had accomplished fully 16 kilometres,
and my poor tired coolies, who came on much slower, did not
reach the bivouac till after sunset. The solid rock is but seldom
seen in this part of the Měnkoetoei, it is invariably either sandstone
or fine quartz-conglomerate, and the boulders are almost entirely
composed of sandstone. Only at Noesa Marienpoeng I found
amongst them a fair proportion of porphyrite.

Being now very desirous to reach the southern division of Borneo October 12.
as soon as possible, I was on the march again before sunrise. But
our progress was slow at first, as the road was very bad. We followed
the banks of the Měnkoetoei river for some little distance passing
right across an enormous boulder-island, Noesa Marienpoeng, and
then leaving the river to our right, we went due south by or
through the little stream of the same name. Sand- and clay-
stone in horizontal position or with a feeble northern dip form
the solid rock all the way. At 8.15 we left the river and followed
a track through the forest on the slope of the water-parting. At
the beginning of this track are two enormous heaps of stones,
piled up as the result of a superstitious custom. Every Dyak when

passing this way, throws a stone upon each heap, fully convinced that by so doing he keeps all evil spirits off his path. My Dyaks also adhered to this custom; of course I followed their example, and I may state that no evil spirits happened to cross our path. A gradual ascent in a south or south-easterly direction, led us, over sandy forest soil, past a small brooklet, the last of the feeders of the Kapoewas river-system which I was to meet with on my journey across Borneo. Fine sandstone was still very prominent and the whole area showed the usual northerly incline. At 9.15, sooner than I had expected, we came on the water-parting at the crest of Mt. Boenjau, and South Borneo lay at my feet. We had now reached an altitude of about 588 metres above sea-level, according to the combined barometrical and boiling-point observations. This approximately agrees with the estimated height of 584 metres, registered for this spot, by the topographical survey. South Borneo, at this first introduction, did not seem disposed to reveal her charms, but kept herself modestly veiled in thick clouds. The wind now and then lifted a corner of the veil and gave me a peep into the mysteries beyond, and I was thus able to verify that there are no high mountains in the south, in fact I only saw one peak Mt. Oro Ohan, which I sighted S. 8° W., and which seemed slightly to exceed Mt. Boenjau in height. In front of me lay the deep valley of the Těmangoei flowing close to the base of the precipice which abruptly terminates the dividing range, towards the south. The nearest ranges on its opposite bank seem to trend from east to west, parallel to the ranges of the water-parting. Behind these rose hills and mountains of irregular height. For about half an hour we continued along the mountain ridge keeping to the E.—S.E., until we reached the place where we could make the descent into the valley of the Těmangoei. On a flat part of the ridge we passed some little enclosures, each about 1.50 metre in diameter, and surrounded by posts. Within each enclosure I noticed a great variety of plants many of which I had not seen growing anywhere else, and the ground was strewn

TOEMBANG KARANG.

with various queer articles covered by tikars and pieces of woven
stuff to protect them against wind and weather. These were
evidently offerings of the Dyaks, upon crossing the divide. I
had a great desire to examine these things more closely, but
the Dyaks were so horrified at the notion that I gave it up,
and I am therefore unable to throw any further light upon the
interior and the purpose of these little gardens. On the divide
the rock *in situ* is fine sandstone. We went down-hill by a very
steep, slippery path, occasionally interrupted by small terraces,
which are generally the outcrops of beds of sandstone. The entire
dividing range is, as I could state undoubtedly on the descent towards the valley of the Těmangooi river, composed of a complex
of sandstone alternating with clay-stone, the strata shewing a feeble
northerly dip. In our descent we crossed the track of a whirlwind.
In the very heart of the forest, a space, which I estimated at
about 40 metres wide, was entirely devastated. Strong trees
snapped or twisted off near the roots lay pell-mell upon the
ground. Trunks of more than 2 metres in diameter testified to
the fury with which they had been wrenched off their bases by
the long fibrous fringes which hung helpless from the fractured
limbs. I mention this because personally I have not witnessed any
hurricanes in Borneo, severe enough to produce anything like such
utter devastation as I saw here, but Schwaner[1]), in the well known
narrative of his travels, tells us of hurricanes which uproot and
demolish the biggest forest trees. I learned afterwards from trustworthy natives that in South Borneo and as far as the boundary
range, squalls of short duration but intense severity frequently
occur during the months of November and December, and during
the latter part of my stay in South Borneo, I witnessed thunder-
storms, such as herald the west monsoon in these parts, which
were accompanied by the most awful blasts of wind.

The distance from the top of the ridge to the bed of the
Těmangooi river is not very great. An hour and a half of steady

---

1) C. A. L. M. Schwaner 55, II, p. 127.

walking brought me to the pangkalan where I found the people
whom I had sent from N⁴ Mĕroeboei ¹). There were only five men
however, as the others, some five days before, had taken a small
boat down to Toembang Habangooi (the nearest settlement on
the Samba river), in order to procure two good sampans for
our transport. Before doing this they had wasted three days in
a futile search for boats at the neighbouring pangkalans.

The five remaining men were in very low spirits; they said

the Poenans harassed them
at night, and that they had
been compelled to shoot to
keep these prowling fiends
at a safe distance from their
camp. Although I knew that
all the current Poenan tales
were grossly exaggerated, I
quite believed that there were
Poenans lurking in the jungle.
Even on the Mĕlawi I had
heard from different quarters
of the head-hunting raid not
long since, in which six Poenans had been killed by Dyaks
of the Djoloi in the Siang-Moeroeng district, also belonging
to the great tribe of the Ot-
Danoms. It was said that the

Fig. 66. Refuge-pondok near Pangkalan
Oeloe-Tĕmangooi.

Poenans had withdrawn into the jungle on the mountain slopes
which separate the valley of the Mĕlawi from the rivers of
South Borneo, and that they were watching their opportunity to
avenge this murder. The pangkalan Tĕmangooi had a reputation

---

1) The nine Dyaks from Boentoet-riam who had accompanied me so far, declared that
they must at once continue their journey. As arranged, they each received 5 florins and their
chief 5 dollars, and I gave them moreover a few small presents as parting gifts.

for being unsafe, in consequence of which the Dyaks passing
to and fro on their journeys from the Samba to the Mĕlawi, had
built a refuge-pondok on very high posts, from which they could
more easily be on their guard against nocturnal attacks (fig. 66).

Of course my men had taken possession of this pondok. As
for the sampan, it was not nearly finished; the tree was felled
and the section of the trunk selected, had begun to assume the
shape of a boat, but it was evidently a far longer business than
the men had given me to understand, and it would certainly
take several more days to finish the work. I had to resign myself to the inevitable, told my men to build a small pondok by the
side of the big one, and waited, with what patience I could muster,
to see the end of it, curiously wondering whether our own make-
shift or the boats from the Samba river would finally conduct
me back to the sea. I made several small excursions in the
neighbourhood, and discovered that there was abundance of game
in this place. At the foot of the precipitous slope of the dividing
range, a few kilometres west of my bivouac, was a pond, the
borders of which were deeply furrowed with the spoor of rhinoceros. It was evidently their bathing and drinking place. One could
trace their foot-marks far into the thick jungle, forming clear,
well-beaten, muddy tracks, where all vegetation was trampled
to the ground. One night I lay in ambush near this pond. After
waiting for several hours a magnificent stag appeared upon the
scene with antlers uncommonly large for Borneo. I could not
resist the temptation to secure it and by my firing evidently
spoiled all chance of further sport. The rhinoceros is extremely
shy and at all events I saw no trace of one that night.

On October 15th my coolies returned from Toembang Haban- October 15.
gooi, but alas with only one sampan, far too small to hold all my
things. What was to be done? I knew that a longer stay would
soon convert our bivouac into a starvation camp, and to make
matters worse, the coolies brought word that there was a rice
famine on the Samba, so I determined to reduce my personal
needs to a minimum, and, at any rate for a short time, to

manage with the one boat. Ten coolies were sent back to
Měroeboei with everything that could possibly be dispensed with.
They were to load all the luggage in our own boat which had
been left behind there and take it down to Sintang; while I
retained only six men besides my own two servants. But in
spite of all these curtailments the boat would not hold us all,
and when we started at 7.45 next morning, 4 of the coolies
had to follow on foot with part of the luggage, while the sampan
was laden to overflowing. At first all went well, the banks were
low and the sandstone rocks protruding here and there offered
no special difficulties, but very soon, scarcely a kilometre below
the pangkalan, the aspect of things changed. Big sandstone banks
almost completely obstructed the bed of the river, reducing it
to two narrow clefts that one could jump across, and through
which the water ran down with great velocity forming a succession of small cataracts.

Fortunately I had by this time gained some experience in
the management of a boat amongst the rapids. There was no
room for idle hands, and it was all we could do to save the
boat from being dashed to pieces. After an hour's hard work
we came into quieter waters, and at 9.30 our boat drifted out
of the Těmangooi[1]) into the Rassahooi, a somewhat wider stream
coming from the east. High sandstone cliffs obstruct the mouth
of the Těmangooi, and a little lower on the left bank is a
boulder-bank containing good specimens of the kinds of rock carried by the Rassahooi. Besides sandstone I saw several pieces
of granite and tonalite, which must therefore occur as solid rock
further east in the immediate neighbourhood of the boundary
mountains. The Rassahooi is here 18 metres wide; beds of
sand- and clay-stone, either horizontal or dipping slightly to the
north, form the bottom rock until, about 2 kilometres further
down-stream, we came to what was evidently the lowest limit
of the sandstone area, and granite now appeared from underneath

---

1) At the pangkalan the Těmangooi is about 12 metres wide.

it. In these lower strata the sandstone is full of bleached flakes of biotite and there is no well marked limit between it and the underlying decomposed granite. It is an amphibole-biotite-granite, containing plagioclase, passing sometimes into tonalite, where plagioclase greatly predominates as the feldspathic constituent. The quantity of amphibole also varies considerably.

Navigation was rather awkward but never really dangerous in these waters. We were however very slow in our progress, because, not being familiar with the currents and rapids, I always took the precaution whenever we came near to a fall to send two of my coolies forward to explore the position of the rocks, and then we selected the best place to run our boat down. Notwithstanding all our care however, our poor overloaded sampan got many serious bumps and knocks and it became at last so full of holes that I knew it could not keep together much longer unless we could relieve it of part of its freight. At this critical moment, I hailed with joy the appearance of a boat full of Dyaks. This was about 3 p. m. They were from Méroeboei and homeward-bound, intending to leave their sampan at the pangkalan Témangooi. I negotiated with them on the spot and it was soon arranged that I should have the use of their sampan in exchange for a present of rice and tobacco. Some of the cargo was shifted into the second boat and all my men could now also be accommodated.

Thus we proceeded comfortably for another 200 metres when a sudden curve of the river to westward brought us face to face with three thickly wooded islands, which completely obstructed the bed of the stream so that the water had to force its way between high granite cliffs. It was a most difficult channel to pass through, and our new acquisition was well-timed, for my battered sampan had received the finishing stroke and was beginning to sink. I went on shore as soon as I could. All that night we were busy patching and calking the boat with bark and fibrous vegetable materials.

October 17. Thus we were enabled to continue our voyage next day. The scenery remained unaltered, high granite banks, occasionally varied with low sandy shores. At the bends, landslips were of frequent occurrence, and several huge blocks of granite had been hurled down into the stream. I saw big strips of the ground which rests on the granite covered with a thick growth of trees, shrubs and plants, removed bodily from the cliff and hurled down into the current where they formed a chaotic mass of wood and foliage, totally barring the passage, and we simply had to cut or dig our way through as best we could. The half-faded foliage and everything else connected with it, pointed plainly to a recent flood, for nothing short of a regular bandjir could have worked such ravages as we here beheld. I learned afterwards that in the last days of May 1894, about five months previously therefore, a terrific flood had swept over the place, and left it thus devastated. We did not meet with any further serious difficulties that day. At 10.30 we reached the place where another stream, 15 metres wide, which we afterwards learned was called the Němangooi, joined the Rassahooi, and half an hour after that, the place where the Rassahooi and the Karang meet and then go by the name of Měnjoekoei. The Karang is nearly as wide as the Rassahooi, and appears to carry an equal volume of water. The feeders of the Karang rise on Mt. Kait Boeran and on Mt. Raja.

The Karang also serves as a means of communication between South and West Borneo, and from the head-waters of that stream a path runs between the Kait Boeran and the Pintoe Běnoewang across the boundary mountains, up to the highest settlements on the Lěkawai and the Sěrawai. The entrance to the Karang (see Pl. XLIV) is extremely picturesque. Pl. XLIII gives the view up the Rassahooi, as seen from the mouth of the Karang. It is a typical river of the highlands of Borneo, where the virgin forest, as yet untouched by human hand, spans the water in untrammelled arches, seemingly resting on gigantic green columns, the garlanded trunks of the colossal forest trees. I spent the greater part of the day in examining the boulders carried by the Karang. Most of them consist of granitite, amphibole-granitite and tonalite, these being also the bottom rocks at the mouth of the Karang, together with some other kindred types, such as quartz-diorite and varieties of granite rich in plagioclase, all strikingly poor in dark con-. stituents, sometimes rather finely grained, sometimes pegmatitic, in fact of exactly the same types that I found in the Upper-Lĕkawai, carried down by the Bassi. This latter river rises on the north side of the boundary mountains at about the same altitude as that of the more important affluents of the Karang on the south side, so that it is obvious that this light coloured granite, so strikingly like arkose when weathered, must be found in the neighbourhood of Mt. Kait Boeran and Mt. Pintoe Bénoewang within the dividing range.

. But my attention was not so much arrested by these already familiar types as by fragments of contact rocks of the group of hornfels, from amongst which I will only mention one type, which looks like a muscovite-biotite-gneiss with numerous accessory minerals, and several varieties of andalusite-mica-schist. These fragments prove that in the boundary mountains or at least on their south slope in the river-basin of the Karang, crystalline rocks occur, which we may reckon to belong to the contact area, lying between the granite and allied deep-seated intrusive rocks and a system of stratified rocks of unknown age. Later on it will be proved that these contact rocks are of great importance in the boundary mountains west of Mt. Raja.

So far the course of the Rassahooi had been almost due south, but below Toembang Karang the south-eastern course of the Karang seems to exercise a decisive influence upon the direction of the united stream which is called Mĕnjoekoei, for from Toembang Karang to Toembang Lamihooi the general direction of the Mĕnjoekoei is south-south-east. Below Toembang Karang granite continues to be the rock *in situ*, in which, however, fully 2 kilometres down-stream (see map X^), some striking modifications become apparent. At first only sporadic concretions occur, in which the dark constituents accumulate, together with an increase of plagioclase in proportion to the orthoclase. But before long these concretions increase in quantity and trend in one direction, which gives them a flattened lens-shaped appearance. The flakes of mica and the prismatic crystals of amphibole then run parallel to each other in the same direction, and thus the rock passes into a kind of gneiss-granite. The concretions become more and more elongated and assume the appearance of flat or undulating layers or sheets which are wedged in between the gneiss-granite and alternate with it. Finally the entire rock becomes more

N.E.   S.W.

🟫	Amphibole-granitite and tonalite.
🟦	Gneissic tonalite.
⬛	Modifications having the same composition as plagioclase-amphibolite.

Fig. 67. SECTION ACROSS THE ROCKS OF GRANITE ON THE LEFT BANK OF THE MĔNJOEKOEI RIVER.

or less cleavable and slaty in the same direction and thus according to the relative proportion of the gneiss-granite and the concretions and their mineralogical constitution, the rock sometimes resembles biotite- or tonalite-gneiss, at other times biotite-schist or amphibolite. It is my opinion that the above described concretions owe their origin to the differentiation of the magma at the outer margin of the intrusive granite massive. To arrive at this conclusion I have in the first place to assume the intrusive character of the granites in the Mĕnjoekoei and the Samba, and I do so with confidence because of the numerous pebbles of contact-rocks (most of them containing andalusite) which I found in the neighbourhood of the granite. In some places on the Samba, as will appear presently, I discovered this same contact formation *in situ*, and in connection with the granite, but in the Mĕnjoekoei

Sandong with sependoks, Roemah Toembang Mantikei.

I found only large, loose, very slightly rounded pieces of the contact rocks, which led me to surmise that solid rock of this formation presumably crops out at no great distance. It is not surprising that although according to my calculations I was now on the extreme edge of the granite massive, I nevertheless did not find the contact rocks *in situ* in the bed of the Menjoekoei, as it should be kept in mind that granite yields more readily to erosive forces than the extremely hard andalusite-hornfels, so that it is quite possible for the river to have cut its bed in the granite, while in the adjoining hills contact rocks may appear *in situ*.

Unfortunately I was not able further to investigate this matter in the open field. Anyone at all acquainted with the difficulties involved in the solving of similar geological problems in hilly districts, thickly covered with tropical virgin forest, will readily understand that with the limited time and means at my disposal, I was obliged to confine myself to the examination of the bed of the river. This phenomenon however, the occurrence of similar modifications close to the contact-zone, is by no means uncommon amongst granites. It calls to mind the contact between the granites of the Cape peninsula and the greywacke and clay-slate surrounding this massive, which is exposed near Sea-point in a section of classical beauty. There basic concretions, abounding with biotite, and occasionally also with tourmaline and amphibole, are found in the granite at a long distance from the contact. These concretions are likewise much flattened and form rough layers or sheets, which in most cases run parallel with the plane of contact between the intrusive granite and the surrounding slate. Only in part are they the result of differentation in the granitic magma, some are caused by blocks of the surrounding rock, which have been absorbed in the granite and thus metamorphosed. So also in the present case lumps of the surrounding rocks, may have been absorbed in the intrusive granite, and possibly have contributed to the remarkable modifications just described.

About 2¹/₂ kilometres below Toembang Lamihooi, olivine-hypersthene-norite appears in the gneissic granite, and still further down-stream low hills of olivine-gabbro break the monotony of the granite area.

I may mention that below Toembang Karang the ravages of the flood just described, became still more evident. The lofty trees, a striking feature of the islands in this river, were almost all broken and beaten down in a direction parallel with the stream, and the lower brushwood and bushes were utterly crushed and demolished, and covered with gravel, sand and mud. The general effect produced was that of a huge cornfield battered and bemired ·by storms and heavy rains. The accumulation of deposit in the bends of the river gave me some idea of the material swept along by the flood in its mad fury and of the objects which served as projectiles to cause further mischief. I saw large heaps of some dozen big trees, with foliage still clinging to them, jammed into the wooded banks, 20 metres above the present water-level, and with such force had they been swept against the rock walls, that the standing trees had been broken down by them, thus forming a truly chaotic pile of living and dead wood.

Fortunately for us the Mĕnjoekoei river was now in a calmer mood, and kept its terrific forces under control. The water remained at a low level, and we passed without accident the numerous rapids below Toembang Lamihooi many of which are far from harmless. At 4.30 we came to the confluence of the Mĕnjoekoei and the Samba, the latter is by far the broader of the two rivers. Down to this point the banks of the Mĕnjoekoei had been rather steep, and the channel narrow, but now the view widened and I could look around more freely, for the Samba valley, although lined with hills, is very wide, and the hills very seldom come down to the waters brink. I passed the night on the narrow strip of land which separates the two rivers, and at no great distance from the big boulder-bank which narrows the Samba considerably just below its confluence with the Mĕnjoekoei.

Obctoer 18. For several days running I had been registering thunder-storms in the early morning before sunrise. For Borneo this is a most unusual time, as 75 % of the storms there take place between 3 p. m. and 9 p. m. On this day again a thunder-storm accompanied by torrents of rain commenced at 6 a. m. and continued all the forenoon. First the Mĕnjoekoei began to rise, the water became muddy and discharged with tremendous force into the, as yet clear, waters of the Samba. Presently the Samba rose and became boisterous. Both rivers were now yellow and muddy and carried large quantities of wood; at 9 a. m. the high karangan was entirely submerged. Meanwhile the rain abated and finally stopped at noon; shortly after, the Mĕnjoekoei fell as rapidly as it had risen; the water cleared, and for a considerable distance beyond the confluence the clear waters of the Mĕnjoekoei could be seen, like a black band, against the muddy waters of the Samba, until they finally amalgamated. I rowed up the Samba until I came to a boulder-island about $1\frac{1}{2}$ kilometres above Toembang Mĕnjoekoei. On this island I became acquainted with the kinds of rocks carried down by the Samba. I noticed amphibole-granitite and amphibole-granite with its gneissic and amphibolitic varieties[1]) in large quantities, besides pegmatite, diorite and different kinds of hornfels, amongst which andalusite-biotite-hornfels was particularly well represented.

The natives informed me, that the Samba arises on the southern slope of Mt. Raja and is navigable for small sampans several day's journey above Toembang Mĕnjoekoei. They said that the natives often used the Samba as the means of communication between the southern division and the river-basin of the Sĕrawai. Their route would then lie past the western slope of Mt. Raja. At Toembang Mĕnjoekoei the Samba is 62 metres across, while the Mĕnjoekoei is only 35 metres wide at the mouth. The valley of the Samba, like the lower course of the Mĕnjoekoei, is cut in granite, while the same gneissic and amphibolitic characters prevail that I noticed in the Mĕnjoekoei.

At 1 p. m. we returned by the same way. About 3 kilometres

1) Some of these latter contain much green spinel and are in all probability contact-rocks.

below Toembang Ménjoekoei the Samba is checked by a range ot hills trending east and west, and its course is changed from south-east to north-east. At the bend the slope of the hill has been washed away so that a sheer precipice has been formed, giving an exposure of a most exquisite rock, a cordierite-andalusite-biotite-hornfels. This belongs probably to the contact zone which encloses the granite-area. Further down-stream gneissic granite remains the component rock on the left, and the contact rock just mentioned continues to crop out on the right bank. Here and there the wooded slopes began to give place to rice-fields (ladangs), a sure sign that human habitations were not far off. In sight of our goal for that day, I added to my collection a very beautiful granitite containing garnets, which I found just above Noesa Poelang. At 4 p. m. I stopped at the Dyak house Toembang Habangooi, situated on the left bank, at the confluence of the Habangooi and the Samba, and thus I was back again in the inhabited if not in the civilised world. The abode was far from inviting. The house, evidently intended originally to accommodate several families, was now for the greater part in ruins and only a small portion of it was inhabited, but even that was dilapidated and very dirty. Round the house were several toras and tĕmadoks, amongst the latter a few carefully executed female figures, which showed that the establishment had known better days. The house is inhabited by Ot-Danoms, and one Malay called Djaja, married to the sister of Raden Kĕramah, the chief of the house. As the chief was absent this man received me, and he told me that he was appointed by the Dutch Government, to represent Dutch autho-rity in these parts. From the first, however, I had very little confidence in this man, and it transpired afterwards that he had established himself as the representative of the Dutch Go-vernment entirely on his own authority, in order that he might the easier swindle and oppress the Dyaks from whom he was in the habit of buying forest produce. All the same, he was of great service to me as interpreter during my stay at Toembang Habangooi, as neither I nor my companions understood the dialect of these Ot-Danoms. South-east of Toembang Habangooi and within easy reach I noticed one mountain considerably higher than the surrounding hills. This peak is called Mt. Atoeng and I determined to ascend it as it seemed to me a desirable point from whence to fix the position of the various mountains of the boundary range between West and South Borneo, in order to enable me to complete a sketch map of the Mēnjoekoei and the Samba.

A party of Dyaks from Roemah Solien (a house not far from Mt. Atoeng) just happened to be coming up the river to see the white man, rumours of whose approach had reached them. I at once availed myself of this opportunity to ensure their services, and by the help of the Malay interpreter I engaged them for the next day to act as guides up the mountain and, if need be, as carriers.

So we started on the 19th at 7 a. m., passed the beautifully October 19. wooded boulder-island Solien, and at 7.30 reached the house of that name, situated 1¹/₂ kilometre below Toembang Habangooi on the right bank of the Samba. A horizontal plan of this house, which is inhabited by five families, is given in fig. 68. In some respects it differs from the general type of Borneo houses. For instance the tree ladder which gives access to the house is here fixed to the outside wall near the centre of the public hall and it is inclosed in a sort of outbuilding which is used as a store-room. Joined to this is another little projecting room in which there is a large fireplace, and serves as the common kitchen of the establishment, while in the private apartments (lawangs) no fireplaces are found. In the common or general room I saw two infants hanging in slings fixed against the ceiling. The sling is held open by means of a cross-bar. The child sits in the sling in a squatting position, with

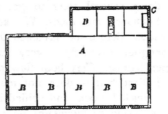

Fig. 68. PLAN OF THE HOUSE SOLIEN ON THE SAMBA RIVER, INHABITED BY OT-DANOM DYAKS.
A Hall in common use.
B Lawangs, private apartments.
C Hearth.
D Storeroom.

a second smaller sling tied round its neck to keep the head steady. Fastened to the top of the sling is a bunch of amulets, such as a bear's tooth, a mais-club, shells of land- and river-snails, bits of rattan and other curios, all of course intended to protect the little one against the evil spirits. The inmates of this house were most friendly and anxious to give me a pleasant reception, but their good intentions were mostly lost upon me as I could not understand a syllable of their language. I did not stay very long, but as soon as my men had refreshed themselves with a good meal of rice we left at about 8.45, fifteen minutes later we reached the pangkalan, and a little before ten o'clock we stood on the top or rather the brow of Mt. Atoeng, 306 metres high. This hill consists entirely of amphibole-granitite. The summit is thickly wooded so that I was obliged to cut down several trees to get anything like a decent view of the boundary mountains. We accomplished the clearing in the afternoon. Before me stretched the valley of the Samba and the Tĕloengooi, evidently a very important right tributary of the Upper Samba. They both flow through a hilly district in which the loftiest height does not much exceed that of Mt. Atoeng upon which I stood, and behind it, like a gigantic wall, rose the line of boundary mountains veiled in clouds. I sighted Mt. Kĕpĕnjahoe, N. 42° W., and the top of Mt. Raja, partly lost in the cloud-group, W. 37° N. Shortly before sunset we returned to Toembang Habangooi drenched through by a short but violent shower of rain.

October 20.   Next morning I ventured a little way up the Habangooi. The solid rock along the banks is first granite, then andalusite-hornfels, and finally gneiss-granite. About $1^1/_2$ kilometre above the mouth of the river are some boulder-islands, particularly rich in varieties of granite with gneissic and amphibolitic concretions, also quartzitic rocks and several varieties of diabase and diabase-porphyrite. The Habangooi rises on Mt. Banjang pali, and an important left tributary rises on Mt. Moehoet, on the other side of which the waters flow towards the Bĕrahooi. The Habangooi is important as the means of communication with the river-basin of the Kahajan; after a three day's journey up this river a short land-route leads to the Lĕbojoe, an affluent of the Bĕrahooi. From the Upper-Bĕrahooi another short land-route leads to the Dangooi, a branch of the Upper Kahajan.

Higher up on the Habangooi are two more Dyak houses; these two settlements are the nearest of all to the mountains in the Samba valley.

I was informed that many years ago, between 1845 and 1880, the highlands in the head-waters of the Samba and its principal tributaries were much more thickly populated then they are now, but the constant head-hunting and plundering raids of the Poenans, rendered life so uncertain to the Ot-Danoms that they retreated more and more from their mountain homes towards the plains, until in 1880 not a single stationary settlement was to be found on the Upper Samba, as far up as Toembang Bĕrahooi. Since then, however, the influence of the Dutch Government has, to a great extent, checked the boldness of the Poenans, and the Ot-Danoms are gradually recovering their old ground.

These facts agree in every respect with the accounts given by SCHWANER in 1847 and by MICHIELSEN [1]) in 1880. Speaking of the Samba, SCHWANER mentions five kampongs on the Bĕjoekoi (by which is meant the Mĕnjoekoei) and one kampong on the Riang, which he calls a right tributary of the Samba a little above the Mĕnjoekoei. Both these rivers are now uninhabited, and the tropical vegetation has so effectually obliterated all traces of former cultivation, that I could only guess at the possible place where the old ladangs had been by the secondary character of the forest in certain places.

I was obliged to buy another boat to take me further down the Samba and the Katingan, as the one I had used so far had to be left at the pangkalan. The Malay DJAJA availed himself of this opportunity to get ample repayment for his slight

---

1) W. J. M. MICHIELSEN 59, p. 72.

services. There was only one boat to be had, a rough dirty old sampan for which he asked the exorbitant price of 30 florins. I left at 10 a. m. having engaged a Dyak guide from the house Solien, to accompany me as far as Toembang Samba. A strong current helped us rapidly along, and very soon we came past Mt. Atoeng at whose foot the river bends to the south, and the rocks of amphibole-granitite, which at this point stand out of the water, create a cataract called Kiham Sĕpan. This type of rock continues to prevail for some distance and the bed of the stream is much narrowed by cliffs rising in mid-stream or jutting out from the banks. We passed a succession of rapids, Kiham Tingang, Kiham Mĕnkĕtjahoe, Kiham Takoei Sandah and Kiham Hoidjera, all of which, no doubt, cause endless delay in going up-stream, but are perfectly harmless when going down the river.

About 1¹/₂ kilometres below Kiham Hoidjera, the granite again acquires that gneissic character which is so general on the Mĕnjoekoei, and here and there andalusite-hornfels reappeared *in situ*. Near Kiham Manding, I again found coarse, dark biotite-amphibole-granite, which seemed to bring us to the southern extremity of this granite area. About 2 kilometres below Kiham Manding, contact rocks take the place of the granite. Nearest to the granite lies muscovite-hornfels containing andalusite, followed by muscovite-hornfels with the character of „Knotenschiefer". A little further on andalusite-clay-slate, then another exposure of andalusite-hornfels, and finally again andalusite-clay-slate occur. Some little way further down-stream the course of the river is checked by a small range of hills, and curves eastward. Just at the curve, the precipitous cliffs on the right shore expose horizontal strata of conglomerate and sandstone, in all probability lying unconformably upon the older formation of which the contact rocks just mentioned form part. On the same bank, a little further on, amphibole-porphyrite with andesitic habitus is exposed. On the left bank the contact-rocks once again make their appearance, in vertical strata striking N.E., after which they disappear under horizontal strata of conglomerate and

Pl. XLVI.

SANDONG, ROEMAH TOEMBANG MENTIKÉH.

sandstone, which continue to form the solid rock on both banks till within $1^1/_2$ kilometre of the mouth of the Pangah. After passing Batoe-Pĕtjoeng we entered upon a hilly district consisting of volcanic rocks, frequently cropping out along the banks. The current here is generally calm and smooth, but occasionally big boulder-islands form obstructions. In the evening we landed on one of these, called Noesa Koetau, situated just in front of the Dyak house of that name. I was told that the chief of this house, TOEMENGGOENG SINGAM, rules over all the neighbouring houses. Noesa Koetau is fully 300 metres long, here and there thickly wooded but for the greater part composed of boulders and devoid of vegetation. I found granite, and amongst it varieties closely allied to gneiss and mica-schist, aplite, diabase, amphibole-porphyrite, some of it with andesitic character, amphibole-andesite, andesite-tuff, a little quartzite and clay-stone, and lastly, pieces of silicified wood. This silicified wood deserves special attention, for neither in the Samba nor in the Mĕnjoekoei, nor in any of the other tributaries, did I come upon any pieces of it. But no sooner had I entered the volcanic area, in which the andesite-tufas play so conspicuous a part, then silicified wood also made its appearance amongst the boulders. The pieces are slightly rounded and none of them bear signs of having been carried for any considerable distance. Beginning from Noesa Koetau, silicified wood is never absent as one of the component parts of the boulder-banks. The conclusion to be drawn is obvious. The silicified wood here originates in the andesite-tufas, the same as in the Müller mountains in the river-basin of the Mandai in West Borneo.

I bivouacked on the island and next morning, shortly before October 21. sunrise, we continued our voyage. Below Noesa Koetau are two more large boulder-islands. The banks are almost entirely composed of gravel covered with sand and loam. About 5 kilometres lower down, another important stream, the Kĕrĕtan, joins the Samba on the right. In a high cliff on the left bank an excellent exposure of andesite-tuff is exhibited. Three kilometres

below Toembang Kĕrĕtan the Samba curves eastward and for
a distance of 28 kilometres from here the current runs east-south-
east. At the bend, a beautiful group of rocks on the right bank
demands special attention. It is called Batoe Bĕsarro and consists
of amphibole-andesite and andesite-breccia. This same formation
is repeated lower down in a pretty group of rocky islets, the biggest
of which goes by the name of Noesa Bĕsarro. Directly before us
we noticed on the left bank the counterpart of the Batoe Bĕsarro,
called Batoe Tĕmarra composed of weathered amphibole-andesite
with brecciated structure. The character of the river and its immediate surroundings continued the same over a considerable
distance. Navigation was quite easy and only very occasionally
the current became somewhat disturbed by the presence of fair-
sized boulder-islands, such as Noesa Tĕmoetjoek, Noesa Hiboeng,
Noesa Mĕnarieng, Noesa Karangan Tapi and Noesa Ingĕh. The
banks are for the most part low, only here and there rocks of amphibole-andesite, augite-andesite, andesite-tuff or andesite-breccia
generally situated on the bends of the river, make a pleasant
variety in the scenery. Below Noesa Tĕmoetjoek and looking
north-east, one has a beautiful view of Mt. Lanai, a lofty, solitary, conical mountain, and below Noesa Hiboeng the exquisitely
slender Poeroek [1]) Tandok comes within sight. For the rest the
river is lined on either side with slightly broken hilly ground,
which as far as one can judge, overlooking it from the river, is
almost entirely brought under cultivation.

The passage through this truly charming landscape was somewhat marred by milliards of tiny gnats (mĕroetoes or agas) which
troubled us considerably in our work, especially in the shady
places close to the shore where I was often obliged to stop to
collect rock-specimens and to draw sections. These agas are much
smaller than mosquitoes and it is astonishing how quickly such
tiny midges manage to pierce the human skin. Their sting causes
a burning irritation. In the lower course of the rivers they are

1) Poeroek (Ot-Danom dialect) = Mountain, peak.

scarce, but in the hilly parts, there are legions of them and especially here on the Samba hundreds of them always covered my face and hands. About noon I arrived at the house Tandok where TOEMENGGOENG BAHÊH is acknowledged as head. I at once made arrangements for ascending Mt. Tandok, and at 2 p. m. I rowed back to the place on the opposite shore nearest the foot of the mountain. Mt. Tandok is a finely tapered cone-shaped mountain surrounded on all sides by very steep rock-walls, almost entirely devoid of vegetation. Seen from a distance the mountain appears to stand quite isolated, but from my stand-point I could see distinctly that it is connected both on the south-east and on the north-west with lower ranges of hills. Like the hills at its foot, Mt. Tandok is entirely built up of andesite, containing olivine and large phenocrysts of feldspar. It is not more than 304 metres high but very difficult of ascent. As a matter of fact it is only scalable on the north-west side and even there one has to face an almost perpendicular rock-wall 30 metres high. In this respect the ascent reminded me of my experiences on Mt. Kĕlam near Sintang, but the rattan ladder which had there been provided by the natives was missing here. Close to the sheer rock-face strong aerial roots hung down from the trees above, and by clutching hold of these, we managed to pull ourselves up to a height of nearly 30 metres, without once being able to get a really good foothold.

The view from the top is extensive in all directions and very surprising, part of the panorama is given in Plate XIV on page 22 of the atlas. To the north the foreground is slightly undulating, thickly wooded, and with a few scattered hills, amongst which Poeroek Lanai, close to Poeroek Tandok and of about equal height, is the most conspicuous. Without any doubt Mt. Lanai is an andesitic cone. More to the north there follows a low hilly country, the granite area of the Upper Samba, which possibly here and there rises as high as Mt. Tandok, but looks much lower because of the high mountains of the dividing range which limit the horizon in the back-ground. These boundary mountains, from our point of view, appeared like a massive sheer rock-wall, the highest part of which, the Raja district, remains hidden in clouds. I took some bearings, most valuable for my map of the Samba river; with my fieldglass I could clearly distinguish the tops of Mt. Kĕpenjahoe and Mt. Damar, the position of which has been exactly ascertained by the topographical survey of West Borneo. I sighted Mt. Kĕpenjahoe at N. 31° W. and Mt. Damar at W. 36° N. East of Mt. Raja the height of the boundary mountains hardly varies and they are shortly beyond concealed from view by other mountains more in the foreground, such as Mt. Tĕmiting (N. 15° W.). Still further east Mt. Moehoet (N. 15° E.) is most conspicuous. One of the right affluents of the Bérahooi river rises here. The top of Mt. Moehoet like that of Mt. Raja, was enveloped in clouds; I estimate its height at 1400 metres at least; it is without doubt the highest mountain in the entire river-basin of the Samba. To the south, as far as the eye reaches, there is nothing but slightly undulating ground, and the monotony is only broken by a few narrow, sharply outlined, ranges of hills trending E.N.E. to W.S.W. The two nearest and most important of these I have called the Tapi range and the Taroeng range, after the principal heights, known by these names. From the top of Mt. Tandok, I could notice how little ground has as yet been brought under cultivation here. The ladangs form only a narrow, light-green strip along the banks of the Samba, and are backed on either side by dark-hued virgin forest. On the top of Mt. Tandok are some places of sacrifice, where the Dyaks offer objects of various kinds to the hantoes.

We accomplished the return journey without accidents of any kind and I spent the evening in the house of TOEMENGGOENG BAHĔK. In the large public room, fantastically illuminated by the fires on the hearths, the greater portion of the inmates squatted themselves a round me. Many a curious piece of furniture, valuable from an ethnographical point of view passed then into my possession. The chief could understand and speak some Malay and I conversed with him about the Poenans. He told me how they had decimated his tribe to such an extent, that many of his kindred had been forced to vacate the Upper Samba district, as life there had become simply untenable. Of late years, however, the Poenans had visibly decreased in numbers, and head-hunting raids had not been heard of for some time past. He declared to me that the grudge which his tribe bore against the Poenans was so strongly rooted, that nothing would deter them from retaliation if any chance offered. It was now their turn, and it would be some time before they should consider their wrongs to be sufficiently avenged.

Worthy of notice is the plan of architecture adopted by the Ot-Danoms of this part of the Samba river, which, as shown in fig. 71, differs in many respects from the plan generally followed in the houses of West Borneo. To mark the differences

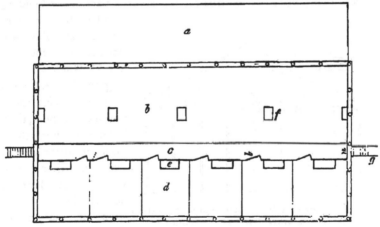

Fig. 69. PLAN OF A DYAK HOUSE, AS ADOPTED BY THE BATANG LOEPAR AND OTHER TRIBES IN THE KAPOEWAS-BASIN IN WEST BORNEO.

a. Open veranda.	*b*. Large public hall.	*c*. Passage.
d. Private apartments.	*e*. Private Hearths.	*f*. Public fireplaces.
g. Ladder.	*l*. Swinging doors.	*n*. Entrance.

I will shortly describe the principal types of building adopted in the different districts of Borneo, visited by me.

374

I commence with the type generally used on the Kapoewas and the Mĕlawi rivers, figs. 69 and 70. A Dyak house is built entirely of materials supplied by the forests. Roughly hewn tree-trunks form the posts and the principal cross-beams upon which the house rests. Strong trees of hard wood, deeply notched on one side at equal distances serve as ladders. Smaller trees are used for the lighter cross-beams, and flat slabs of wood chipped out of the thick part of the trunk constitute the roof coverings (siraps). Bark is used to make partitions, doors and roofing. The delicate lathwork, the bale-bale, which forms the floors, is made of bamboo, rattan or very thin, straight stems of trees. The whole structure is fastened together by ropes of rattan, which nothing can equal for strength.

The Dyak house is oblong in shape. The width does not vary much, being about 10 metres for the covered part of the building.

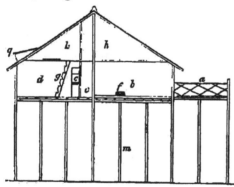

Fig. 70. SECTION OF A DYAK HOUSE USED BY THE BATANG LOEPAR AND OTHER TRIBES IN THE KAPOEWAS BASIN IN WEST BORNEO.

a. Open veranda.   *b*. Large public hall.   *c*. Passage. *d*. Private apartments.   *e*. Private hearths.   *f*. Public fireplaces.   *g*. Ladder.   *h*. General attic. *k*. Private attic.   *m*. Posts.   *q*. Skylight.

The length however differs considerably, and depends upon the number of families inhabiting the house. On one side, that facing the river, if one is near, there is an open veranda or platform where all such work as can not conveniently be done indoors, for instance, winnowing rice, drying clothes, etc., is carried on. The covered portion of the house is divided lengthways into two parts, almost equal in size. One of these, generally the largest adjoining the open-air gallery, is the public room. It is what we should call the living and drawing room, and the greater part of the customary daily business is also done here. It is moreover the sleeping place for guests and the unmarried men of the community. Here and there are small fireplaces of stone or hard clay for general use. The most precious of their possessions, the skulls of their enemies, are stored here, they are hung from the ceiling in bunches like chandeliers, directly above the door of the private apartment of the head of the family to which these trophies belong[1]).

The second part has been subdivided by permanent partitions into a certain number of private apartments each inhabited by one family. Here a variety of precious articles amongst which the famous tĕmpajans stand against or hang on the walls; the principal piece of furniture, standing against the inside wall, is the hearth, and above it is a rack in which firewood is always drying. In these apartments, husband, wife, and children sleep altogether; the family cooking is also done, and the women in particular spend most of their time in them. Between the private apartments and the public room runs a passage, the floor of which is made of very strong boards and is generally a little lower than that of the public room. This passage connects the two doors of the house, access to which from the outside is obtained by the ladders placed underneath. Amongst some of the Dyak tribes certain domestic occupations take place in this passage. Thus, for instance, the Dyaks on Mt. Kĕlam pound their rice in it, but several other tribes, for instance the Oeloe-Ajers and the Dyaks on the Sibau, perform this operation underneath the house. All the family apartments open directly into the passage, generally by a swinging door, but occasionally by a sash-door. The doors and the beams which support the ceiling are often ornamented with carved figures of animals, generally crocodiles.

Above these apartments are the attics, divided similarly into

---

1) There was one remarkable exception to this custom in the house of the Ot-Danoms of Mĕrakau on the Jĕkawai (see page 324) where there was a separate shed built for the skulls quite close to the house. I know of no other instance where this is done.

one general and several private rooms. The general attic is the store-room; there one sees the rice-barrels made of bark, also weapons, fishing and hunting implements. Each family uses that part of the general attic adjoining their private apartment; there are no partitions to keep these stores separate. Ladders lead from the different private quarters below, through holes in the ceiling into the carefully partitioned private attics immediately above each dwelling. These serve as bed-rooms for the young unmarried women and sometimes for the young men. The roof projects about 2 feet beyond the walls, and is chiefly made of palm leaves, siraps and bark. The house receives its light from a few low sash-doors or windows through which one can creep from the public room to the open-air veranda, and the private apartments have little sky-lights cut out in the roof, which can be covered up. According to European notions however the light in the houses is wholly inadequate. In the evening the inside of the house is lighted up by the wood fires and dammar torches, but during the last ten years very simple paraffin lamps have been introduced amongst several of the Dyak tribes.

The houses are always built on posts of heights varying according to the district. On the Mandai river I saw houses on posts from 4 to 5 metres high and in the Batang Loepar districts the houses often stand only about a man's height above the ground. The space under the house is utilized for different purposes. With the Batang Loepars, Kantoeks, and several other tribes, it is one large dunghill, where the pigs and the dogs live; sometimes also the fowls roost there. Practically all the refuse is thrown down through the crevices of the balé-balé, which is covered with thick mats to keep out the bad smells. Other tribes, for instance the Dyaks on the Sibau river, and also, though not so generally, those on the Mandai, use the greater part of the space under their houses for storage or for carrying out certain domestic duties, such as rice-pounding, etc., and the remaining space is then used for refuse. On each side of the house is a steep ladder, or rather a tree-trunk with notches, by

New Sandong, Kasoengan.

which one enters the house. On the top of the ladder either a face or a female figure is generally carved. Sometimes, at least in the smaller houses on the Kapoewas, the ladder and the entrance hole are placed near the middle of the open-air veranda.

An exception to this type of building is the fortified house, the most typical specimen of which I saw on the Mĕlawi. The space under the house is enclosed by palisades, armed on the outside by sharp-pointed, bent, rattan tops. The stairs are under-neath the house, and a small close-fitting door in the fence leads to a tree-trunk which takes one across the dunghill the house to the ladder. In other respects the arrangement and

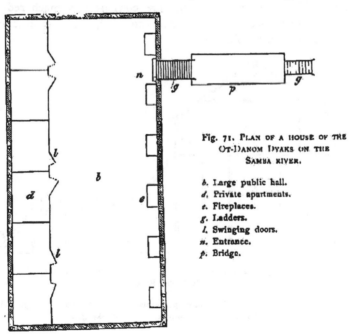

Fig. 71. PLAN OF A HOUSE OF THE OT-DANOM DYAKS ON THE SAMBA RIVER.

b. Large public hall.
d. Private apartments.
e. Fireplaces.
g. Ladders.
l. Swinging doors.
n. Entrance.
p. Bridge.

division of the house are as described above. The houses of the Ot-Danoms on the Upper Mĕlawi and its tributaries are almost all built in this style (Fig. 68). The Oeloe-Ajer-Dyaks belonging to the same tribe but living north of the Madi plateau have done away with the fortifications and enclosure of the space underneath their houses, but the stairs in the large house, (at N⁺ Raoen there were three) are in the same position under-neath the house. The passage along the private apartments is also preserved.

The houses of the Ot-Danom Dyaks on the Samba river are altogether different to the type of building described above. Figs. 71 and 72 give plan and section of the house Tandok, close to the mountain of·that name, which may serve as a typical speci-men. The public room occupies almost two-thirds of the entire space. The private apartments are consequently much reduced in size. They are directly under the roof, and the attic only extends over the public room.

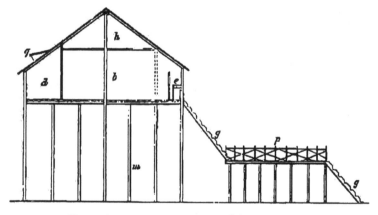

Fig. 72. SECTION OF A HOUSE OF THE OT-DANOM DYAKS ON
THE SAMBA RIVER.

b. Large public hall.	*d*. Private apartments.	*c*. Hearths.	*g*. Ladders.
h. Attics.	*m*. Posts.	*p*. Bridge.	*q*. Skylight.

The domestic hearth is not in the private apartment but in the general room, and every family has a fireplace which can be locked. This is placed opposite to the door of their private room, which is used exclusively as bed-room and for storing their valuables.

This arrangement necessitates the performance of all household duties in public, and the big room is from morning till night and often far into the night, the scene of much bustle and activity. The passage in the centre is done away with, and there is only one exit at the front of the house. The staircase leading down from it, does not go direct to the ground but rests on a sort of bridge or platform, from which a second staircase lands one on the ground close to the river. There is no open-air veranda, and the domestic duties which in West Borneo are performed on the open-air veranda, are here done on the bridge between the two staircases. Three types slightly deviating from this plan of architecture, are illustrated in figs. 63, 64 and 68. They are taken from the house Nᵃ Bawai on the Mělawi, the house Nᵃ Měroeboei on the Lěkawai, and the house Solien on the Samba, not far from Toembang Habangooi.

On this day we passed several hills of the Mt. Tandok type, October 22. which line the meandering course of the Samba. At the bends of the river the banks were generally steep and rocky and the stream-bed full of boulders. These karangans were particularly rich in petrified wood. The solid rock on the banks was of volcanic origin, principally andesite or andesite-tuff. The most beautiful spot seemed to me the place where the river cuts through the Tapi range of hills, so called after Mt. Tapi, 227 metres high, on the left shore of the Samba. The rocks on the bank are here 7 metres high and are composed of doleritic basalt. The hills of the Tapi range are conical in shape but much less steep than Mt. Tandok. They are without any doubt of volcanic origin, those nearest to the Samba consist of doleritic basalt. Below the Tapi hills andesite and andesite-tuff again prevail along the banks, they are covered in many places with loam and gravel and a surface layer of sand. At 11 a. m. I reached the place where the Běrahooi joins the Samba. It is the most important tributary of the Samba, carries very nearly the same volume of water,

and at the confluence is almost as wide as the Samba itself.
The Běrahooi is supposed to rise on the mountains north of
Mt. Moehoet, and is largely fed from Mt. Moehoet itself. It
is an important water-way between the Samba and the Upper
Kahajan. Schwaner[1]) tells us, upon native authority, that by
following the Běrahooi for half a day's journey up-stream one
comes to Toembang Papo on the left bank. Following the
Papo for half an hour longer one can land and walk across
the water-parting, which is so narrow that the boat has only
to be dragged, for about 20 paces across it to the Patangoi
river, by which one can then reach the Kahajan in a day and
a half. I was unable to procure any further reliable information
concerning this route.

There had evidently been much rain in the basin of the Běrahooi for the river was swollen and muddy. Above Toembang
Běrahooi the Samba flows, for a distance of 28 kilometres, in
a due south-easterly direction, but at its confluence with the
Běrahooi it curves to the south and follows the current of that
stream which is at first south-south-west, and then changes to
due south, within $4^1/_2$ kilometres below Toembang Běrahooi. At
first the rock *in situ* continues to be of a volcanic nature. The
rocks, Batoe Amboe and Batoe Lěbang Pohien, consist of ande-
site-tuff-breccia, and lower down, amphibole-andesite comes to
the surface, forming steep cliffs at Noesa-Běkakkie on the right
shore, and elsewhere. Still lower down, about 7 kilometres below
Toembang Běrahooi, I had reached the southern limit of the
volcanic area on the Samba.

Here begins again a large granite area which extends as far
as the mouth of the Samba or possibly still further. Exactly $7^1/_2$
kilometres below Toembang Běrahooi and just above the island
Hamoessang, the first granitic rocks were exposed in the bed of

1) C. A. L. M. Schwaner 55, II, p. 122.

This statement does not strictly agree with what is said in the same book II, p. 70.
There we are told that the Papo does not flow into the Běrahooi itself but into the Tan-
gooi which discharges into the Běrahooi.

the river, where they create a rapid. Together with the rocks on the right bank they consist of amphibole-granitite, and one kilometre further on tonalite is shown in the cliffs. After that the river flows due south through a hilly granite district. Rocks are of frequent occurrence in the stream, creating a succession of falls, the principal of which is Kiham Pĕnĕkilloe. Although the current is very rapid, none of these falls are dangerous. The principal rock-groups in this granite area have separate names, such as Batoe Barabai, Batoe Embak and Batoe Himba, near the rapids of the same name. Fully a kilometre below the entrance to the Pĕnĕkilloe, the river bends to the east, and navigation is made almost impossible by large diorite rocks which project from both shores while in midstream detached rocks form a kind of dam leaving only a very narrow channel for the water to flow through, which it does with furious velocity. This point is called Kiham Habida and is full of danger when going up-stream. Next we passed the mouth of the Tĕlangan Koedjoe opposite a group of rocks composed of granite with dioritic concretions. The Samba then resumes its course southward, the hills retreat on either side, and reveal a wide plain, which, as far as one can see, has been entirely brought under cultivation. We soon caught sight of the principal settlement on the Samba, Toembang Mentikéh. Small floating houses, and bathing rafts on the left shore came first within view. They stamped the part at once as a Malay settlement, but in reality the population of Mentikéh is a mixed one. The Dyaks inhabit a large house on the neck of land at the confluence of the Mentikéh and the Samba. The Malays live partly on the left, partly on the right bank, near the Balei [1]). I was pleasantly surprised to see the Dutch tricolor fly from the Balei; it gave me the comfortable assurance that the unknown interior lay behind me, and that I was once more within the domain of Dutch power. The chief of this settlement, who bears the title of toemenggoeng, is the

---

1) The Balei is the place where travellers and merchants can lodge, also the place of judicature.

representative of the Dutch Government on the Samba river. He is under the district head at Kasoengan on the Katingan, who in his turn is subject to the controller of Sampit. I took up my abode in the Balei where I met a Malay merchant who had a bĕlanga ¹) worth 4000 florins, which he was trying to sell to one of the Dyak chiefs in the neighbourhood.

This Dyak house, the residence of the toemoenggoeng MAKKOO is a most interesting specimen of architecture (see Pl. XLV). In the open space between the house and the river there are eight memorial poles ranged in a row; they are here called sĕpandoks and they resemble the poles in front of the house Mĕrakau, described on page 325. Exactly in front of the house is a sandong raoen (Pl. XLVI) ²).

The Mentikéh is the most important of the right tributaries of the Samba. It rises in the hilly district south of the Raja mountains. This river is difficult of navigation and only very small sampans can go up it, as the rapids commence close to the mouth. The first of these, Kiham Pĕning, one metre high, is caused by rocks of amphibole-biotite-granite, standing out above the surface of the water. At a place three day's journey up-stream beds of iron ore are found, and in time past the Dyaks used to smelt the ore and make very excellent parangs of it, the fame of which spread over the whole of South Borneo. But since large quantities began to be brought over from Bandjermassin into the higher districts, this industry has flagged considerably. Yet even now parangs made of the Mentikéh ore fetch very high prices

October 23. At 8 in the morning I continued my course down the river; the water was fairly high and we ran down at a good speed. The banks consist of sand and gravel and they are generally wooded to the water's edge. Here and there the solid rock crops out through these fluviatile deposits, forming groups or islets in the stream-bed and cliffs along the banks. Rapids are

---

1) Bĕlanga (Mal.) = kind of tĕmpajan.
2) Sandang raoen: Shed of curved wood, in which the corpse of a chief is buried.

numerous in such places, but they offer no great difficulty for a well-managed sampan. Amongst them are the Kiham Hĕbilit, not far from Toembang Mentikéh, and the Kiham Habangang, just above Toembang Tĕloenai near the island of that name, and further on the Kiham Haboi, 1½ kilometre below the Tĕloenai, and the Kiham Liboeng, near the island Liboeng, another kilometre lower down. The rocks in all these places are amphibole-granitite rich in plagioclase. The Tĕloenai forms part of a line of communication between the Kahajan and the Samba rivers. From SCHWANER [1]) we learn that: "From the Kahajan it is one day's journey up the Maratja to a point on the water-parting where it is only 40 fathoms wide. There the boats are lifted out of the water, and launched again on the other side in the Tahoijan river, leading into the Troenai (= Tĕloenai) and eventually into the Sampa (= Samba). The travelling down the Tĕloenai takes two day's more."

Up to the present time this appears to have been used as the regular route of communication as may be gathered from the following account of the missionary TROMP [2]): "About 20 day's rowing from Pangkok (near the mouth of the Kahajan) is the entrance to the little river Maraja. Some way up this streamlet, one has to go ashore and drag the boats for a short distance over land to the river Tĕloenai, leading to the Samba and the Katingan."

On the banks of the Samba, below Toembang Mentikéh, are several deserted and often ruined houses, built on the slopes of the low hills close to the river side. These houses are generally surrounded by a large number of toras and memorial pillars. A wide margin along the bank shows evident signs of former cultivation. Below Toembang Tĕloenai, and on the right bank are a few Dyak graves, and on the left bank is a place for cremation. Below Noesa Liboeng, amphibole-porphyrite is exposed; this has

1) C. A. L. M. SCHWANER 55, p. 70.
2) H. TROMP 55, p. 68.

probably forced its way through the granite, because a little lower down, the same kind of granite as that at Noesa Liboeng, is again seen. From this latter island to the mouth of the Pattinnéh river, the general direction of the river is to the west, but at this point it bends round and resumes its course to the S.S.E. The long and fairly straight piece of river below Toembang Pattinnéh runs through especially flat ground, shut in towards the south by a low range of hills which we reached two kilometres further down. At Sěbanja, where the Samba breaks through these hills, both the banks and the river-bed consist of granitite containing much amphibole. As far as Toembang Pěssil the river then runs almost due south, the hills retreat more and more from the waters edge and just below Toembang Pěsseer or Pěssil, $38^1/_2$ metres above sea-level, the Samba widens rapidly and flows perceptibly slower. Here also on the convex sides of the bends in the river, long rushes begin to line the banks, a sure sign that the upper course is past.

Another way to reach the Kahajan is by the Pěssil. TROMP tells us [1]): "From Toembang Talaken on the Kahajan one can reach the little village of Gahis in two short day's journeys per boat, and from there a footpath leads in one day to the river Pěssil, close to the village of that name. This little river flows into the Samba." SCHWANER also mentions this same route (l. c. II, p. 122).

Below Toembang Pěssil the solid granitite only once crops out through the alluvium, near the house Měngkiang, where the river breaks through a low line of hills. For the rest the ground is quite flat, and the banks are composed of gravel, covered with loamy sand.

About three kilometres below Měngkiang, the Samba describes a large loop-shaped curve, on the east side of which the Saki is supposed to have its outlet. It was by this latter river that SCHWANER, in December 1847, came from the Kahajan down the

---

1) II. TROMP *58*, p. 69.

Pl. XLVIII.

Old Sandong, Kampong Kasoengan.

Samba, and finally into the Katingan. In the thickly wooded banks I have possibly overlooked this outlet, I have therefore marked it on map X^c in the same place where it is indicated on SCHWANER's map. Fully 10 kilometres further down, is the confluence of the Samba and Katingan. This point we reached on Oct. 23^d at three o'clock in the afternoon. In many respects this beautiful spot reminded me of Sintang in West Borneo, where the Melawi joins the Kapoewas. On map X^c the position of this place, Toembang Samba, is given as $1°25'30''$ S. Lat. and $113°3'30''$ E. Long. from Greenwich. Formerly [1]), at any rate for the latitude, I had followed SCHWANER's [2]) indications viz., $1°30'$, but when compiling my maps I found that the figures given by MICHIELSEN [3]), corresponded more closely with my own observations, viz., $1°25'30''$. As a matter of fact the true astronomical position of Toembang Samba is as yet unknown.

The Samba is the most important affluent of the Katingan and the volume of water it carries is about half of that of the principal stream. The width of the Katingan before the Samba joins it, is about 250 metres, and the Samba is about 100 metres wide. The actual length of the Samba from Toembang Mènjoekoei down to the mouth is 130 kilometres, and its entire actual length, I estimate at 158 kilometres. This river rises on the southern slopes of the Raja mountains. For steam-boats of slight draught it would be navigable up to the island Sèbanja, but for small rowing boats, as shown in my journal, it is possible to get past Toembang Mènjoekoei without meeting with any serious difficulties. And finally the Mènjoekoei and its tributary the Rassahooi are navigable for canoes up to the foot of the boundary mountains of West Borneo, although the difficulties are there of a somewhat serious nature.

The Samba is one of the principal means of communication between South and West Borneo, and several forest tracks lead

1) G. A. F. MOLENGRAAFF *41*, p. 201.
2) C. A. L. M. SCHWANER *55*, p. 119.
3) W. J. M. MICHIELSEN *39*, p. 76.

from the head-waters of the Samba to those of the Kahajan
and the southern tributaries of the Mĕlawi. Most of these tracks
were known to SCHWANER[1]), on native authority. Thus, for instance,
the path which I followed from Mĕroeboei on the Lĕkawai
(SCHWANER calls it Rakauwi) past Liang Kĕnangka (SCHWANER's
Liang Nangka) is carefully mentioned by him.

It is a fact well worthy of notice, that a track which might
be called an important road of communication, is not traversed
oftener than perhaps by about 100 individuals during the course
of twelve months, and that in consequence each time it is used
it has to be opened up afresh. Moreover this same track may
not perhaps by any means be the most direct or the best route
to a given place, but in spite of this it continues to be used,
until some very urgent reasons compel the Dyaks to alter their
route. I noticed this more than once in Borneo. This procedure
coincides with their generally conservative notions and their great
respect for the works of their elders and ancestors.

The Samba has been but little known up to this time. SCHWANER
only visited the portion of the river between Nangah Saki and
the mouth, and MICHIELSEN[2]) and HENDRICH[3]) did not travel much
beyond that point; the former reached as far as the Kampong
Asem Koemboeng, a few kilometres beyond Toembang Péssil,
the latter, as far as Tampak which is supposed to be just below
Toembang Péssil. But no European appears to have advanced
any further up the Samba [4]).

Toembang Samba is a very important settlement. The left
shore, close to the confluence of the two rivers, is chiefly inhabited
by Mohammedans, most of them natives from Bĕkoempai (Lower

1) C. A. L. M. SCHWANER *55*, II, p. 154.
2) W. J. M. MICHIELSEN *39*, 1880.
3) HENDRICH *21*, p. 372.
4) The natives certainly did tell me that some years ago a European had travelled in these
regions; that he went up the Samba and by the Bĕrahooi to the Kahajan. I heard afterwards from the missionary TROMP, that an English bible-seller, IRVING by name, had visited
the Samba, no doubt this was the same man of whom the natives had spoken. I could,
however, obtain no very clear information as to his movements.

Barito). Higher up on the right shore of the Katingan lies the
very extensive Dyak kampong, occupied by Katingan Dyaks and
a large contingent of Dyaks from the Lower Kapoewas Moeroeng (South Borneo).

Below Toembang Samba the Katingan becomes a majestic
stream about 300 metres wide and carrying a volume of water
equal to the Kapoewas above Sintang. I did not visit the Katingan up-stream from Toembang Samba, but we have a fair
knowledge of it from the elaborate and trustworthy accounts of
SCHWANER[1]) and the circumstantial, unadorned description of
MICHIELSEN[2]). Both accounts state that the Katingan, beyond
Toembang Samba, flows alternately in an eastern and a south-
eastern direction, and that there are several dangerous rapids,
marking the spots where the river breaks through granite hills.

I did not tarry long at Toembang Samba and after dismissing
my escort from the house Solien, I continued that same day to
travel down the Katingan. The shores immediately below Toembang Samba are low, but not far from the present bank are the remains of an old one indicating a much less sinuous course of the
river than it now describes. The old bank is more distinct on
the right than on the left. At Toembang Pělawai, almost opposite the kampong Giring, the river flows for some little distance
along one of its old banks, which is here composed of a bed of
bluish clay (turning yellow when decomposed) $1^1/_2$ metre thick,
upon which rests a layer of coarse gravel one metre in depth,
containing much silicified wood. Covering the gravel is a layer
of loamy sand, 2 metres thick. Opposite the large kampong
Pěndahara much further down-stream, the old bank once again
approaches to within 35 metres of the river. It is there still $5^1/_2$
metres high and consists of fine gravel mixed with sand and
clay. The components of these ancient banks have a decided
fluviatile character. We may conclude therefrom that, formerly,

---

1) C. A. L. M. SCHWANER 55, II, p. 120 and following.
2) W. J. M. MICHIELSEN 59, p. 34 and following.

the Katingan could carry gravel and sand in places where at present its action is merely of an eroding nature and its carrying power confined to fine sand and mud. This may date from the time when the mountains of the Upper Katingan, for instance, were less denuded than they are now, and the general fall therefore must have been considerably greater. The existence of these fluviatile deposits pleads against the theory of recent submersion by the sea, and consequently of any negative shifting of the sea shore in recent times; theories which are often accepted for South Borneo. The remainder of the voyage down the Katingan is for the geologist most uninteresting. With many meanderings the river flows down to the Java Sea, over flat, low and occasionally marshy ground, and the six days which I spent in my sampan are amongst the dullest and most unpleasant of my Borneo recollections. I had nothing to do, and was unable to leave my boat. There I sat, waging a constant but unsuccessful war against the milliards of mosquitoes for which this river is justly notorious. I had one little diversion at Kasoengan where I paid a short visit to the chief of the district DĔMANG ANOEM TJAKRA DALAM.

The kampong Kasoengan is situated on a narrow strip of land, shut in by a gigantic bend of the river. Although on foot one can cross this neck of land in a few minutes it takes quite half an hour's hard rowing to get round it by boat. The Malay occupy the smaller, the Dyaks the larger part of the kampong. The Dyak kampong Kasoengan, is the largest and most flourishing on the whole of the Katingan. The ladders leading up to the different houses are frequently ornamented with elaborate carving. Here and there beautiful sandongs stand in front of the houses flanked by one or two rows of pantars. Several particulars are shown on the plates XLVII, XLVIII, XLIX, L and fig. 73 in which different views of the kampong are illustrated.

The district chief of Kasoengan is a Christian and by the side of his house is a small school-house where a goeroe (native teacher) opens school every morning with the singing of a psalm. This school was founded by the Rheinish Mission, which has its head-quarters at Bandjermassin. The hospitable chief offered me his commodious bidar to continue my journey in, and I gladly accepted his offer. The bidars of South Borneo are much more elegant in shape then those used in West Borneo, and are not unlike Venetian Gondolas, (see plate LI). They are invariably built of ironwood, hence of a beautiful colour and quick

Fig. 73. View of the kampong Kasoengan close to the office of the chief of the district.

of motion. They have one drawback however. The great weight of the ironwood makes them float low on the water, and they are easily swamped. Moreover the ironwood, although very hard and strong, is neither pliable nor tough, and the bidars are not proof against serious shocks and bumps. They are only suitable in the lower courses of the river, which are free from rapids, and in case of a heavy gale the greatest caution has to be observed.

On Oct. 28th I arrived at Mëndawei, 28 kilometres distant from the mouth of the Katingan, and the next day we reached Pëgattan close to the estuary. The kampongs here are extremely dirty, and the inhabitants, mostly Malays, seemed to take a pride to rival with one another in the slovenliness of their surroundings. I delayed here as short a time as I possibly could, the more so as a malignant epidemic beri beri was raging, which had already carried off one-third of the population. The progress of the disease was in this instance so rapid that the victims often succumbed within 24 hours of the attack.

There was no great choice of vessels to carry me from here to Bandjermassin, and I had to content myself with a miserable unseaworthy sailing boat, in which I left Pëgattan on October 30. We proceeded by the strait between Dammar island and the mainland, and safely passed the shallows at the mouth of the estuary, which prevent ships of more than six feet draught from entering the Katingan. We were now in the large bay of Së-bangan, which is only separated from the wide, deep bay of Bandjermassin, by a rocky promontory, cape Melatajor. In this latter bay the Kahajan, Kapoewas Moeroeng and Barito, all discharge. Past cape Melatajor a terrific storm, such as heralds the approach of the west monsoon, brought us in great danger, but it turned out to my advantage in the end, for the storm drove us in the right direction and I reached Bandjermassin early in the morning of November 1. Under unfavourable circumstances these little native boats often take ten days to go from the Katingan to the mouth of the Barito.

I left Bandjermassin again the next day and sailed for Soerabaja, then travelled to Batavia across Java and afterwards back to West Borneo, in order to arrange my collections previous to my return to Amsterdam, where I arrived without further incidents in January 1895, with all my collections in splendid condition.

CHAPTER XI.

GEOLOGICAL DESCRIPTION OF THE SECTION ACROSS DUTCH
BORNEO, FROM NORTH TO SOUTH.

(Compare Atlas page 18. Sections AA', DD', EE' and FF').

Our starting point will be from (section AA') the top of Mt.
Tjondong (1242 metres) near the frontier between West Borneo
and Sarawak. This is one of the highest peaks of the various
chains which compose the Upper Kapoewas mountain range. These
mountains consist of clay-slate, mostly phyllitic clay-slate with
well marked silky lustre, quartzite, quartzitic sandstone and greywacke-slate, the strata of which are strongly folded, plicated,
disturbed and intersected by quartz-veins. I have called this formation the old slate-formation. Its age is uncertain because of
the absence of fossils, but it is probably older than the cherts
with Radiolaria, the pre-Cretaceous, possibly Jurassic, character
of which, has been proved.

To the north this formation probably extends for some distance
into Sarawak. But if we follow our line of section we find that
44 kilometres more to the south the Upper Kapoewas mountain
chain, and, simultaneously with it the old slate-formation are
suddenly cut off. Almost immediately the mountains here pass
into the Upper Kapoewas plain. Where our line of section cuts
through this plain it is almost entirely covered with recent fluviatile sediments, but we may conclude from observations made

in the lake district on the one side, and at the sources of the Kapoewas on the other side, that these river deposits rest on strongly denuded strata of the pre-Cretaceous Danau formation.

In the Kapoewas plain the surface only shows alluvial mud and fine sand, but 68 kilometres more southward, along the banks of the Boenoet, the Lower Tébaoeng, and the Sèbilit, we come again upon solid rock, sandstone and clay-slate, the strata of which dip feebly to the north. (Sections AA', DD' and EE'). Seams of coal are interstratified in the youngest, upper part of this formation on the Lower Tébaoeng and the Sèbilit. This formation rests unconformably on the Danau formation and was probably formed in the Tertiary period. Nine kilometres south of the Boenoet river, andesite has broken through this sandstone. A range of cone-shaped andesite hills stretches in an east to west direction, to which Mt. Oendau on the Sèbilit and Mt. Loeboek on the Tébaoeng belong, while in the vicinity of these mountains numerous dykes of andesite were seen in the river-beds cutting through the formation. This andesite, which almost invariably contains quartz, has an ancient type, and might perhaps with equal right be determined as amphibole-porphyrite. On geological grounds, however, I surmise that these andesite hills are the remains of more extensive volcanic formations, of which they form the deep-seated parts which until now have escaped denudation. This explains why they resemble the deep-seated plutonic rocks in some of their peculiarities, and in other respects bear the character of dyke rocks.

At a distance of 5 kilometres south of Mt. Loeboek (see Section EE'), an older formation appears from underneath the Tertiary sandstone, with strongly folded, nearly vertical strata. On the Tébaoeng its consists of silicified diabase-tuff, and diabase-porphyrite which I classify with the pre-Cretaceous formation because their petrographic character bears a strong resemblance to the rocks of this formation found in the Lake district. The trend of the strata is west-north-west. On the Sèbilit near Mt. Oendau (see section DD') an older formation also appears, namely,

Pl. XLIX.

HOUSE AT KÄSOENGAN.

strata of amphibolite and glaucophane-amphibolite, the beds standing almost vertical with a west-north-west strike. They closely resemble the amphibolite in the hills near Sěmitau. Provisionally I have considered these rocks to be archaean crystalline slates, although on purely geological grounds they might, together with the rocks of Sěmitau, be regarded as eruptive rocks belonging to the family of the diabases which have been uralitized and altered by mountain pressure, and thus belonging to the pre-Cretaceous Danau formation.

The pre-Cretaceous deposits on the Tebaoeng are immediately followed by strata of coarse, greywacke sandstone, standing perpendicular with exactly east to west trend, but apparently less altered by mountain pressure than the layers of the Danau formation. Their striking petrographical resemblance to the greywacke sandstone along the Upper Kapoewas and the Sěberoewang, whose Cretaceous age is ascertained, compels me, provisionally, to accept a Cretaceous age for these beds also. The sudden abutting of the strata of the Danau formation with a west-north-west trend against the banks of sandstone which trend from east to west, (compare separate plan on Map IX⁴) could then be explained by admitting that these latter, even before their upheaval, rested unconformably on the, at that time, already disturbed pre-Cretaceous strata.

On the Sěbilit as well as on the Tébaoeng the tract where the Cretaceous and pre-Cretaceous deposits come to the surface is the region of the dangerous rapids. As soon as, above goeroeng Hěnap, the older formations again disappear under the younger sandstone with feeble northerly dip, resting unconformably upon them, the water becomes calmer, and where the sandstone lies undisturbed, no further serious difficulties present themselves.

Following the section it appears that the position of the Tertiary sandstone remains the same as far as Riam Pandjang. Here, however, strong distortions occur, causing the sandstone to stand either vertical as in the Riam Pandjang, or else to have a steep incline as in the Riam Bĕnoewang. Disturbances quite as strong as these occur further south where the line of section cuts through Mt. Bĕransa. Here also, the banks of sandstone are generally vertical till close to the river Tĕbaoeng Tĕnang, where the vertical position quickly passes into a feeble south or south-east slope, then into a horizontal position, and shortly after changes into the feeble northerly incline prevailing in this district. The strike remains almost exactly east and west. As indicated in the section I do not separate this strongly tilted sandstone of Mt. Bĕransa from the undisturbed sandstone which I have called Tertiary although I consider the vertical beds of sandstone near Riam Nĕkan to be of older date. The petrographical characteristics of the sandstone of Mt. Bĕransa have induced me to do so. The rock is coarse grained, friable, with a very poorly developed cement of clay between the quartz grains. This characteristic, and also the presence of pink quartz-grains, it has in common with the sandstone to the north of the Riam Pandjang and with the sandstone of the Madi plateau, both of which lie in an almost undisturbed position. The sandstone of the goeroeng Nĕkan, on the other hand, does not show these pink quartz-grains, and it is further characterized by a strongly developed, ferruginous clay-cement, which I also found in other parts of Borneo in sandstones of undoubted Cretaceous age. I am fully aware that these arguments are far from conclusive, but it stands to reason that in a first journey of exploration the classification of sedimentary formations must of necessity be somewhat vague, especially where fossils are absent. The highly inclined position of the beds of this Tertiary sandstone near Mt. Bĕransa is, in my opinion, caused by the same series of disturbances, which produced the great faults trending from east to west, by which large masses have been let down to a much lower level than in the territories adjoining to the north.

The Madi plateau, which we now reach, like the sandstone and the clay-slate of which it is composed, has a slight incline to the north. The highest part forms a marshy, trough-shaped plateau, situated between Mt. Babas Hantoe and Mt. Madi, which is covered with a tropical peat formation. The Madi plateau ends suddenly to the south forming sheer precipices, and it appears as if the block of sandstone with a northern incline is here abruptly broken off. I believe that this precipice is the flank of a great fault, the Madi fault, which has been shifted northward by erosion and denudation. The continuation of the sandstone of the Madi plateau must now be looked for much further to the south at a lower level. On the steep southern slope of Mt. Madi we find along the Midih river underneath the Tertiary sandstone, older sediments, disturbed and often strongly folded, of the same character as those found along the Panai and Kĕremoei rivers. This formation consists of sandstone, arkose, quartzite and clay-stone; the strike of the strata is about east to west. These rocks bear a great resemblance to the deposits on the Sĕbĕroewang, of Cretaceous age, and I have therefore considered them as belonging to the same period.

Near Nᵃ Gilang these disturbed deposits suddenly give place to beds of sandstone and clay-stone lying almost horizontal, which bear a striking resemblance both in character and position to the sandstone of Mt. Madi, only that the level is about 1000 metres lower. I am of opinion that the fault, for which I have adopted the name of the great Madi fault, which has thrown the sandstone down to this level, and which therefore, in principle, originated the southern declivity of Mt. Madi, which is now shifted much more to the north, runs near Nᵃ Gilang in an east to west direction along the Mĕlawi valley.

Following the line of section past Nᵃ Gilang along the Mĕlawi we find Tertiary sandstone and clay-stone lying horizontally, or with a feeble southward incline. Further south, along the Lĕkawai, these beds slope to the north, in consequence of which the Tertiary formation in the Mĕlawi valley has a trough-shaped position. Near riam Batoe Tossan badly preserved fossils were found in the clay-stone, which KRAUSE considers to be of Tertiary age. This clay-stone here rests on diabase-porphyrite and olivine-diabase, closely allied to basalt. A little further on, near Liang Bohies, clay-shale with numerous shells appears under the sandstone. The palaeontological examinations of KRAUSE[1]) and MARTIN[2]), proved these to be brackish-water deposits of old-Tertiary age. At first KRAUSE believed them to be of upper Cretaceous age, as he wrote to me in a letter dated October 21, 1896. Accordingly on maps I and IX, and in the section EE', these deposits were separated from the Tertiary formations in the Schwaner mountains, and in order to explain their position I was obliged to infer the existence of important dislocations in the Lekawai district. After the maps were printed in this manner KRAUSE, as the result of further research, altered his opinion. Therefore, when studying maps I and IX and section EE', we must look upon the deposits of the Lekawai and of the Schwaner mountains as belonging together, ignoring the faults there indicated, as given below in a simpler form.

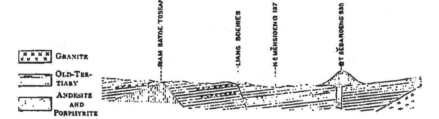

A little more to the south, near Meroeboei, these deposits are once more replaced by sandstone and clay-stone, which, with the exception of a few local disturbances, have a northerly incline and are most likely also of Tertiary age. From here to the watershed there stretches a plateau feebly sloping to the north-north-east, which in many respects is a repetition of the Madi plateau. There are

1). P. K. KRAUSE, 37, p. 169—220.
2) K. MARTIN, 38, p. 257—315.

however a great many dislocations trending from east to west, of the same type as those observed in the section near Liang Bohies (map IX^c, section H). Some of these lines of fracture seem to be of exceptional importance, eruptive rocks having forced their way up through them. These are porphyrites, which I think we may take to be deep-seated representatives of andesites. A range of conical hills trending almost due east and west shows the direction of these fractures. To these hills belong Mt. Sĕdaroeng (930 metres), Mt. Lĕmoekoet (1001 metres), Mt. Mĕmboeloe or Moeloe (1268 metres). In some places this range of andesite hills is duplicated, in other places there are great gaps. It is not yet known how far they extend to east and west.

The sandstone plateau terminates in the south in a very steep flank. The average height of the southern edge of the plateau is about 800 metres. Then follows a steep incline averaging 600 metres in height, strongly reminding one of the southern slope of the Madi plateau, and which brings us into the hilly district of South Borneo. Near Mt. Boenjau (Section FF'), this declivity consists entirely of sandstone and clay-stone, and not until some distance further south, does granite appear below the sandstone. The structure of the Raja mountain-range is a little more complicated (Section EE'). This mountain consists of granite broken through in many places by porphyritic rocks of which, amongst others, some of the highest peaks of Mt. Raja are formed. This porphyrite shows different types, which, provisionally, I have included under the common name of Raja-porphyrite. On the north flank of this range lies the sandstone with a gentle northern incline, which causes the northern slope of the mountain, from a height of about 1600 metres, to be very feeble and gradual. On the south side there is no trace of sandstone, and the steep granitic mountains rise abruptly from the hilly country of South Borneo. It is remarkable that on the Raja mountains and in their immediate vicinity, sandstone is found at a much higher level (up to 1650 metres) than in a more easterly direction (± 800 metres), and that the declivity which terminates the sandstone plateau towards the south, here projects considerably southward. It is fair to suppose that the sandstone plateau originally stretched much further southward, and that the steep southern incline has been shifted through erosion northward, to the place where it now exists. We may also conclude that the granite and porphyrite mass of the Raja mountains has offered serious resistance to the eroding influences, and, as a consequence, the sandstone plateau, where it rests against these mountains and in their immediate vicinity has suffered much less from erosion and has not been shifted northward to nearly so great an extent.

Summing up our observations on the vicinity of Mt. Boenjau and the Raja mountains, we may say that the boundary mountains between South and West Borneo have a typical monoclinal build. The main part of this mountain range is a plateau built up of sandstone and clay-stone forming a system of strata of about 600 metres in thickness. This plateau feebly inclines to the north but terminates to the south in a very steep slope. This difference of slope is the origin of the monoclinal structure of the range. The sandstone and clay-stone rest on granite of which the hilly district of South Borneo to the south of this range is principally composed. In most places the granite occupies a lower level than the sandstone, but in the Raja mountains granite and porphyrite rise above the highest tops of the sandstone plateau and where this occurs they have protected the plateau against the eroding influences of the streams running southward.

This monoclinal boundary range between West and South Borneo I have called the Schwaner mountains [1]) after Dr. C. A. L. M. Schwaner, the first European, who coming from the Katingan river, in January 1848, crossed this range, entering it by the sources of the Sěnamang, and afterwards reaching the Kapoewas river by the Sěrawai and the Mělawi.

About the geological structure of the Schwaner mountains we possess little knowledge but Schwaner has given us a few particu-

---

1) G. A. F. Molengraaff 41, p. 207.

lars, and a collection of rocks from the mountains west of the Sĕra-wai has also been put at my disposal [1]). From these we gather that granite and numerous varieties of hornfels, chiefly andalusite-biotite-hornfels, that is to say sedimentary rocks of unknown age altered by contact with the granite, play there a more important part than further eastward, and also that sandstone and clay-stone rest almost horizontally on these formations.

Summarising from SCHWANER's description of these mountains, he says: "The bulk of the geological formations consists of coarse grained granite. In the north and the south, sandstone formations of little cohesion, alternating with clay-stone, rest on the granite, which up-heaved them. The highland slopes to the south in terraces following each other in a series of steps, which succeed one another more rapidly than on the north side, where the transition from the highland to the plain is more gradual. Steep mountain slopes terminate on high plains, on the north side of which terrace follows upon terrace until they finally lose themselves in the plains below. These terraces are called by the natives tata (= tatai) i. e. steps. Bogs and extensive bamboo forests are character-istic features of the tatas. The regularity of this terrace formation is often interrupted however, by isolated heights or groups of peaks, and the highest points to which the ground rises on the summits of the mountains, are as usual not found within the line of the watershed, but rather in the mountainous districts to the north of it. This is the cause of the much interrupted flow of the waters and their apparently contrary course [2]).

SCHWANER's description which, like the whole of his work, bears the stamp of perfect truthfulness, would agree entirely with my own observations of the Schwaner mountains, only that SCHWANER must have found the tatas more distinctly developed than I did; nor can I agree with his statement that the overlying sandstone and clay-

---

[1] This collection was made for me by the surveyor WERBATA of the topographic ser-vice. The authenticity of the material and the correct indication of the places where they were found, are beyond dispute.

[2] C. A. L. M. SCHWANER 55, II, p. 171, 172.

stone have been lifted by the granite from their horizontal position.

I guess that the existence of the southern declivity of the Schwaner mountains is due to the fracture in the sandstone plateau caused by a fault trending east and west, and that the portion of the plateau which formerly extended to the south side was thrown down. I believe however that, as in the case of the Madi plateau, the fault which caused the dislocation in the sandstone plateau must be looked for in a much more southerly direction, and therefore that its flank was originally also situated there. The eroding action of the rivers flowing towards the Java sea have gradually lowered the steep incline and at the same time pushed it back a considerably distance northward. This fault, which we might call the great boundary fault, cannot as yet be located with absolute certainty, and I dare not even say decidedly that there is any connection between this great fault and the faults along which, further southward on the Samba river, volcanic formations have been found.

Returning to the section near Mt. Boenjau we find that in the hilly country of the Upper Samba, along the banks of the Rassahooi, the Mĕnjoekoei and a part of Samba the bottom is everywhere composed of granite. It is biotite-granite and amphibole-granitite, which, as already described in detail elsewhere, in many places (compare Map X) gradually pass into rocks which, according to their petrographic characteristics, might be respectively called tonalite-gneiss, mica-diorite and amphibolite. The phenomena agree entirely with observations made elsewhere, where the granite near the boundary of enveloping sedimentary rocks has sustained similar, strong endomorphic transformations, caused probably, partly by differentiation in the magma and separation of the more basic elements near the outside of the intrusive masses, partly by modifications in its chemical composition due to the influence of large quantities of enveloping rocks which during the intrusion have been absorbed by the ascending granites. It is principally owing to BRÖGGER's writings that the nature and the significance of such differentations of

Pl. 77

Sᵘ Katingan near Kasoengan.

intrusive magmas has become more generally known. We may moreover draw attention to the fact that the alterations sustained by the granites and tonalites on the Samba agree, even in details, with the observations made by SALOMON[1]) during his examination of the tonalite massive of Monte Adamello.

The suggestion that the olivine-gabbro and the olivine-hypersthene-norite found in the granite district of the Samba, may also be basic products of the differentiation of a granite or tonalite-magma, cannot "a priori" be rejected as improbable, since LACROIX'[2]) latest investigations.

In some places below Toembang Mĕnjoekoei the contact may be observed between these gneissic granites, altered by endomorphic metamorphism, and certain sedimentary formations which in their turn, by their contact with the granite, have been altered into andalusite-hornfels, andalusite-biotite-hornfels, "garbenschiefer", spotted clay-slate and andalusite-clay-slate. The age of these rocks altered by contact-metamorphism is not known. They appear in a few places along the Samba, but I never saw them in an unaltered state and no fossils were found in them. The strike of the steeply inclined strata could only be observed in a few places, where it is from north to south, totally deviating from the directions prevailing in West Borneo. Near Nᵃ Pangah, the granite formation with its accompanying contact rocks ends abruptly, and in its place we find horizontal layers of sandstone and clay-stone, which are soon overlain, and in their turn replaced along the river-banks by volcanic rocks, especially andesite-tuff, tuff-breccia, and dolerite. Numerous hills, built up of volcanic rocks, which group themselves into single ranges trending east-north-east—west-south-west, break here the monotony of the hilly granite area.

It seems to me not improbable that there may be some connection between the great boundary fault and this dislocation, where the granite stops suddenly and gives place to the hori-

---

1) W. SALOMON *49*, p. 409.
2) A. LACROIX *29*, pag. 1021 et seq.

zontal sandstone and clay formation in which I see a continuation of the presumably Tertiary sandstone formation of which the Schwaner plateau and the Madi plateau are composed. These strata of sandstone and clay, as shown in the section, lie here on a downthrown block between two faults in the granite district, and the faults have enabled the volcanic rocks to rise to the surface.

The volcanic rocks seem now almost exclusively to be found on this sunken ground, but originally they doubtless covered also a part of the granite district north of the line of fracture. It is clear however that whilst the sliding in the same direction along the line of fracture continued, the eruptive rocks on the northern wing of the fault became ever increasingly exposed to the eroding and denuding forces, while these same formations on the south side of the fracture, on the downthrown block, came gradually into a more and more favorable position for escaping the influence of these forces. The strip of volcanic territory is about 15 kilometres wide, and to the south of it amphibole-granitite, tonalite, and diorite again make their appearance, all these rocks bear a close resemblance to the predominating rocks in the granite district north of the volcanic strip. Although it is hardly probable that the granite would lie at the same level on both sides of the volcanic territory [1]) it is quite certain that we are justified to accept here a "Grabenversenkung" in the granite district. To the south of the volcanic ground the granite forms a hilly district which gradually becomes lower and lower, and is more and more covered by recent and diluvial fluviatile deposits, and finally, at any rate in the neighbourhood of the Samba river, about 10 kilometres above Toembang Samba, it disappears altogether below these fluviatile deposits.

From SCHWANER's travelling account we gather the following

---

[1]) Of course I am quite prepared to admit that the granite to the south of the volcanic ground occupies a lower level, i. e. has been shifted lower down, than the granite district to the north of the volcanic tract.

particulars about the continuation of the formation here described, in the hilly districts of South Borneo. SCHWANER found, below Toembang Mirih on the Upper Kahajan, several cone-shaped hills (he estimates the height of Mt. Pohon Batoe close to the river, at 400 feet). These hills consist of trachyte (probably andesite), and from his description they seem to resemble Mt. Tandok and the other andesite hills along the Samba. Prolonging the volcanic strip (with which we are familiar from the Samba valley) in an easterly direction, it would intersect the Kahajan in such a manner as to bring Mt. Pohon Batoe within its boundaries. It is highly probable that the volcanic strip actually continues persistently in an east to west or east-north-east— west-south-west direction and that SCHWANER struck it at Mt. Pohon Batoe.

Presuming the volcanic tract to extend still further eastward it would intersect the Kapoewas Moeroeng, above the kampong Habatong, and at that very place SCHWANER found eruptive rocks, which therefore possibly belong to this same volcanic territory. Positive proofs of the continuation of the volcanic strip of ground west of the Samba, we unfortunately do not possess.

With regard to the granite territory, we may take for granted that it forms a large portion of the ground of the hilly district of South Borneo north of 1°3′ S. L. Neither have I any doubts as to the correctness of SCHWANER's statements, (after what I saw myself on the Samba) concerning the numerous notable rapids in the Katingan above Toembang Samba, which, he says, are caused by granite rocks in the river-bed [1]).

Over a distance of 200 kilometres along the Katingan below Toembang Samba our section only shows diluvial and more recent fluviatile deposits. As a matter of fact no more ancient deposits were met with along this river. SCHWANER however points out that the wide alluvial plain, through which the Katingan flows is surrounded by hilly country. Between the Katingan and the

---

1) C. A. L. M. SCHWANER 55, II, p. 141.

Kahajan this hilly district, partly at any rate, consists of sandstone and clay-stone, interstratified with coal-seams. Schwaner calls this formation Tertiary. Gaffron [1]) regarded these formations as consisting of Secondary and Tertiary deposits, but did not specify them. Martin [2]) however showed that Gaffron merely classified a portion of these formation under the Secondary period because in some places they contained limestone full of shells, in which he recognized "Muschelkalk", and that he classified old fluviatile (diluvial) deposits as Tertiary. Martin therefore suggests, that for Gaffron's Secondary and Tertiary formations, it would be better to read: "Tertiary and Quaternary formations". On Gaffron's map [3]) these deposits cover the greater part of the tracts of high ground which separate the broad valleys of the rivers Pĕmboeang, Sampit, Katingan and Kahajan. Between the Pĕmboeang and the Sampit they are supposed to extend as far as the coast, and between the Katingan and the Kahajan to within a short distance from the sea. All things considered however, we must come to the conclusion that our knowledge of the Tertiary formation is very scant and very uncertain. It would seem that in South Borneo the Tertiary formation occupies the territory between the granite hills and the Java Sea, that it is not an unbroken stretch of country however, but divided into tracts (strips) by the broad valleys covered with more recent fluviatile deposits, which separate the great rivers of South Borneo, and that these tracts of Tertiary formation extend generally from north to south in the same direction as the rivers.

On the section Mt. Kaki is marked, near the Java Sea, this being the only height to be seen between Toembang Samba and the sea. Mt. Kaki lies 18 kilometres to the east of Méndawei between the Katingan and the Sebangan. It is entirely isolated and surrounded by marshland. According to Gaffron it is 800 feet high and consists of greenstone (diabase).

1) H. von Gaffron, vide J. Pijnappel 46, p. 143.
2) K. Martin 33, page 131 and 34, page 16.
3) This map has been published with the two above mentioned essays of Martin.

CHAPTER XII.

———

I. FORMATIONS AND ROCKS.

In the greater part of the sedimentary deposits met with in Central Borneo no fossils have hitherto been found, so that an absolute determination of their age is in most cases impossible, and even a relative determination is generally doubtful. For where the ground is covered by thick forests, as is the case in Borneo, and rock *in situ* only appears here and there and often at long intervals, the relation between the strata in one place and those in another can rarely be determined with indubitable certainty.

In the first place then there is always a certain amount of doubt as to the relations between the deposits of one given place and those of another, and an equal uncertainty prevails as to what complex of strata constitutes one greater unity, a formation. Then again the absence of fossils leaves the age of these formations and in consequence their mutual relation, a matter of doubt.

As a matter of fact in West Borneo, to the east of Sintang, there exists only one system of strata the age of which is known. These are the beds with *Orbitolina concava*, Lam., the Cretaceous (Cenomanian) age of which is ascertained. These therefore formed the starting point for the consideration of the relative ages of the sedimentary formations of Central Borneo. Moreover it is proved that one of the older formations, the chert and tuff with Radiolaria, belongs to the pre-Cretaceous, probably Jurassic period. Finally the brackish-water deposits along the Pinoh and the Lĕkawai river, and elsewhere in the Mĕlawi valley, are known to be of old-Tertiary age (see page 396). For fixing the age and the mutual relation between the other formations we have no palaeontological data whatever, and we are, therefore, entirely dependent upon general stratigraphical and petrographical characteristics. Amongst the stratigraphical distinguishing marks, the undisturbed or folded condition of the strata, and their steep or feeble incline, deserve primary consideration. It stands to reason however, that this can only exceptionally lead to a complete and exact definition of the succession of the different systems, while the petrographical characteristics, which in some regions lead to almost as well defined partitions as those based on palaeontological evidence, do not hold good in Borneo. In the formations here grouped together as Cretaceous and Tertiary, a series of strata of sandstone, quartzitic sandstone and clay-stone follow one another and resemble each other as a rule so closely that it is impossible to distinguish any well defined horizon in them by their petrographic characters. It is therefore not only possible, but even probable, that time will show, that other formations besides those of Cretaceous and Tertiary age are hidden in this uniform series of strata.

Not long ago a happy find in the district of Sambas in West Borneo has revealed that a portion a series of beds of sandstone, clay-stone, and marls, of which one could only suspect the Mesozoïc age, belongs to the Lias [1]), whereas elsewhere in a similar series, Upper Jurassic [2]) fossils have been found, and it is supposed that even a part of this series may prove to be of Cretaceous age.

The following summary of the formations and rocks which take

1) P. G. KRAUSE 26, p. 154—168.
2) K. MARTIN 36, p. 23—51 (p. 31) and F. VOGEL 65, p. 127—153.

part in the structure of that portion of Central Borneo which has been investigated, must therefore of necessity be provisional and to a certain extent uncertain.

A. *Crystalline schists, Archaean period.*

Rocks which, judging only by their petrographical characters, should be considered to belong to the crystalline schists, are found in several places. Mica-schist is found in the Schwaner mountains between the Ella and the Sĕrawai rivers [1]), on Mt. Boenoet, Mt. Sĕmoenga, Mt. Dammar, and a few other places near the sources of the Mĕntatai river. I also discovered fairly large quantities of this rock amongst the boulders of the Karang and the Samba. This biotite-schist however, generally contains andalusite, and it is connected by several transition-types with the andalusite-biotite-hornfels (cornubianite), and spotted slate, which, are sedimentary rocks altered by contact with granite. Such indubitably contact rocks are found a. o. on Mt. Batoe, Mt. Liang Djĕlai, Mt. Noewa, Mt. Boelai, and at several other places along the Samba river in the vicinity of the granite bosses.

In consideration of these facts I am inclined to look upon the biotite-schist of the Schwaner mountains, not as a true crystalline schist, but as a slate which, through contact with the granite, has become crystalline attaining the type "quartz-biotite-slate". Neither can I look upon the biotite-amphibole-schist, which I found in various places along the Samba river, as a true crystalline schist. For I could prove that these schists, have probably originated from basic segregations in the granite, as they are lying close to the periphery, and therefore in the zone of endomorphic metamorphism of the intrusive masses.

It is therefore doubtful whether the mica-schists and the gneissic rocks, found in such abundance in the portion of the Schwaner

---

1) I have a collection of rocks from this mountain land, which has been carefully collected by Wĕrbata, surveyor, of the topographical service, and SCHWANER also gives geological notes about this part of the district.

mountains east of the Ella river, in contact with the granite
which there is the predominating rock, should be reckoned to
belong to the formation of crystalline schists (the Archaean period).
They are more probably to be regarded partly as modifications
of the granite itself near the line of contact with the sedimentary rocks, partly as rocks of unknown age, but older than the
granite, as they are metamorphosed through contact with it and
possibly also through mountain pressure. This explanation would
agree with observations made in the mountain districts of Soekadana and Matan[1]) which may be considered to be the western
continuation of the Schwaner mountains. There amphibolite and
mica-schist are only mentioned as existing on the Upper Pen-
djawan, while the sedimentary deposits of unknown age have,
in various ways, been strongly metamorphosed through contact
with granite, the predominating rock.

In the hilly country of Sémitau, amphibolite, chlorite-schist, and
quartzitic slate play an important part. These strata alternate with
quartzites and cherty rocks, which can scarcely be distinguished
from the cherts belonging to the Danau formation. Their intensive folding shows, moreover, that they have been exposed to
strong mountain pressure. It is therefore not improbable that these
amphibolitic schists are, in reality, rocks of the diabase family which
have been uralitized and altered through mountain pressure. If so
those diabases, must here have formed beds between the strata of the
Danau formation, the same as they have done in the lake district.

But it is also possible that these amphibolites are in reality
Archaean rocks, and that originally the cherts and the quartzites
of the Danau formation were deposited unconformably upon them,
and afterwards the entire complex has been crushed up again
so that these later beds were folded in amongst the crystalline
schists.

This explanation has been adopted in section KK', and on
the maps.

---

1) R. EVERWIJN *II*, p. 58 and following.

GRAUWACHE WITH NUMMILITES AND ORBITOIDES FROM THE VALLEY
OF THE SOENGEI EMBALOEH.

The hills of Sémitau are intersected more eastward by the Embahoe river and along the bed of this river gneissic rocks are exposed. Macroscopically these give the impression of being granites altered through mountain pressure, and even on microscopical examination they show granitic characters [1]).

The amphibolite and glaucophane-amphibolite described from the Sébilit river, may be considered to form the eastern continuation of these Sémitau schists. Probably they are true crystallines schists, but it may be that these strongly crumpled rocks should be looked upon as metamorphosed partly by mountain pressure, and perhaps also to some extent by contact with the very basic eruptive rocks, which, altered into serpentine, are found in their vicinity.

In my opinion it is quite certain that the amphibolite and talc-schist, exposed in different places on the Upper Kapoewas between Poetoes Sibau and N^a Kéryau, are not old crystalline schists (comp. page 175). Most likely they are sedimentary rocks altered by contact with basic eruptive rocks, picrite and harzburgite, (which are for the greater part altered into serpentine). They occur in close connection with the chert with Radiolaria and belong presumably to the pre-Cretaceous Danau formation.

If this suggestion should prove to be the correct one, then this occurrence should be comparable to the finds on the Coast Range of California. RANSOME says with regard to these: "These ranges consist mainly of (probably Cretaceous) San Francisco sandstone, with intercalated beds of Radiolarian chert; but intruded into these are sills of basic eruptive rocks (fourchite) and a dyke of serpentine. The bedded rocks, both the sandstone and the Radiolarian chert are at the contact with the basic eruptive rocks converted into glaucophane-schists [2])."

---

1) In one place a beautiful piece of ruck was found, apparently consisting of alternate layers of biotite-gneiss and muscovite-gneiss. I presume that it was a biotite-granite injected with numerous, thin veins of muscovite-granite, and that through mountain pressure it has obtained a gneissic structure.

2) F. J. RANSOME 47, pp. 193—240.

Summarising the above, it appears likely, but not quite certain, that the amphibolite, glaucophane-schist and chlorite-schist exposed in the Sĕmitau hills and along the Sĕbilit are old crystalline schists and may therefore be reckoned to belong to the Archaean period. With regard to the gneiss, mica-schist, amphibolite and talc-schist found in other localities in Central Borneo it is fairly certain in some instances, and probable in others, that they are rocks of more recent age, which have been metamorphosed by various influences such as mountain pressure and contact with intrusive rocks, and have thus obtained petrographically the character of crystalline schists.

In the neighbourhood of Sémitau the crystalline schists form the centre of a strongly denuded, hilly district (see maps III and IV) hardly ever rising to more than 150 metres above sea-level. The strata stand for the greater part on edge and have evidently been exposed repeatedly to the action of mountain pressure. The same remark applies to the gneissic rocks along the Embahoe and the glaucophane-schists and amphibolites on the Sébilit.

Very little is known concerning the occurrence of the mica-schists in the Schwaner mountains. We only know from Schwaner's notes and Wĕrbata's collection that some of the mountain tops are composed of these rocks, thus, for instance, mica-schist is found on Mt. Dammar at a height of 1100 metres.

B. *The old slate-formation.*

Under the name of old slate-formation I wish to comprise a system of deposits, which, with the exception of the above mentioned crystalline schists, I consider to be the oldest sedimentary formations of Borneo and which are characterized by the frequent occurrence of an easily recognizable phyllitic clay-slate with silky lustre. The strata of this clay-slate alternate with beds of sandstone, greywacke, greywacke-slate and quartzite. The clay-slate, which predominates over all the other rocks is sometimes true drawing slate, but as a rule it is characterized by strong and clear silky lustre; in that condition it resembles phyllite, and I have therefore called it phyllitic clay-slate. Its petrographical characteristics agree with those of the old slate-formation [1]) of the Dutch state mining-engineers in the coast district of West Borneo and in Sumatra. WING EASTON [2]) says with regard to the complex of strata in the districts of Sambas and Landak, which he reckons to belong to the old slate-formation:

"The old slates are in Sambas most probably the most ancient rocks. As compared with the younger slates, mentioned later on, they possess the following characteristics:

1. A well marked uniform character connected with constancy of composition.

2. Fairly great hardness.

3. Excellent silky lustre, well recognizable, although of course in a lesser degree, even in fairly weathered pieces.

4. Perfect fissility with very smooth planes of stratification.

5. General imperfect development of several directions of cleavage, so that these slates can easily be procured in large, thin, slabs.

The colour is generally dark blue, sometimes a little lighter, probably due to bleaching by ulterior alteration.

With a few exceptions the strike is east to west, the dip being from 70° to 90° north or south. This strong upheaval and folding is due, for the greater part, to the later granite eruptions.

So far no fossils have been found anywhere, the age is therefore doubtful.

The intersection with numerous quartz veins is also characteristic. Thousands of fragments of these quartz veins lie scattered about in the area where this slate formation prevails."

This description shows that in the petrographical character and the colour of the rocks, in the position of the strata, in fact, in all particulars, the old slate-formation of Sambas agrees entirely

---

1) R. D. M. VERBEEK 62,. p. 153. Even now the name of old slate-formation is used by the Dutch Indian mining-engineers, with exactly the same meaning attached. R. D. M. VERBEEK and R. FENNEMA 63, p. 886.

2) N. WING EASTON 69.

with my old slate-formation of the Embaloeh valley. But of course the folding of the strata in the Upper Kapoewas range is not due to granite eruptions. For although it is an ascertained fact that the upward intrusion of igneous rocks, may cause the tilting of overlying beds, few geologists will positively admit that granite eruptions (sic) can cause the folding of surrounding rocks on a large scale.

This old slate-formation used to be looked upon as Palæozoic, and although it became probable later on that a portion of this formation in Sumatra is of Mesozoic age [1]), this has merely had the result of restricting the limits of the old slate-formation while on the other hand the term *old" has received a more concise meaning. I use the term old slate-formation in the same meaning as originally attached to it by the colonial mining-engineers, viz., a formation in which clay-slate, of unknown although probably of Palæozoic age is the ruling constituent. In the Upper Kapoewas range the beds of this formation are steeply tilted, strongly plicated and folded, and intersected in various directions by veins of varying thickness. The predominating strike of the strata is east to west. The area of this formation is marked by ranges of hills trending east to west, their sharp ridges remaining uniform in height over considerable distances. Going from west to east, however, the general elevation of this mountain land increases. The ranges are separated from each other by deep, narrow, longitudinal valleys. This mountain land reaches its greatest height near the boundary of Sarawak, and the general average height of the ridges is there about 1000 metres in its western part, and about 1400 metres in its eastern part.

I found this formation developed in the territory of the Upper Embaloeh and the Tékëlan rivers, BÜTTIKOFER [2]) found it on the Upper Sibau, and van SCHELLE [3]) on the Lébojan. The boulders

1) N. WING EASTON 70, p. 152.
2) From verbal communications to the author, and from specimens collected by him in the bed of Sibau river.
3) C. J. VAN SCHELLE 51, II, p. 37.

which I found in the Upper Kapoewas, indicate as very probable
that the mountains near the sources of the Kapoewas, above
N° Tandjan, also belong to the same formation, and the general
configuration of this region justifies us in accepting that the
Upper Kapoewas mountain range, built up principally of rocks
of the old slate-formation, stretches without interruption from the
boundaries of Sarawak, near the sources of the Lĕbojan, to the
boundaries between West and East Borneo, near the sources of
the Kapoewas, a distance of 225 kilometres. The trend of this
mountain range, is in the western part about W. 10° N.—E. 10° S ,
then direct E.—W., and near the sources of the Kapoewas
probably W. 5° S.—E. 5° N. Thus the Upper Kapoewas mountain range describes, as far as is known, a feeble curve the convex
side of which is turned southward. Moreover the character of
the landscape which I viewed from Mt. Pan, Mt. Tjondong, and
Mt. Lĕkoedjan, justifies one in supposing, that this range extends
still further both westwards and eastwards. As regards its continuation eastwards, this must perhaps be looked for in the high
mountain district of Bĕrau and Bĕloengan.

Up to the present no fossils[1]) have been found in this
old slate-formation and its absolute age is therefore unknown.
As to its relative age, I consider that it is older than any
of the other sedimentary formations in Borneo. I hold this
view in consideration of the fact that the strata of this formation are more disturbed, plicated, and overfolded than those
of any of the others, while they likewise bear evident marks
of having been subject more than once to the influence of
mountain pressure. Again, the petrographical character of the
rocks indicates an old formation, although we should not attach
so much importance to this in a case like that under consideration, where the beds are so strongly disturbed. In summing
up we come to the conclusion that the age of the old slate-
formation is unknown, but that there are reasons for supposing

---

1) See however pages 238 and 239.

that it is probably older than the other sedimentary formations hitherto known in Central Borneo.

C. *The Danau formation.*
Formation of the great lakes.

Under the name of Danau formation I comprise a system of deposits, which I saw for the first time typically developed in the area of the great lakes and in the hilly district bordering the north side of these lakes. The constituent rocks are diabase-tuff, diabase, diabase-porphyrite, quartzite, chert, jasper, horn-stone, clay-slate, and sandstone. The most characteristic, the leading rocks, in this formation are a silicified and partly serpent-inized diabase-tuff, which I have called Poelau Mĕlaïoe rock, and jasper. In all places where they were found, the jasper and horn-stone proved to contain many Radiolaria, and these organisms were also found in many places in the cherts and occasionally in the diabase-tuff.

The examination of these Radiolaria has revealed that these deposits have a pre-Cretaceous, probably Jurassic age.

The strata of this formation are almost everywhere folded and consequently tilted, but hardly ever so strongly disturbed as the strata of the old slate-formation. Neither are they, like these latter, intersected with numberless quartz veins.

Without any doubt the most interesting rocks in this formation are the chert, jasper, hornstone and diabase-tuff with Radiolaria [1]). The jasper and hornstone in particular consist in many places almost exclusively of closely packed tests of Radiolaria, between which a few spicula of sponges may be seen, all joined to-gether by a siliceous cement. The amount of silica contained in such a hornstone with Radiolaria from the Boengan river was not less than 97%. I found jasper with Radiolaria in various places in the hilly country to the north of the lakes, and later on again, much further eastward, in the extension of the line of strike of

---

1) See the appendix to this work.

the above named strata along the Upper Kapoewas, the Boengan, and even as far as the boundary between West and East Borneo, altogether over a distance of fully 230 kilometres. The character of the rock remains unchanged over that distance and the beds are always folded with a constant east to west strike. I feel justified in maintaining that these jaspers and hornstones with Radiolaria are deep-sea deposits, formed most probably at a considerable distance from a mainland of any importance. They consist almost exclusively of the tests of pelagic organisms. Now this in itself is no conclusive proof that they must be oceanic deposits, for we know that in certain localities close to the coast where, through peculiar conditions, the deposition of terrigenous materials on the ocean floor was entirely or almost entirely prevented, even at a short distance from the shore, deposits are found which may be composed exclusively of remains of pelagic living organisms. In the case here under consideration, however, the peculiar conditions just mentioned would have to extend along the coast for a distance of at least 230 kilometres which would be alltogether too unlikely to happen. The fact that these cherts and jaspers consist almost exclusively of tests of Radiolaria, stamps them as deep-sea deposits; their great horizontal extension with perfectly uniform petrographical and palaeontological characters stamps them as oceanic deposits. We may therefore conclude from them that in the pre-Cretaceous age, when these deposits were formed, this portion of Borneo was the floor of a deep sea, far away from any land, and this in itself is a important fact. But these deposits become particularly valuable as they offer a very convenient starting point for studying the geological history, not only of Borneo but probably also of other regions in the East Indian Archipelago. For in attempting to disentangle the geological history of a shifting portion of the earth's crust, to which the East Indian archipelago no doubt belongs, a portion therefore that has past through all the different conditions of existence (from forming part of a mainland to being buried deep down under the ocean) it becomes of the greatest importance, if only approximately, to note the different periods in that history in which the most extreme conditions have been realized. The most extreme conditions are evidently, *mainland* and *deep-sea bottom*. Geological documents[1]) regarding the condition as land are, as a rule, almost entirely wanting, because a land period is more a period of denudation than of sedimentation, and can therefore as a rule leave only negative vestiges. But if that same region has at one time been deep-sea bottom the true deep-sea deposits bear witness of it. Moreover however great variations in level we may allow for a shifting portion of the earth's crust, it would be difficult to conceive of a comparatively limited area, like the East Indian archipelago or a portion of it, to have formed repeatedly, during a limited part of the earth's history, a part of the floor of the deep sea.

On the other hand, apart from all other grounds which justify our calling the East Indian archipelago a shifting portion of the earth's crust, the fact alone that deep-sea deposits are found there, where land now exists, entitles us to do so.

One of these extreme conditions in the history of the development of the East Indian archipelago is thus indicated by our Radiolarian rocks. But because they are not only deep-sea deposits but also oceanic deposits we must accept as probable that the adjoining areas of the Indian archipelago were at one time, partially at any rate, covered by the sea[2]). In other parts of the East Indian archipelago similar beds of jasper with Radiolaria have recently been found;

1) Geological documents i. e. remains, from which a tradition or document of the condition of things then existing can be made clear.

2) If our Radiolarian rocks were only deep-sea deposits and not oceanic deposits as well, we could not come to this conclusion. The present condition of the East Indian archipelago shows that various deep basins, as the Banda Sea, the Celebes Sea, and the Soeloe Sea can exist between and near islands of considerable size. But the soundings of the Challenger have proved that the floor of the Banda-sea consists for the greater part of terrigenous elements, and because they are deep local basins in the neighbourhood of the land no pure oceanic deposits can be found in them.

The soundings made by the Siboga-expedition in the East Indian Archipelago (1899—1900) under the direction of Prof. MAX WEBER have fully confirmed these facts and have provided abundance of additional evidence. Compare Siboga Expeditie. I. MAX WEBER. Introduction et description de l'expedition p. 29 et 129—134. Leiden 1902.

at Celebes on the Posso river [1]), and in South East Borneo [2]), while tuff with Radiolaria occurs in North-East Borneo near Batoe Tjinagat [3]), and VERBEEK found chert with Radiolaria on the island of Billiton. There is every chance that elsewhere in the East Indian archipelago, rocks with Radiolaria may be discovered, because only a comparatively small area of this territory has as yet been geologically investigated, and also because the Radiolaria often remain unnoticed as in true occanic deposits rarely any macroscopically visible organisms occur in the same rock with them. We have therefore some ground for expecting that in course of time all deposits with Radiolaria in the East Indian archipelago will be proved to date back to that period when this area formed partly, if not entirely, the floor of a deep sea. As regards the period in the geological chronology when this condition was realized, we only know that it must be looked for before the Cretaceous period.

I quite expect it will be shown later on that this formation of Radiolarian rocks represents, a very considerable space of time, possibly more than one geological period. The thickness of the Radiolarian "ooze" in the Upper Kapoewas and the Boengan I estimate at 100 metres [4]) at the very least, and when we take into consideration that the investigations of the Challenger have revealed that in many parts of the ocean floor, during the whole space of time from the end of the Tertiary period up to the present day, only a layer of a few centimetres thickness has been deposited, it becomes evident that for the formation of strata of Radiolarian "ooze" of more than a hundred metres in thickness very long periods of time must have been required. There is reason to suggest that the deposition of the

---

1) A. WICHMANN 67, p. 164.
2) The same, p. 164.
3) According to personal information from Prof. WICHMANN to the author.
4) It is very difficult to give an exact estimate of the thickness of the chert deposits, because they are strongly folded and much disturbed; but 100 metres is at any rate a minimum estimate.

deep sea sediments, if it did not take place altogether in the Jurassic period, at all events extended far into it.

The researches of WICHMANN and ROTHPLETZ in Rotti and Timor, MARTIN in Boeroe, MARTIN, VOGEL and KRAUSE in West Borneo, have proved that in Jurassic times various parts of the East Indian archipelago were covered by the sea.

The greater part of the fauna described in these papers, does not bear the character of a littoral fauna, and the limestone with Aptychi, found by MARTIN on the island of Boeroe, has unquestionably been formed in a deep open sea.

For the interior of Borneo I have taken the area covered by the Danau-formation [1]) (see maps II and III) to be as follows:

1. In the lake district and from there in a continuous strip of ground, 280 kilometres in length, to the other side of the divide right into the basin of the Mahakkam in East Borneo.

This entire area has probably been sunk along a great line of fracture (red line on map II) running almost due east and west, and it is almost entirely covered over by recent fluviatile deposits of the Kapoewas and its affluents. In the western part of the territory we find the Danau formation most fully developed in the hilly district north of the lakes, forming the divide between the Kapoewas basin and the basin of the Batang Loepar in Sarawak. More southward these strata are covered by alluvial deposits and they are also partly hidden under the water-level of the lakes. Only in a few places, as for instance in the Sěmběroewang range, in the islands of Sépandau and Mélaioe, along the northern edge

---

1) It is evident that the Danau formation as a whole covers a still greater period of time than the Radiolarian deposits. For the formations which we have included in the Danau formation are not all deep-sea deposits. The diabase-tuffs may owe their origin to the loose ejectamenta from volcanic eruptions on oceanic islands of that time, and may have been deposited on the bottom of the sea. The diabase may have originated from submarine eruptions. The sandy shales, the clay-slates and sandstone which form an inferior part of the Danau-formation are certainly not deep-sea deposits, and consequently must have been formed when the relation between land and sea was quite different to what it was during the formation of the Radiolarian rocks. These rocks are too much folded and disturbed to say with certainty whether the sandy shales and clay-slates lie on the top or below the chert; they appeared to me to rest on the top of it.

of Mt. Lémpai, Mt. Séligi, etc., do the strata of this formation crop out again from under the alluvial plains. Probably the contact rocks situated round the granite boss of Mt. Kénépai also belong to this formation. To the east of the Sémbéroewang the Danau formation disappears altogether under the alluvium of the Kapoewas river, and not until a good way further on above Poetoes Sibau, where the Upper Kapoewas plain terminates and gradually passes into the hilly district of Boengan, does our formation reappear once more, with the same trend and the same characteristics as in the lake district. Here also the strata are folded and tilted. As in the lake district, the Poelau Mélaioe rock, jasper and hornstone with Radiolaria, are abundantly and characteristically represented in this formation, and the only variation of any importance consists in a few beds or very flat lenses of limestone which appear in the valley of the Boelit conformably intercalated between the folded strata of this formation. Whereas in the lake district it may be a matter of speculation whether the granite bosses are intrusive rocks which have forced their way through the strata of this formation, on the Upper Kapoewas above the island Lolong, and on the Boelit above pangkalan Mahakkam, there can be no doubt whatever about this fact. Basic igneous rocks, now partially altered into serpentine, also occur in this formation; they are partly intrusive, and therefore younger; partly, however, they must together with the diabase, be considered as of equal age with the Danau formation.

2. South of the Upper Kapoewas plain the Danau formation has only been recognised in a few places. Thus the cherty quartzites in the Sémitau region must probably be considered as strata of the Danau formation folded in the underlying amphibolite and chlorite-schist. Moreover, as shown on page 408, it is as yet uncertain whether the amphibolite, chlorite-schist, and slaty quartzite, which mainly compose the hilly district of Sémitau, are true crystalline schists, or as far as the amphibolites and chlorite-schists are concerned must be looked upon as highly altered rocks of the diabase-family, and as far as the slaty quartzites are concerned as

much altered cherts and sandy shales. This supposition is supported by the fact that these crystalline rocks bear evident marks of having been exposed to intense mountain pressure. The same may be said of the glaucophane-amphibolites and the amphibolites on the Sébilit, where moreover contact-metamorphism may have played its part by the intrusion of basic igneous rocks which, altered into serpentine, are found in the immediate vicinity.

3. With more certainty we may reckon as belonging to the Danau formation, the serpentinised tuff-breccia, tuff and chert, which are the predominating rocks in the hills bordering the north side of the Sĕbĕroewang valley, and more eastward in the dividing area between the Gaäng and the Tĕpoewai rivers. For in several places Radiolaria have been discovered in the chert and especially in the tuff, sufficiently closely resembling the Radiolaria of the chert and jaspers of the Upper Kapoewas to justify our associating these deposits with those of the Danau formation. From the Sébĕroewang valley I followed this formation as far as the Pyaboeng mountains the base of which is composed of it. The beds stand almost vertical but appeared less disturbed than on the Upper Kapoewas; the predominating trend was east and west. In the same way the presence of this formation was recognised at several points along the Embahoe.

4. And, finally, I realised its presence once more in the appearance of rocks of the Poelau Mĕlaioe type on the Tébaoeng, where, amongst others, the rocks in the dangerous Riam Pĕlai are composed of it. The strike here verges to W. 30° N. and the strata are vertical. From other regions of Borneo which have been geologically investigated, I only know of some deposits in the Sambas district which, judging from the description, appear to be in any way related to my Danau formation. I refer to that part of the mesozoic sediments of Sambas which WING EASTON [1]) considers to be either pre-Cretaceous or else the base of the Cretaceous. The resemblance lies in the fact that, like the Danau formation, they trend in the same direction as the old slate-formation, that they

---

1) N. WING EASTON 69.

are folded like the latter, though in a lesser degree, and that a complex of cherts is found in their upper division. The characteristic diabase-tuffs however, are not found in Sambas. I do not attach much importance to the fact that no mention is made of Radiolaria in the Sambas rocks, as in the open field they are not conspicuous. Should the cherts of Sambas prove to be of the same age as those of the Danau formation, Radiolaria will doubtless be found in them if searched for.

D. *Cretaceous deposits.*

The age of the Cretaceous deposits, or rather of part of these, was established earlier than that of any of the other sedimentary deposits. The investigations of MARTIN proved that the Foraminifera, collected in marls along the Sèbèroewang river, first by EVERWIJN, then by SCHNEIDER and finally by VAN SCHELLE, belong to the species *Orbitolina concava*, Lam. It is therefore unquestionably certain that these marls belong to the Cretaceous formation, and more particularly to the Cenomanian stage. As a matter of fact the beds with *Orbitolina concava* were in 1894 the only strata in West Borneo, the age of which was accurately known, and consequently they formed an important starting-point for the determination of the chronological order of the sediments. I discovered *Orbitolina concava*, Lam. not only on the Sèbèroewang but also in a marly sandstone along the Upper Kapoewas near the goeroeng Dèlapan. On the Sèbèroewang I found *Orbitolina concava* not only in the marl but also in the sandstone [1]). Although it was thus easy enough to furnish proofs of the existence of the Cretaceous deposits in

1) Other shells besides *Orbitolina* have been found in the neighbourhood of Sajor on the Sèbèroewang river, probably in the same strata, but these seem to have been lost, so that the identity of the material can no longer be determined. Besides *Orbitolina* I found near Sajor only fragments of tests and spines of Echinoderms, and a few shells and Ammonites, all however too imperfect, or too badly preserved to be of value for the determination of the age of the strata.

the above named neighbourhoods, it was much more difficult to ascertain which deposits in other parts of Central Borneo might be reckoned as belonging to the same formation with these marls and sandstones. Both on the Sĕbĕroewang and on the Upper Kapoewas these beds with *Orbitolina* form part of a system mainly of moderately tilted and folded sandy deposits, amongst which varieties akin to greywacke and arkose play an important part. Indeed it seemed to me that arkose and allied varieties of sandstone were characteristic for the Cretaceous deposits, and consequently each complex of strata in which arkose plays an important part, if it were also tilted, I have provisionally reckoned to belong to the Cretaceous formation.

Whereas the pre-Cretaceous strata bear, on the whole, the stamp of deep-sea deposits, the character of the fauna in the Cenomanian strata shows that they were formed not far from the shore, and their character of coast-deposits is further proved by the appearance of numerous vegetable remains together with the tests of *Orbitolina*. This I noticed especially on the Sĕbĕroewang not far from pangkalan Piang. The distribution of the Cretaceous deposits (in the area concerned) is indicated on maps II and III. Cenomanian deposits appear all along the Sĕbĕroewang river and more eastwards in the valley of the Gaäng and the Tĕpoewai, also in the Boengan hills where along the Kapoewas and the Boengan the Cretaceous strata have been folded with the older deposits and have filled up the troughs bordered by pre-Cretaceous formations. Judging exclusively by their petrographic characters the arkose and sandstone of the Tĕbaoeng, near the Goeroeng Nekan, are classed with the Cretaceous formation, while for the Bojan, Cretaceous limestone is indicated on the authority of van Schelle and Martin. Finally, I presume that the fairly strongly folded deposits which I found south of the Madi plateau along the rivers Midih, Panai and Kĕrĕmoei Mĕlawi, amongst which arkose is very well represented, form part of the Cretaceous formation.

E. *The Eogene formation.*

The existence of deposits of Eogene age in Central Borneo has so far only been identified in boulders of a grit, resembling greywacke with *Nummulites* and *Orbitoides* which I discovered in great quantities in the valley gravels of the Embaloeh and the Tékélan rivers, and which Büttikofer found in the boulder-banks of the Sibau river. It is much to be regretted that these Eogene rocks have not been met with anywhere *in situ*, for we are now left in absolute doubt with regard to their position in relation to other formations. Their discovery, however, enables us to draw some interesting conclusions.

Originally these boulders were examined by C. Schlumberger, who considered them to be of Eocene age. Later the same boulders together with the sections for microscopic study made by Schlumberger, were reexamined by R. Bullen Newton [1]) and compared with similar rocks from Borneo and Sumatra. He determined in them the presence of *Nummulites Djokjokartae*, K. Martin and *Orbitoides* (*Discocyclina*) and came to the conclusion, that the rocks are of *Oligocene* age.

Pl. LII exhibits sections of this rock, showing both *Nummulites* and *Orbitoides*.

At present we know for certain that the Upper Kapoewas range through which the Tékélan takes its entire course, and through which the Embaloeh flows until it enters the Kapoewas plain, is built up exclusively of one formation, the old slate-formation.

To explain the presence of these Eogene rocks we are therefore bound to accept one of the following possibilities:

a. The greywacke with *Nummulites* belongs itself to the old slate-formation and both are therefore of the same age, in which case we are compelled to accept an Eogene age for the old slate-formation.

---

[1]) R. Bullen Newton and R. Holland. On some tertiary Foraminifera from Borneo collected by Professor Molengraaff and the late Mr. A. H. Everett, and the comparison with similar forms from Sumatra. Annals and Mag. of Nat. History Ser. 7, Vol. II, p. 245 seq. and Pls. IX, X, 1899.
For all further particulars about these fossils may be referred to this treatise.

This opinion would fit in very well with the facts observed, and with the petrographic character of the rock containing the *Nummulites*, which shows a great resemblance to the greywacke-slate which plays such an important part amongst the rocks of the Upper Kapoewas range. Against this opinion however the following objections arise:

1. In accordance with this view the Eogene rocks would have been folded and disturbed to a much greater extent than the older pre-Cretaceous and Cretaceous formations. This argument, it must be admitted, is weakened when we consider that the predominating rocks of the old slate-formation (clay- and greywacke-slate) were easily moulded by pressure and would therefore show those indications, such as strong folding, isoclinal stratification, overfolding and cleavage, which are considered to be the result of severe pressure, under circumstances where a system of less pliable rocks, such as chert, hornstone, sandstone, quartzite, would only be moderately folded, but, on the other hand, cracked, fissured and faulted. The more intense pressure to which the Eogene formation seems to have been subjected, may not therefore, according to this suggestion, have been actually greater than that which has acted on the pre-Cretaceous sediments.

2. From what is known concerning the geological structure of other parts of the East Indian archipelago, we learn that the strata of the Eogene formation, although often disturbed and locally somewhat folded, are, as a rule, not highly tilted, and are nowhere strongly folded so as to form a mountain chain.

3. It is very difficult to imagine how the *Nummulites* and *Orbitoides* could be preserved in such good condition, and comparatively so little compresed and distorted, if the rock in which they are found, had been in reality exposed to the same mountain pressure as that by which the Upper Kapoewas range has been upheaved.

It is principally the force of this latter argument which has induced me rather to favour the second possibility, namely:

b. The greywacke and greywacke-breccia with *Nummulites* is younger than the old slate-formation, and was deposited after the strata of this latter had been tilted.

This view is also quite compatible with the petrographic character of the rocks. For when they were deposited on the floor of the shallow sea in which the *Nummulites* lived, their material was derived from the nearest coast (both the petrographic character and the fauna stamp the rock as a shore — or shallow sea — deposit) which consisted of clay-slate and greywacke-slate; thus the rocks were composed of fragments of slate and greywacke-slate, united by a cement of the same rock in a finely triturated condition, together with the skeletons of the buried *Nummulites* and *Orbitoides*, i. e., a kind of grit resembling greywacke.

If we hold this opinion, we are bound to accept that these Eogene deposits lie unconformably on and against the old slate-formation, and although as a rule such overlapping beds show very distinct and prominent in the landscape, I noticed nothing of the sort in the mountains along the Embaloeh and the Tĕkĕlan. This is indeed a very serious difficulty, which forces us to suppose, either that these Eogene deposits have already, for the greater part, been carried away by erosion and are no longer conspicuous because of their slight extent, or else that they in their turn have been disturbed by post-Eogene movements, perhaps even have been folded into the older, already disturbed, rocks.

c. Finally a third possibility might be suggested, viz., that the presence of *Nummulites* need not necessarily be a proof of the Eogene age of the rock, but that the rock might belong to an older period. In that case we might regard the old slate-formation and the greywacke-breccia with *Nummulites* as contemporary without being forced to consider the old slate-formation as Eogene. Palaeontological evidence is, however, as already mentioned, so strongly against an interpretation of this kind, that it may safely be discarded.

F. *Tertiary sandstone formation.*

Old Tertiary.

An enormous tract of ground in the area investigated by us, consists of a formation characterized everywhere by the predominance of sandstone and quartzitic sandstone with intercalated layers of clay-stone, and by an almost horizontal position of the strata which have been very little disturbed. Coal-seams and strata of clay-stone with fossil shells occur in various parts of it.

To this sandstone formation belongs in the first place the greater part of the Schwaner mountains and of the Madi plateau, from where it stretches westward into the striking and far extending ridges in the neighbourhood of the Silat river. In the Schwaner mountains and in the Madi plateau the sandstone strata have a feeble northern dip, and more towards the west, both in the Silat mountains and in the lake district, they dip somewhat more steeply to the south. This sandstone also forms the summit of some of the hills in the lake district, and lastly the horizontal or almost horizontal sandstones and shales, which are exposed on the Embahoe, the Boenoet, the Tĕbaoeng, the Mandai and, more eastwards, on the Boewang and in the boundary mountains between East and West Borneo, are reckoned to belong to this formation. Possibly it covers also a large area in South Borneo. It is true that I found it there only in the downthrown area, together with the volcanic formations, on the Samba river, but we may gather from the accounts of SCHWANER and GAFFRON that in South Borneo it occupies a considerable part of the hilly country which separates the great alluvial plains through which the rivers flow down into the Java Sea. The strata of the sandstone formation generally are horizontal or but slightly tilted, and they are only locally here and there tilted and strongly disturbed and sometimes even vertical, as is the case on the Upper Tĕbaoeng. Nowhere have they been folded, but the area of the sandstone formation is divided into separate blocks by dislocations

(faults) striking in an E.—W. direction. I have stated in some
earlier chapters that the probable age of this formation is Tertiary, but I must now add that up to a short time ago the proofs
for this Tertiary age were very insufficient. All there was to go
upon consisted in the discovery of fossils of Tertiary age in beds
of clay and sandstone, on the Kapoewas near Télok-Dah above
Sintang, and on the Lower-Mélawi near Bantok [1]). It is only a
short time since, thanks to the studies of KRAUSE and also more
particularly to those of MARTIN [2]), that the Tertiary or rather
old-Tertiary age of the fossils in this formation has been finally
fixed. MARTIN showed that the fossils which I had found on
the Pinoh river and at different points along the Lĕkawai river,
and those which WING EASTON found afterwards at different points
on the Mélawi river above Nᵘ Pinoh, and on the Kajan and
the Tébidah rivers, also at a few places in the lower course of
and at the mouth of the Mélawi, and finally at the point where
the Tĕmpoenak flows into the Kapoewas, all belong to the old-
Tertiary formation. He further drew the conclusion that all the
deposits with fossils at these several points form one geological
whole and represent "eine zusammengehörige, im Wesentlichen
gleichalterige Ablagerung."

The great majority of the fossils are brackish-water forms,
but mixed up with them are a good quantity of fresh-water
shells. KRAUSE and MARTIN conclude from this that these deposits

---

1) P. VAN DIJK 8, page 147.   2) P. G. KRAUSE 27, pages 196—220.
3) K. MARTIN, Brakwatervormingen van de Mélawi in het binnenland van Borneo. Versl.
en Med. Kon. Akad. v. Wetenschappen Amsterdam, 26 Jan 1899 and 38.

It must be specially mentioned here that KRAUSE's report was published when the maps
of the present work and a portion of the text were already printed, and MARTIN's account
did not come out until both maps and text were almost ready for publication. The consequence is that I have not been able either to mention or to make use of the results of
either of these investigations in Chapt. VI (pp. 141—143), Chapt. IX (pp. 280 and 281
and Chapt. Xᴬ.

On page 396 I mentioned that owing to some private information received from
KRAUSE I had made on the maps a distinction between the strata with fossils on the Upper
Lĕkawai, and the Tertiary sandstone formation, the former being most likely older, probably
young-Cretaceous deposits. I further mentioned that KRAUSE's final researches had proved
that after all this distinction did not exist.

were formed into estuaries, in which rivers emptied themselves.

MARTIN comprises this entire old-Tertiary brackish-water formation under the name of the Mĕlawi-group.

KRAUSE also examined some shells which I had collected on the Embahoe river near Poelau Boedoeng (see page 428 and Map VIII) and considered them to belong to the stage α-Eocene (Verbeek). But as there is a striking resemblance both in structure, in position and in geological occurrence, between this formation on the Embahoe and the above named deposits of the Mĕlawi-group, — and especially as KRAUSE seems as yet doubtful whether the α-Eocene (of Verbeek), is really Eocene — I do not feel at liberty to separate these deposits, but include them all in the Tertiary sandstone formation.

In this same formation near Sĕlimbau on the Kĕnĕpai river, on the Lower-Tĕbaoeng and the Sĕbilit, on the Mandai and on the Pinoh, coal-beds have been discovered, which, at the last named place, rest immediately on strata containing brackish-water shells (see fig. 35, p. 142).

Future examinations may yet reveal to us that the deposits in which coal-beds and old-Tertiary molluscan shells [1]) appear, are of younger date than the sandstones of the Madi plateau, the Schwaner mountains, the hills of Silat and the lake district (the Lĕmpai-sandstone) and may lie unconformably upon them [2]). This would be not in contradiction to my stratigraphical observations, but the data so far known are too imperfect to allow of my carrying through the separation. For the present, therefore, I have included all the sandstones and shales with an almost horizontal position into one formation, the Tertiary sandstone formation.

1) In many places the sandstone has a distinctly diagonal stratification, and where this is the case, it is certainly of fluviatile origin. The coal-beds occurring in this formation must have originated from vegetable matter carried down by the streams and formed in the manner described on page 44.

2) The sandstone, which in the Madi plateau and in the Schwaner mountains is found tilted to a height of 1000 metres above sea-level, will then perhaps prove to be of young-Cretaceous, or possibly, of old-Eocene age.

G. *Young fluviatile and lacustrine deposits.*

(*Quaternary*).

The alluvial fluviatile and lacustrine deposits, consisting for the greater part of fine sand and mud, largely mixed with vegetable débris, are very strongly developed in Borneo. The horizontal extent of the alluvial area in the Kapoewas basin, may be estimated at 20000 square kilometres (including the danaus, here taken as submerged alluvial ground), 6850 square kilometres of which belong to the Upper Kapoewas plain and 5400 square kilometres to the Kapoewas delta, while the remainder lies principally on the side rivers, and on the Lower Kapoewas. When we include the alluvial areas of other river-basins it would not be extravagant to estimate the total area of alluvial ground in West Borneo, at 30000 square kilometres. The total area of West Borneo being 145000 square kilometres; more than one-fifth therefore is taken up by alluvial deposits.

Between the hills and the flat alluvial grounds near the principal rivers and their tributaries, we find in many places older fluviatile deposits which, besides sand and mud, also contain pebbles. These deposits lie partially or entirely above the present average water-level in these rivers. They have evidently originally been deposited by the rivers, and their greater and smaller branches, and afterwards the rivers and streams have again cut out their beds in these deposits. It is most difficult to draw a line between the young and the old fluviatile deposits, nor is it easy to ascertain whether these are formations of the present period, or of the Quaternary period, or whether perhaps they should partially be counted as young Tertiary deposits. Personally I am of opinion that in tropical regions generally, and therefore also in the present case, the characteristic features which might help us to come to a definite conclusion in the matter are wanting. I therefore distinguish only between older and younger fluviatile deposits leaving their actual geological age undetermined. Exactly the same thing occurs in South Borneo. There indeed the fluviatile deposits on the borders of the great rivers occupy proportionately a far larger area than in West Borneo.

The old gravelly , river-deposits in Borneo, and especially in West Borneo generally contain gold. In certain rich spots these alluvial gravels are worked by the Chinese in their ordinary primitive style (see pp. 130 and 269). The yield of gold is not very great; I heard that, relatively speaking, the richest deposits were found on the Embahoe and the Sĕrawai. Up to now they have never yet been systematically prospected and it has not yet been proved whether it would be worth while to work these extensive gravel-deposits which everywhere, in the hilly district of Borneo, skirt the valley-slopes of the great rivers. The most rational method of working them would be by hydraulic machinery. Personally I think there would be a fair chance of success, although I hope the day may yet be far distant when Borneo will be invaded by gold-hunters; for this must of necessity lead to the destruction of the beauties of nature, and to the moral, if not total, ruin of the Dyak population.

H. *Intrusive and eruptive rocks.*

1. Granite and Tonalite.

Granite plays an important part throughout West Borneo. The characteristic type is an amphibole-biotite-granite carrying a fair proportion of plagioclase, so that transitions into rocks of a tonalite type are of frequent occurrence. In many places in West Borneo the intrusive character of the granite is proved by the intense alteration of the surrounding rocks near the zone of contact.

In the territory here concerned, east of Sintang, the following granite areas occur:

a. *The granite area of the Schwaner mountains.*

When speaking of the Schwaner mountains I refer here to the entire range, extending, as far as known at present, from

the sea coast near and to the north of Soekadana, as far as the
Oeloe-Samba, and further eastward to an unknown distance into
East Borneo. The base of these mountains consists, as far as
I can judge, principally of granite, but this rock predominates
less in the mountains than in the hilly districts of South Borneo
bordering the Schwaner mountains. Probably there are quite a
number of granite bosses more or less connected with each other,
which follow one after the other in a line from east to west.
As far as known, amphibole-biotite-granite and biotite-granite
are the principal rocks. Both varieties, but more especially the
first, owing to its larger amount of plagioclase, are found at different spots passing into tonalite.

The granite massives in the Schwaner mountains appear to
have altered the sedimentary rocks, which enveloped them during
their intrusion, to a great extent. Although this envelop of contact
rocks appears to have been removed in many places, remains of it
are found in the river-basins of the Samba, the Méntatai, and
the Ella, also in the mountain districts of Soekadana and Matan,
sometimes *in situ*, or at various places in boulders which undeniably proclaim their origin from the contact zone. The contact-
minerals found are:

α. In the rocks, enveloping the intrusive granite and within
the zone of contact, andalusite, biotite, actinolite, cordiërite and
muscovite;

β. In the endomorphically altered outer part of the granite,
tourmaline and muscovite.

The most characteristic contact-rock in the neighbourhood of
the granite is an entirely crystalline schist, not very perfectly
foliated, composed of quartz, biotite and andalusite (with, generally, a little cordiërite), and which might perhaps appropriately
be called andalusite-cornubianite.

The granite and the tonalite in the basin of the Samba are
gneissic in several places, and there are also numerous flat
basic segregations near the borders of these massives, and the
enveloping sedimentary rocks have been transformed into strata

of biotite-gneiss, tonalite-gneiss, mica-diorite, or amphibolite.

The age of the rocks of the Schwaner mountains metamorphosed by contact with the granite, and the age of the granite itself, are so far unknown.

b. *The granite of the Sémitau hills.*

VAN SCHELLE found granite to the south-east of Sémitau near Mt. Séboeloeh, and I found the same in several places on the Embahoe above goeroeng Oelak. On the Embahoe, granite, amphibole-granite, and tonalite are exposed; dykes of muscovite-pegmatite break through them. There the granite has a gneissic structure through dynamo-metamorphism, and in more than one place it passes into a type of rock which cannot be distinguished from true gneiss. No observations have as yet been made with regard to the metamorphosing influence of this granite on the surrounding rocks. Its age is unknown.

c. *The granite of the lake district and of the Boengan hills.*

In the lake district, tourmaline-granite, tonalite and augite-tonalite appear in the Kénépai mountains; tourmaline-granite and tourmaline-tonalite in Mt. Měnjoekoeng, and biotite-granite in Mt. Sap and its nearest surroundings. The tourmaline-granite itself here passes, in the contact-zone, into tourmaline-fels. Contact-rocks, such as quartz-muscovite-tourmaline-rock, andalusite-hornfels, and muscovite-hornfels were found skirting the augite-tonalite and the tourmaline-granite of Mt. Kěněpai. This granite is in part certainly, and in part probably, intrusive, and younger than the pre-Cretaceous Danau formation.

More eastward small granite massives were found in this same formation, one of which was a granitite situated on the Upper Kapoewas near Poelau Lolong, and one a tonalite on the Boelit river just above pangkalan Mahakkam. In the Boelit the forces which folded the strata of the pre-Cretaceous formation have put their stamp on the tonalite also and have given it the appearance of tonalite gneiss which here and there resembles amphibolite.

The numerous boulders of granitite in the valley of the Médjoewai lead one to expect the existence of more bosses of this kind in the basin of that river. The numerous granite boulders, rich' in varieties, which are scattered in the valley of the Mandai, above N⁰ Kalis and also in the Kalis itself, have probably originated from a similar source.

As regards the origin of the boulders of granite and pegmatite found in great quantities in a conglomerate on the Séběroewang river, probably of Cretaceous age, and on the slopes of Mt. Oejan, nothing is known.

2. DIORITE.

Diorite is found *in situ* on the Samba river near Kiham Habida, and on Mt. Bakah and Mt. Méraboe on the north side of the Schwaner range. I also found diorite amongst the boulders in the Tébaoeng, in the Samba above Toembang Ménjoekoei, and in the Karang river. Nothing is known with regard to its mode of occurrence and its connection with other rocks. As regards the quartz-diorite found in various places amongst the boulders of the Lékawai and its tributaries, e. g., the Bassi, and others, we know that it must have originated from the granite district of the Schwaner mountains, where this rock forms segregations in amphibole-granite, and where it also probably occurs as a variety of tonalite, exceptionally poor in mica.

Quartz-diorite of unknown origin is also found in the conglomerates on the Séběroewang river probably of Cretaceous age, and on the slope of Mt. Oejan and in the breccia in the Gaäng river probably of the same age, and abundantly in the pebbles of the Séběroewang river.

3. GABBRO AND NORITE.

A small massive of olivine-gabbro is found in the granite district on the Ménjoekoei in South Borneo at no great distance

28

from the spot where olivine-hypersthene-norite is exposed, the position of which is indicated on map X^A.

A few dykes and small bosses of gabbro were found on the Upper Kapoewas, above Poetoes Sibau, within the area of the pre-Cretaceous Danau formation (see map VII), and the many boulders of gabbro in the valley of the Mědjoewai, prove that this rock must occur there also. It is almost certain that the gabbro is younger then the Danau formation.

4. PERIDOTITE AND SERPENTINE.

In the area of the Danau formation on the Upper Kapoewas, at and above Poelau Těngkidoe, a few small massives are found two of which are composed of pikrite, one of Harzburgite and one of Lherzolite, which rocks have almost entirely been altered into serpentine. In this same region I found numerous boulders of serpentine, principally derived from Harzburgites, carried down by the Kěryau and the Mědjoewai.

In the Sěbilit serpentine occurs both *in situ* near N^a Sabat and in boulder-banks. To some extent it is serpentinized Harzburgite.

I also found serpentine, derived from Lherzolite, in Sarawak near to the Dutch frontier, on the path from N^a Badau to Loeboek Hantoe.

In all these instances — except on the Sěbilit river where it is somewhat uncertain — the serpentine appears either as intrusive bosses, or or dykes in the Danau formation, and must therefore be of later date than this.

5. ROCKS OF THE DIABASE FAMILY.

It has already been remarked, when describing the Danau formation (p. 414) that it is largely composed of beds of diabase-porphyrite, amygdaloidal diabase and diabase-tuff, and that some varieties, as for instance the silicified diabase-tuff which I have called Poelau Mélaioe rock, are quite as characteristic of this

formation as the chert with Radiolaria. Most of these diabase
rocks are of the same age as the other rocks in the Danau
formation, a few however are probably of more recent date.
We know this for certain of some dykes of diabase exposed on
the Boengan which cut through the chert.

Apart from the Danau formation, rocks of the diabase family
are seldom found in Central Borneo. The following varieties
only remain to be mentioned.

a. Diabase-porphyrite and labrador-porphyrite, found in dykes
in the granite of the Kénépai mountains, and also composing
the highest cone of the same mountains.

b. Diabase-porphyrite near riam Batoe Tossan, and olivine-
diabase, remarkably like dolerite, near riam Pandjang on the
Lèkawai. With regard to the age of these rocks we only know
that they are overlain by Tertiary deposits.

c. Diabase-porphyrite composing the highest tops of the Raja
mountains.

This diabase-porphyrite is very variable in its composition; it
generally contains quartz and then gradually passes into quartz-
porphyrite, while at other places it shows such a decided andesitic
character, that it might almost be determined as augite-andesite.

All these diverging types are classed together on maps II
and IX^c under the name of Raja-porphyrite. Dykes of this porphyrite cut through the granite and the tonalite of which the
base of Mt. Raja is composed.

Diabase-porphyrite appears moreover as solid rock on the
northern slope of Mt. Damar in the Schwaner range, and also
in boulders in the Lèkawai and other rivers which have their
source in the Schwaner mountains.

d. Diabase of Mt. Kaki near the Java Sea.

6. QUARTZ-PORPHYRITE AND QUARTZ-PORPHYRY.

The most remarkable occurrence of quartz-porphyrite in Central Borneo, is without any doubt at Mt. Kélam. This mountain
rises abruptly from the alluvial plain as a gigantic rock mass

of microgranitic quartz-porphyrite. The neighbouring lesser
heights of the same shape, as for instance Mt. Rèntap, consist
probably of the same material. Its connection with other rocks
is unknown, but we may conclude that it forms part of one or
more dykes of quartz-porphyrite which to a great extent have
been destroyed by erosion.

In other places quartz-porphyrite is only found very occasionally. It appears in a few places on the Embahoe river just
below N⁸ Tèpoewai, in dykes cutting through the rocks of the
pre-Cretaceous formation, and on Mt. Raja in dykes through
the granite.

Quartz-porphyrite was observed in the boulders carried down by
the Sèbèroewang, the Embahoe, the Mandai and the Kalis; also
in the Lèkawai and the Samba and their tributaries, so far as
these have their source in the Schwaner mountains. Quartz-porphyry appears in the boulders of the Tèpoewai and the Mandai.

And, finally, there are various rocks in Central Borneo which
looked at separately (individually) would be called quartz-porphyrite but which are proved by careful examination in the field
to be merely local modifications of other rocks, to which they
are related through various stages of transition. Thus, in the
Pyaboeng range, I found quartz-porphyrite as a local modification
of quartzose amphibole-porphyrite.

In the eastern portion of the Müller mountains and in the
valley of the Kèryau I found quartz-porphyrite related, by various
transitions, to dacite and mica-dacite, it is doubtless a somewhat altered variety of the dacite and mica-dacite originally having
been solidified at some depth below the surface.

7. AMPHIBOLE-PORPHYRITE.

Amphibole-porphyrite abounds in Central Borneo. In its geological occurrence it is so nearly related to volcanic formations,
that I have considered it best not to separate the two; its prop-
erties will therefore be discussed together with those of the volcanic formations.

8. Volcanic rocks.

Under the name of "volcanic" rocks a great variety of types have been classed together (on map II); as they are hard to separate on geological grounds. The following regions of past volcanic activity exist in Central Borneo.

a. *The Müller mountains.*

The Müller mountains, forming the southern boundary of the Upper Kapoewas plain, extend in an E. 10° N. direction from the Embahoe river right across the central part of West Borneo, and further eastward to an unknown distance. The whole of this range, or more correctly the higher elevations which make this a mountainous district, consist of volcanic rocks. Orographically and geologically these Müller mountains can be divided into three parts.

α. The western division to the west of the Soeroek valley.

This division is characterized by the appearance of numerous conical mountains, either detached or grouped together in larger or smaller groups. They consist of rocks comprising transitions between amphibole-andesite, amphibole-dacite, amphibole-porphyrite and quartz-amphibole-porphyrite. As regards their habitus they have a decided andesitic character, so that the determination in the field is for most of them 'andesite", while a microscopic examination brings a large portion of them under "porphyrite". Almost all these rocks are characterized by the presence of a fair amount of quartz in the ground-mass. Most probably they are modifications of andesite and dacite now exposed by erosion, but which originally, during their solidification, were at some depth below the surface.

β. The middle division.

The middle division of the Müller mountains is bounded on one side by the Soeroek valley, but the boundary on the other side is not accurately known — it must be looked for in the Kéryau basin. The mountains here have a very striking character. They are table-mountains with clearly defined terrace structure, and separated from each other by deep valleys. The table-mountains consist of thick beds of volcanic tuff, occasionally alternating with horizontal or almost horizontal flows of andesite or basalt. The volcanic rocks here are less acid than those in the western, and also, as we shall see presently, than those in the eastern division. The predominating types are: hypersthene-andesite, enstatite-andesite, augite-andesite and basalt. Slaggy modifications occur frequently. The fine material of the tuffs, the volcanic ashes, consists principally of hypersthene-andesitic material. The total thickness of these tuff-deposits may be estimated, for the Mandai district, at 1200 metres, but it was doubtless considerably thicker originally. Amongst the highest peaks in this part of the Müller mountains are; Mt. Tiloeng 1112 metres, Mt. Toengoen, the highest top of the Gĕgare-massif, 1390 metres, Mt. Liang Koeboeng 1332 metres, and Mt. Pali 1338 metres.

γ. The eastern division.

This is a continuation of the middle division in an eastern direction and it extends to an unknown distance on the other side of the boundary between West and East Borneo. The scenery here is more in keeping with that of the western than with that of the eastern division. Many volcanic mountains, either isolated or united in groups, occur here, and by their greater height and grotesque shape they more strongly attract one's attention than those in the western division. In some places they are surrounded and perhaps mutually connected by tuffs, which however are never so strongly developed as to govern the character of the landscape.

In this eastern part of the Müller mountains acid rocks have the predominance. They are of the rhyolitic and dacite type, which is also represented in various, mostly vitreous, modifications, such as rhyolite-pitchstone, dacite-obsidian, etc. As far as I could see all the tops of the volcanic mountains consist of these acid rocks, chiefly dacite and mica-dacite, while a little lower down their slopes andesite predominates, of which, in many cases the base of the actual volcanic mountain seems to be built up. Varieties of mica-andesite and mica-dacite of an ancient character, are of frequent occurrence, and if only microscopic characteristics were to be considered some of them ought to be called kersantite.

Some of the highest peaks of the Müller mountains fall into this eastern division, Mt. Lĕkoedjan 1190 metres, Mt. Tĕrata 1467 metres, Mt. Pémeloewan 1340 metres, and Mt. Sara 1317 metres.

As to the age of the volcanic Müller mountains, we may remark as follows:

The Müller mountains are the result of very prolonged volcanic action. The Beloewang mountains, including Mt. Pyaboeng and Mt. Oejan [1]), appear to me to be the oldest part, and I could not say with certainty that here the eruptions did not occur before the Tertiary sandstone formation had been deposited, although in section LL′ I accepted the opposite view, as the more probably correct one.

The andesite mountains of the Embahoe, Sĕbilit and Tĕbaoeng districts, in fact all that remains of the western part of the Müller mountains cannot be of a much more recent age than the Béloewang range.

I believe the eastern part of the Müller mountains to be younger than the western part, for here the volcanic formations distinctly overlie the sandstone formation, whereas fragments of the latter are found in the lavas [2]) of the eastern division. When the volcanic forces began their work here, the shape and division of the valleys were practically the same as they are now, and this might induce us to consider these eruptions as comparatively

---

1) On map II the rocks of these mountains are separated from the volcanic rocks of the Müller mountains as porphyrites of unknown age.

2) It should be remembered that the sandstones, arkoses, and clay-slates which underlie the volcanic formations in the Boewang valley and at the watershed between the Abang and the Pĕoaneh, have merely been reckoned to belong to the Tertiary sandstone formation because of their horizontal position, but as regards their petrographic composition, they bear a greater likeness to the Cretaceous deposits of other parts of Central Borneo.

recent, but, on the other hand, the extensive encroachments made by erosion in the volcanic area prove that, geologically speaking, these eruptions cannot possibly be called recent.

The middle division, the tuff-beds of the Mandai, etc., is unquestionably the youngest. The volcanic formations distinctly overlie the very latest of the coal-beds of the Tertiary formation, and the eruptions must have begun while these latest strata were being formed. We remarked p. 428, that these most recent strata of the sandstone formation, in which coal is found in various places on the border of the Upper Kapoewas plain, are possibly younger than the sandstone and clay-stone of the Madi plateau, which latter ought perhaps to be separated from them as belonging to an older group. The erosive and transporting action of water has not yet succeeded here in carrying away the loose volcanic ejectamenta, which, in the eastern division, have been partially, and in the western, entirely removed. We may conclude that the Upper Kapoewas plain at the time the volcanoes were in action, existed almost exactly as we see it now, because we find that some streams of basalt have flowed from the Müller mountains in the direction of the plain, one of them even reaching the Kapoewas river (see p. 244). The average height of the mountains, i. e., the fragments from a formerly unbroken tuff-plateau, now separated one from another by erosion, is about 1100 metres, but judging by the enormous alterations wrought here by running water we may safely conclude that at one time the tuff-plateau was a great deal higher. Centres of volcanic activity, volcanoes in a narrower sense, have, so far, not been found in this region, a phenomenon which, I think, can only be explained by the fact that as yet only small tracts of the margin of this region have been carefully examined. I estimate the average width of the Müller mountains in this middle division to be not less than 45 kilometres. In the tuff of the Mandai district large quantities of silicified wood [1]) are found, partly

---

1) This material is still waiting for examination by a specialist.

as tree-trunks still standing erect in the tuff-beds. When these pieces of fossil wood have been determined we shall probably be enabled definitely to fix the time in which these volcanic tuffs were formed, but judging from the general habitus of the rocks I do not hesitate to state, even now, that the volcanic action in the Müller mountains has ended long since, and that as a consequence these volcanoes must be of much older date than those of the Sunda islands and of the Moluccas.

In summing up I regard the Müller mountains as the result of prolonged volcanic action, which beginning during, or shortly after the Cretaceous age, continued for a considerable time during the Tertiary period, and terminated long ages ago, probably before, or at the beginning of the Quaternary period. These volcanic mountains are built up on a line or system of lines of fracture the trend of which corresponds with that of the other chief lines of dislocation, which govern the present relief of Central Borneo.

b. *The porphyrite cones on the northern slope of the Schwaner mountains.*

A line of conical hills rise in an almost east and west direction on the feebly sloping, sandstone plateau which forms the northern incline of the Schwaner mountains. Of these hills I examined Mt. Sĕdaroeng, Mt. Moeloe and Mt. Lĕmoekoet in the source area of the Lĕkawai. They consist of a rock resembling in its structure both quartzose amphibole-andesite, and quartz-amphibole-porphyrite. On geological grounds, I am of opinion that these rocks are younger than the sandstones of the Schwaner mountains. On the maps and in the diagrams I have, as a rule, marked them as porphyrite, nevertheless I see in these conical hills denuded centres of volcanoes. The porphyrite cones evidently stand on a line of fracture running almost due east and west, the extent of which is not yet known.

c. *The volcanic region of the Samba river.*

Near Toembang Běrahooi a stretch of volcanic formations, about 18 kilometres in width, breaks through the extensive granite area of the Samba. This volcanic range trends, on an average, E. 10° N.—W. 10° S. It consists of cupolar, conical, and column-shaped hills of amphibole-andesite and augite-andesite; olivine occurs in some of the rocks, which then form transition types into dolerite. Microscopically these andesites show a structure approaching to that of porphyrites, but not to so large an extent as the andesites of West Borneo. I believe that these andesite hills are also the denuded centres of volcanoes. All the loose volcanic material ejected has not yet been carried away by denudation, and the low ground surrounding these conical hills, consists largely of andesite-tuff and tuff-breccia, as we learn from the section on the banks of the Samba. Nevertheless it would be premature to conclude from this, that these volcanic formations are of more recent date than, for instance, those of the Embahoe where no loose volcanic ejectamenta are found. For the volcanoes of the Samba are built up along the edges and in a *Grabenversenkung* in the granite, which probably has increased in depth after the volcanic action had come into operation. The tuffs and breccias accumulated in the *Grabenversenkung* occupied therefore a lower level in comparison with the surrounding granite district, and consequently they were not so much exposed to the denuding and erosive forces. On good grounds we may surmise that these tuffs contain silicified wood, and it is therefore highly probable that a comparative examination of this fossil wood and the similarly silicified wood found in the tuffs of the Mandai in West Borneo, will teach us something concerning the relative age of the volcanic tuff-region of the Müller mountains and the volcanic hills of the Samba. The fault-lines which produced the *Grabenversenkung* in the granite region of the Samba, in which and on the edge of which the volcanoes arose, all trend in an E.N.E.—W.S.W. direction and correspond therefore, as regards their direction, with all the other great lines of dislocation of Central Borneo.

———

Apart from these three principal volcanic regions, volcanic rocks were only found in a few detached localities in Central Borneo. Thus, for instance, on the Panai river, an andesitic quartz-amphibole-porphyrite dyke occurs (see map IX^A), breaking through probably Cretaceous deposits. The boulders of volcanic rocks found in the karangans in various places, and mentioned here and there in previous chapters, can almost all be traced back, with a fair amount of probability, to one of the above named volcanic regions. Exception must be made to a few boulders of andesitic amphibole-porphyrite, and of amphibole-andesite, found in the Panai river, and to a piece of augite-andesite found in the bed of the Tĕbaoeng near riam Nĕkan. The origin of these rocks is unknown. The isolated andesitic porphyrite dyke of the Panai, just referred to, gives us a clue to a possible explanation.

Summarising the principal facts regarding the appearance of volcanic formations in Central Borneo, it is evident that no volcanic rocks are found north of the Kapoewas, that to the south of that river there are three volcanic regions extending in long strips almost due east and west and probably connected with dislocations trending in the same direction. The principal of these three regions of volcanic activity is the Müller range, which is known to extend over 280 kilometres, but probably stretches much further eastwards. The two other volcanic areas, the andesite hills on the northern slope of the Schwaner mountains and the volcanic region of the Samba river, are certainly of less importance than the Müller range, but their extent cannot, as yet, be estimated.

II. THE TECTONICS OF CENTRAL BORNEO.

The principal directions which govern the tectonics of Central Borneo run almost due east and west. Proceeding from north to south therefore, it is comparatively easy to obtain a fairly correct estimate of the orographical importance of the various geological formations and of their relation with regard to one another. For the description of the various subdivisions we refer to previous chapters.

The northern strip of the west part of Central Borneo is occupied by the Upper Kapoewas range. These mountains extend beyond of the frontier into East Borneo for an unknown distance. Orographically they still occupy an important position and form a typical mountain chain, although much worn down by erosion. They are built up of the old slate-formation considered to be the oldest of the sedimentary formations of Borneo. This range terminates abruptly on its south side, probably against a large fault[1]) which I have called the great Upper Kapoewas fault, and which has brought a younger formation down to the same level with the old slate-formation. This younger formation is the Danau formation, which in the hilly district of Boengan, south of the great dislocation, immediately adjoins the rocks of the Upper Kapoewas formation (see section CC'). Further westward, however, it only appears occasionally from below the much more recent fluviatile deposits which cover the greater part of the Upper Kapoewas plain (see sections BB' and HH'). The whole of the mass, which was broken off and sunk along the southern limit of the Upper Kapoewas range (and to which the Upper Kapoewas plain owes its existence), appears to dip slightly northward, and consequently has not sunk so much on the south side. There are no very clear indications as to whether the south side of the Upper Kapoewas plain also terminates in a fault, but we there find the undisturbed upper strata of the sandstone formation lying everywhere unconformably upon the pre-Cretaceous formation, and dipping feebly towards the

---

1) It is evident, that, the strata of the old slate-formation and the adjoining Danau formation both being highly tilted and folded, the abrupt termination of the Upper Kapoewas range to the south and the position of the two formations of different age on the same level could also be explained by differences in resistance to erosive forces without accepting the existence of a fault.

Kapoewas plain. The pre-Cretaceous formation crops up again here and there further south. Continuing in a southerly direction we notice a striking difference between the general structure of the western part of the Dutch territory and the more easterly districts. In this latter region the sunken Upper Kapoewas block terminates southward against another fracture, and a fault of the same type and trend as the great Upper Kapoewas fault has thrown the southern territories once more down to a lower level. Thus we see (section Nᵃ Boelit-Pĕnaneh) how the sandstones of the Müller mountains with a feeble northern dip, terminate near the Boewang river against the folded Cretaceous deposits of the Boengan hills which lie there on the same level. Again, near Mt. Bĕransa the more southern territory has been thrown down when compared with the northern districts, and the movement has given rise to strong disturbance of the strata of the Tertiary formation. The volcanic Müller mountains are built up, either (as near the Boewang) on the northern edge of the sunken tract, or (as near the Tĕbaoeng, see section EE') on the southern edge of the upper limb of the fault; in both cases, however, it seems evident that there is a close connection between the volcanic mountains and this dislocation, and it is more than probable that not merely one but several faults, all trending in the same direction, have given the volcanic material an opportunity of reaching the surface. This connection is further proved by the fact that the general trend of the Müller mountains and the trend of the faults are in perfect accordance.

In the western part of the Dutch territory (the great lakes and the Sĕmitau hills), we find no trace of this second dislocation, of the same type as the great Upper Kapoewas fault (see section HH'). Looking at it merely in broad outline we might imagine here a feeble anticlinal structure, in which the Sĕmitau hills form the axis of the anticline. This structure however is marred by several dislocations, the principal of which have been marked in the sections. The direction of these fault-lines is again almost due east and west. In this western division the Müller mountains break up into separate small groups and soon disappear altogether.

Returning to the more easterly divisions where our investigations much further to the south make one connected whole, we find that this same typical structure repeats itself several times.

Faults trending almost due east and west have divided the territory into tracts, the more southerly of which always occupies a lower level in relation to the one adjoining it on the north side. Each tract is slightly tilted northward, and consequently all the strata in these several tracts have a slight southerly rise. We have noticed how the great Upper Kapoewas tract, consisting chiefly of folded pre-Cretaceous strata, rises feebly to the south. On the south side near the Tĕbaoeng, between the riam Nĕkan and Mt. Béransa, this tract is broken off by faults trending east and west, in consequence of which the beds of the sandstone formation in the Béransa mountains have also been tilted to a vertical position. Then follows — at a lower level — the tract composing the Madi plateau. This tract also gradually rises towards the south and is there broken off by the Mĕlawi fault or great Madi fault F_s. This fault has occasioned the steep southern incline of the Madi range, with an east to west trend, which must once have been considerably higher and situated more to the south (see section EE′), but erosive forces have shifted it gradually to its present more northerly position. The so called Madi mountains (Madi-tentoe) which, seen from the Mĕlawi valley, rise as a commanding rock-wall 1000 metres in height, are therefore, in reality, merely the northerly displaced broken edge of the Madi plateau. Between the Mĕlawi and the Madi plateau the sandstone formation has been entirely carried away by erosion, and the underlying, probably Cretaceous, strata have become exposed.

The same type of structure is again repeated, in the more southerly Schwaner mountains. On the north side this range also consists of a feebly inclined sandstone plateau, and terminates abruptly towards the south in a steep incline. In this range the sandstone overlies the granite which occupies the greater part of the lower ground of the Samba hilly district. There can be little doubt that faults with an east and west trend, have abruptly terminated the Schwaner plateau towards the south, but so far our researches are insufficient to localise these dislocations. And finally there is a remarkable *Grabenversenkung* in the granite-region of the Samba, where the beds of the sandstone formation have once again been brought to the same level with the granite. Volcanic rocks have been brought to the surface along the lines of dislocation and cover the greater part of the sunken tract. South of the *Grabenversenkung*, granite is also exposed, but there it gradually disappears underneath the recent, fluviatile deposits, which from thence as far as the Java Sea, hide all further particulars regarding the tectonics of Borneo.

Summing up, we find that the structure of Central Borneo is governed by lines of dislocation with an almost due east and west trend. Along a strip of country bordering the 113th degree of east longitude from Greenwich, they have given origin to a type of crustal structure which American geologists have termed 'Basin Range' structure [1]).

III. HISTORICAL GEOLOGY OF CENTRAL BORNEO.

Considering our extremely fragmentary knowledge of the geology of Borneo it would certainly be premature to attempt to give a sketch of the origin of this island. In this chapter therefore we will only shortly note the possible chronological order of the different geological events which have made Central Borneo what it now is, as far as we can deduce it from the known facts.

The crystalline schists, the sedimentary deposits altered by metamorphism in contact with the granite of the Schwaner mountains, and the old slate-formation, cannot be included in our summary, because practically nothing is known with regard to their age in relation to other formations.

The only thing perhaps worth mentioning is that the deposits

1) J. W. POWELL, Geology of the Uinta-Mountains 1876, p. 16.

of the old slate-formation were most probably laid down at no great distance from the coast.

Once in pre-Cretaceous times Central Borneo lay buried deep under the sea, with no dry land any where near. The skeletons of pelagic living organisms sank down to the bottom of this sea, and as a rule only those composed of silica, were preserved at this great depth where they formed deposits on the ocean floor consisting almost exclusively of the tests of Radiolaria. Contemporary with this, submarine eruptions poured streams of diabase over the ocean floor, while further sediments of diabase-tuff-material, originating partly from submarine eruptions, and partly perhaps from the sunken ashes of volcanoes existing on islands in the ocean were deposited. This oceanic "ooze" became converted into chert and hornstone with Radiolaria, now found in strata, alternating with altered and silicified varia-ties of diabase and diabase-tuffs. We do not know the geolo-gical period when the upward movement began which was to make dry land of this deep-sea region, we only know for certain that in Cretaceous times (in the Cenomanian, horizon of *Orbitolina concava*, Lam.) a portion of Central Borneo must have been dry land, for the character of the deposits formed at that time shows that they originated close to the coast. This is very clearly proved by the Cretaceous conglomerates exposed in the Sĕbĕroewang district, for instance near Mt. Oejan. These are shore-formations. The land which furnished the material for these conglomerates lay probably in a southern direction, and we may presume that the conglomerates were deposited along the north coast of the then existing land of Borneo. This was probably a high land stretching far to the southward and embracing the Schwaner mountains. Mt. Kĕlam, a gigantic dyke of quartz-porphyrite, which afterwards by erosion was left standing up as a wall out of the denuded country, may be considered as a remnant of it. Mt. Kĕlam indicates to us the enormous tracts of land which have here been carried away through erosion. Simultaneously with the pro-cess of land-forming, the folding of the strata commenced with

its accompanying dynamo-metamorphosis, and I consider it probable — although not proved — that the pre-Cretaceous strata
were already folded in a generally east and west direction at
the time when the Cretaceous sediments were deposited. Granite
generally forced its way into the cracks or folds of the disturbed
rocks; these intrusive masses did not cause eruptions but became
solid rock at a certain depth and strongly metamorphosed the
adjoining rocks. To the granite intrusions of that time I reckon
amongst others, the granite of Mt. Kĕnĕpai, of the Mĕnjoekoeng, of the Embahoe, of the Upper Kapoewas, and of the
Boelit. It is not known whether the great granite massives of
the Schwaner mountains and of the Samba were also forced up
at that time.

During the Cretaceous period, or at least until the close of
the Cenomanian age, the folding process continued steadily by
pressure from the same direction, at any rate in the Upper
Kapoewas district, with the result that there the Cretaceous
deposits were folded into the pre-Cretaceous deposits so as to
form troughs. To the south of the Sĕmitau hills the crumpling
was not so intense.

Towards the end of the Cretaceous age or at the beginning
of the Tertiary period the folding ceased and was followed by
a transgression during which the whole of Central Borneo, with
the exception, probably, of the Upper Kapoewas mountain chain,
sank below the surface of the sea. Unconformably on all older
formations the beds of the Tertiary sandstone-formation were
then deposited.

This however was soon followed by movements in a contrary
direction. The submerged land was upheaved again and in such
a manner that in rising it canted over a little towards the north.
In the north therefore the actual upheaval was very slight,
sometimes even the general level became a little lower, while
more to the south the land was brought up to a considerably
higher level. The great Upper Kapoewas fault dates back to
this early time. It still forms the boundary between the Upper

29

Kapoewas mountain range, and the Upper Kapoewas plain. This boundary became even then one of the chief drainage lines through which the Kapoewas discharged towards the west. The river then, as now, curved southward in the neighbourhood of Sĕmitau, probably stopped in its course further westward by the granite massives of the Kĕnĕpai, the Toetoep, etc., which then formed a much higher and more connected range than they do now. And in proportion as the land immediately adjoining the southern edge of the Upper Kapoewas range, either sank or remained at the same level, and the land in the south continued to rise, the Kapoewas cut its bed deeper down and finally formed a deep valley in a territory situated at a much higher level than the greater part of the area of its upper course.

As might be expected however, the eroding action of the Kapoewas and the changes of level caused by the canting over of the rising tract by no means always counterbalanced one another. This would have been highly improbable and as a matter of fact did not take place, because the tract, during the process of rising and canting over, was broken up into smaller areas, and in the eastern portion of the region we traversed, there was formed a series of step faults with the architectonic character of the Basin Range type. In the western division the upheaval is also marked by several faults. All these dislocations trend in almost due east and west direction, i. e., parallel to the great Upper Kapoewas fault, and therefore parallel also to the turning axis of the upheaved tract.

Owing to these disturbances in the regularity of the great dislocation movement, the Upper Kapoewas plain was sometimes dry land, at other times an immense freshwater lake, with temporary invasions of the sea, leaving behind them brackish-water deposits. These oceanic invasions were certainly not confined to the Upper Kapoewas plain, and the strata of the Mélawi-group (see p. 427) probably owe their formation to one of these invasions of the sea into the low lying district which formed the boundary between the Madi mountains and the Schwaner range, these being the two chief portions into which the rising tract was divided up.

During the long period in which this tract was broken up by a great system of parallel faults, the volcanic forces were also at work along some of the fissures, creating ranges of volcanic mountains with an east and west trend. These volcanic ranges vary considerably in age, and we must assume that, during the breaking up of the great tract, volcanic activity was at work sometimes in one place, sometimes in another.

To some extent, as for instance in the Běloewan range, this volcanic activity was in operation before the above mentioned system of faults began to be developed, and to some extent the volcanic activity continued (as for instance in the tuff-mountains of the Mandai) after the general configuration of the territory was already very nearly if not altogether similar to that of the present day [1]).

Of course during all this time erosion was busy reducing the existing land — for instance the Upper Kapoewas mountain chain — and as the new land rose out of the sea, erosion exerted all its powers in trying to level it again. These persistent attacks have worn down the Upper Kapoewas range to the ruined chain we now see. The steep escarpments which in the broken tract of the Basin Range type would have orographically marked the boundaries of the different blocks, were attacked by the erosive forces as soon as they were formed, and were demolished or displaced towards the side of the upthrow. Thus, for instance, the steep southern incline of the Madi

---

1) This great and last upheaval of Borneo, which, apart from the secondary movements, the canting over and breaking up of the tilted tract, may be called a period of continuous negative coast-displacement, is, I believe, connected with similar movements which have occurred in Tertiary times in various other parts of the Indian Archipelago, Timor, the Kei islands, etc. In spite of this similarity in the results, I consider that our knowledge of the exact time at which these movements took place and of their mutual connection is, as yet, of too fragmentary a nature to justify our drawing the conclusion that these negative coast-displacements date from Tertiary times, and were caused by the receding or flowing away of the water of the ocean.

plateau was shifted more northward, and the steep slope on the south side of the Schwaner mountains is probably the northern displacement of the steep edge of a similar fault the original position of which has not yet been discovered. A large proportion of the newly formed sandstone was eroded immediately after its upheaval, and the sand was probably deposited again at a short distance off in running water, thus forming a fluviatile sandstone which can scarcely, if at all, be distinguished from the sandstone from which it was derived.

Not only erosion, however, but vegetation also seized upon the new land. Forests covered the sandstone plateaus and the rivers then, as now, carried great quantities of vegetable detritus and drift-wood. From the Madi plateau especially large volumes of water flowed northwards towards the Kapoewas plain, and thus along the southern edge of this plain, a fluviatile, recent formation of sandstone and clay-stone was deposited. Petrographically this cannot be distinguished from the sandstone of the sandstone-formation, from which its material was derived, but it contains numerous particles of coal, and thick and thin beds of brown-coal, originating from the large quantities of vegetable débris and drift-wood which have been carried down by the streams from the Madi plateau [1]).

In the same manner we see, even now, that sand and clay with intercalated layers of vegetable material and tree-trunks are being deposited along the Mandai river. Indeed, to my mind, there is no essential difference, but rather a gradual transition from the sandstones and clay-stones with intercalated beds of brown-coal which now, on the Upper Mandai, underlie the tuff-beds of the Müller mountains, to the sand- and clay-beds with interstratified layers of pressed, half-decayed remains of plants and

---

1) This also occurred during the period or periods of invasion of the sea, and may possibly account for the fluviatile sandstones, brackish-water clay-stones with fossil shells and coalbeds, alternating and in close connection with one another, which are exposed on the southern edge of the Upper Kapoewas plain and in the Mēlawi valley.

trees which even now are deposited along the Lower Mandai during the annual floods[1]).

When the crust movements here described, which probably lasted from the latter part of the Cretaceous period till far into and per- haps to the end of the Tertiary period, had subsided, there does not seem to have been any further upheavals of importance up to our time, and all the subsequent changes in the orographical appearance of Central Borneo appear to have been caused by the altering or displacing influences derived from the atmos- phere. These have doubtless greatly diminished the general height of the island. Through them the mountain slopes were covered and the valleys filled with gravel- and sand-deposits, all more or less gold-bearing, while thick beds of fine sand were depo- sited over the plains. The circumference of the island was enlarged through the rapid deposition of the alluvial materials carried by the rivers down to the sea. Thus the delta of the Kapoewas was formed which doubtless has surrounded and annexed some of the coast-islands now rising as small mountain groups from the low lying land. On the south coast of Borneo the acquisition of land through alluvial deposits must have been still more extensive. The Schwaner mountains with the sand- stone plateau overlying them, must have reached much further southward at one time, and were therefore also considerably higher than they are now, and while this range was, as it were, pushed back northward by the erosive forces, the rivers flowing southward deposited enormous quantities of gravel, sand and mud derived from the Müller mountains, by which process much land was acquired and the coast-line of South Borneo was pushed back southward. This process still continues, but naturally with slowly decreasing intensity.

In the whole of Central Borneo I have not found one single fact or phenomenon from which one might deduce that in Borneo during quaternary or recent times, a negative displacement of the

---

[1] On page 44 the origin and the appearance of these layers on the Mandai is described.

coast-line has taken place. For South Borneo, however, we must admit that our knowledge, especially as regards the strips of high land which separate the broad swampy valleys of the great rivers, is as yet very slight, and certainly quite insufficient to enable us, with any chance of success, to draw conclusions concerning a geological question of so vast a compass.

From the above we see that the constant action of erosion during a long period of rest — as far as the crust movements are concerned —, has remodelled the Borneo of the Quaternary age, which we may suppose to have been a high and very broken mountain land, into the Borneo of to-day, a worn, much denuded mountain land surrounded for the greater part by broad tracts of low land covered with fluviatile deposits. The Borneo of to-day is a typical example of what must be the result of prolonged and intense erosion of a mountainous island surrounded by a shallow sea.

With a few words we may finally point out wherein the illustration here given differs from that which may be called the generally accepted idea respecting the latest period in the geological history of Borneo. For, strange to say, notwithstanding our very inadequate geological knowledge of Borneo there exists a fairly well defined theory regarding the origin and growth of the Borneo of the present day. This theory is based in principle upon a supposed homology in structure between Borneo, Celebes and Halmaheira, which finds expression in a conformity of trend and of ramification of the principal mountain chains. HORNER in 1837 [1]) was the first to draw attention to this conformity. He gives in broad outline the theory concerning the latest episodes in the geological history of Borneo. In 1882 MARTIN gave it more in detail [2]). He says: "An ein aus Graniten, Syeniten und Dioriten, Serpentinen und krystallinischen Schiefern bestehendes Grundgebirge schliesst sich eine von

---

[1]) L. HORNER *22*, p. 104. HORNER's report was handed to the government in 1837.
[2]) K. MARTIN *33*, p. 145.

jüngeren Eruptiv-gesteinen durchbrochene Tertiär-formation an, welche aus Sandsteinen, Mergeln und Kalken besteht. Die Letzteren repräsentiren ein tertiäres Korallenriff, welches die Insel in einer Zeit umsäumte, in der ihre allgemeine Form derjenigen von Celebes glich. Nachdem die Tertiär-formation über den Meeresspiegel sich erhoben, wurden die tiefen Buchten Borneo's von den Zerstörungs-producten der genannten Gebirgsarten, unter denen auch Gerölle der Tertiär-formation zahlreich vertreten sind, ausgefüllt."

Later, in 1890, MARTIN [1]) emphasized more especially that during the Quaternary period the upheaval of marine sediments must still have been going on in Borneo as well as in other parts of the Indian Archipelago.

Lately MARTIN [2]) has more strictly formulated this theory, and he has inferred that a negative shifting of the coast-line has taken place in Borneo during the Quaternary age.

POSEWITZ' [3]) representation differs but slightly from this. He says: "Die Gestalt Borneo's bis zum Beginne der Tertiärzeit glich einem ausgedehnten Insel-archipel, in welchem theils kleinere Inselgruppen, theils grössere Eilande von der See umspült wurden." His opinion is that in Tertiary times sediments had been deposited in the sea round these islands and that when these regions were made dry land the islands became for the greater part connected, and Borneo assumed a shape similar to that of Celebes and Halmaheira at the present day. He continues: "Zu Beginn der Diluvialzeit begannen die Seebusen langsam auszutrocknen; es bildete sich ein Streifen flachen Landes längs dem Gebirgsrande und z. Th. auch im Gebirgslande selbst und Gold, Diamanten, Platin hergeschwemmt wurden hier abgelagert. Es entstand die Jetztzeit, indem die Seebusen langsam

1) K. MARTIN. Die Kei-Inseln und ihr Verhältniss zur Australisch-asiatischen Grenzlinie. Tijdschrift van het Kon. Ned. Aardrijkskundig Genootschap VII, pp. 241—280. Comp. especially pp. 262—267.

2) K. MARTIN 37, p. 25.

3) TH. POSEWITZ 45, p. 197 et seq.

an Tiefe verloren und sich zurückzogen. Mächtige zahlreiche Ströme bahnten sich ein Bett in den ausgetrockneten aber noch sumpfigen Niederungen und strömten majestätisch dem weit zurückgedrängten Meere zu." Did POSEWITZ here refer to the negative shifting of the coast-line or to the filling up of the bays by alluvial deposits without the help of shifting of the coast-line? The former is the more likely, but there remains room for some doubt as to POSEWITZ' actual meaning.

And now, since our knowledge of Borneo is considerably greater, the question is how much of this theory can be accepted, and what must be rejected as improbable or incorrect?

It is certainly quite incorrect to say that Borneo has a central mountain range, analogous in shape and build to the central ranges of Celebes and Halmaheira [1]). Nor can we accept that this mountain mass formed one island or an archipelago of islands consisting solely of archaean or palæozoïc rocks, which at the beginning of and throughout the Tertiary age formed a nucleus of land untouched by the sea.

The nearly horizontal sandstone strata which are found up

---

1) WICHMANN [1]) pointed out in 1893, with Celebes for his starting-point, that the supposed homology between Borneo, Celebes, and Halmaheira, does not exist. And, relying upon the latest discoveries made in Central Borneo, I came to the same conclusion in 1895 [2]).

The fact that this theory, the incorrectness of which could be proved so easily, has remained so long in existence, partly finds its explanation in that this idea of homology adapted itself splendidly to schematic deductions, and became a welcome hobby for discussing the origin and mutual relationships of the relief forms on the earth's surface.

To some extent the propagation of the error must be laid to SCHWANER's [3]) charge, whose authority in every other respect is perfectly reliable, and to whom, up to quite lately, we owe almost all our positive knowledge concerning Central Borneo. SCHWANER agrees entirely with HORNER, and when describing the mountain-system of Borneo he speaks of: "a central mountain mass, from which radiate the ranges which separate the different water-sheds and constitute the frame of Borneo, dividing the whole into four principal parts." When we closely study SCHWANER's wanderings we can plainly see how he came to hold this view, but I am convinced that if SCHWANER had travelled in the central part of West Borneo he would never have written these words.

---

1) A. WICHMANN 67, p. 225.
2) G. A. F. MOLENGRAAFF 40, p. 506.
3) C. A. L. M. SCHWANER 55, I. p. 1.

to a very considerable height in the Schwaner mountains prove sufficiently that in Tertiary times, (probably at the beginning of this era) the whole of West Borneo was covered by the sea, with the exception perhaps of the Upper Kapoewas range and a few of the highest tops of what are now known as the Schwaner mountains. Following this, in Tertiary times and possibly during the first part of the Quaternary period, the great upheaval and breaking of the tilted tract took place, as already described. This upheaval appears to have been felt also in South and East Borneo, for along the Teweh, the Barito, in Tanah-Laut, etc., Eocene and recent Tertiary marine sediments occur, and now form a hilly district. But with the exception of Tanah-Laut, which occupies a more detached position, so little is known concerning South and East Borneo that we are unable to decide whether the movements which raised the sandstones of the Schwaner mountains up to their high level, have been at work here also.

One thing is correct in the old theory, viz., that the Quaternary age was an age of considerable acquisition of land by the forming of alluvial deposits, a process which goes on to this day.

Some alterations, however, should be made in the old theory also with regard to this matter. For we do not believe that Borneo has gone through the Celebes-stage, and we cannot accept the theory of the existence of bays deeply encroaching on the land which were supposed to have been filled up in Quaternary times, and thus to have given Borneo its present shape.

We would rather believe that the Schwaner mountains reached much further south at that time, perhaps beyond 1° 30' S. Lat., and that they were higher than now, while on the south side they probably had a steep slope down to the sea. The Java Sea of those times we picture to ourselves as being shallow in the vicinity of Borneo. In Quaternary times the rivers rising in the Schwaner range deposited enormous quantities of gravel sand and mud in that shallow sea, which, in consequence, near the coast had to recede more and more and gradually became dry land. Naturally the Schwaner mountain range suffered from this growth of new land, for it had to provide all the material for it. It was eroded, lowered, and pushed back northward with the result that the Quaternary fluviatile deposits of South Borneo lie in some places where, at the beginning of the Quaternary age the Java Sea still reached, and in other parts they overlie remains of the earlier southern continuation of the Schwaner mountains now worn down and levelled by erosion. And even now the rivers continue to carry enormous quantities of sand and mud from the mountains down to the plains and into the sea, and still the process of land-making by growth of the alluvial materials is continued at the expense of the mountains of Borneo [1]). Although I can well imagine that here and there in Quaternary times the sea periodically encroached upon the land and hence marine sediments may occur locally between the Quaternary fluviatile sediments, these are to my mind only local phenomena and not of general importance [2]). In no case, when travelling in these regions, have I come upon any facts which would support the theory of a negative shifting of the coast-line having occurred in Borneo in Quaternary times.

Thus I cannot accept for West and South Borneo [3]) a nega-

---

1) Interesting examples of this are given by POSEWITZ in the Chapter: "Verlandung Borneo's", POSEWITZ *45*, p. 196.

2) I reckon amongst these the occurrence of marine shells of recent age (at all events shallow water forms, *Ostrea*, *Cardium* etc.) in the gold-bearing gravel near goenoeng Lawak in Tanah-Laut not far from Martapoera. I do not agree with the conclusion drawn from this occurrence that the Quaternary sediments of Borneo „zum Theil bereits als echte Meeresbildungen erkannt worden sind" [1]). Particularly I consider this generalisation too far reaching, as the general geological structure of Tanah-Laut seems to differ considerably from that of the island of Borneo itself. HOOZE [2]), in his exhaustive study of the geology of the Martapoera district, has not made here any deductions concerning the geological structure of other parts of Borneo. It is true, that he accepts, on grounds which I do not consider very plain, a subsidence followed by an upheaval of this district in the Quaternary period.

3) As regards South Borneo this only refers to the part through which I travelled, i. e. the Katingan territory.

---

1) K. MARTIN. Tijdschr. van het Kon. Ned. Aardr. Genootschap 1890, p. 265.

2) J. A. HOOZE *23*.

tive shifting of the coast-line in recent or Quaternary times, although I do not deny the possibility that recent upheavals of land may have occurred locally. On the contrary, I believe that after the movement had stopped which caused the upheaval of a considerable part of Borneo in Tertiary and early Quaternary times, no further raising of the land above the level of the sea has taken place, at least none of importance. The great difficulty in Borneo, which in some parts amounts to an impossibility, is to distinguish between Tertiary, Quaternary and recent formations, and it is therefore all the more necessary sharply to define the differences between my suggestion and the old theory. The principal difference is this: Hitherto it has been thought, that after the formation of the vast sedimentary deposits of Quaternary gold-bearing gravels and sands, an upheaval of land above the surface of the sea has taken place; my opinion is that the important crust movements and the period of upheaval, date from before the formation of these deposits, and that all the Quaternary sediments which I have included in the "old fluviatile deposits" have been formed during a period of stability and still occupy the same level, or nearly so, as they did when first deposited.

CHAPTER XIII.

— - ·· ——

REMARKS ON THE KAPOEWAS AND SOME OTHER RIVERS OF BORNEO.

Atlas, page 20.

In his well-known work, SCHWANER gives short characteristics of the rivers in Borneo. He calls "upper course" the portion of the stream from the sources to the point where it enters the alluvial low lands, and he mentions that this portion of the river is generally characterized by windings and irregularities in its course, by waterfalls and rocks in the river-bed, and particularly by banks and islands built up of boulders. In the middle course the river winds through the low lands and the absence of islands is the striking characteristic of this middle division. The lower course is that portion of the stream in which the influence of the tides is felt. Islands again form a notable feature in this portion of the course. This scheme is somewhat different to the usual one. Generally the upper course of a river is known as that portion of it where erosion and power of transport largely predominate over accumulation and sedimentation; the middle course is that portion where the erosive and sedimentary forces locally alternate, but, generally speaking, balance one another; the lower course is that where the sedimentary process rules alone. The most serviceable characteristics by which to distinguish these three divisions, are the following: In the upper course the fall is variable; there are still obstacles to surmount and the normal decline has not yet been reached.

Hence rapids and waterfalls, boulder-islands and boulder-banks (karangans) are formed. In the middle course the fall is gradual, but still great enough to prevent any accumulations of sediment which might cause the river to leave its old bed in search of a new one. The lower course has also a gradual but very slight fall and great beds of accumulated sediments give occasion to the river to repeatedly change its bed. Islands, lakes as portions of deserted river-beds, and delta-formations, produced as consequences of these shiftings of the river-bed, form the principal characteristics of the lower course. I wish now to trace, how far and with what alterations these general divisions may be applied to the best known rivers of Borneo, and in particular to those which I have personally visited.

In the first place then we note the curious fact that this simple scheme is in no way applicable to the two largest rivers of Borneo, the Kapoewas and the Mahakkam, as will be shown presently. To the rivers which flow from the Schwaner mountains in a southern direction towards the Java Sea, the Katingan, the Kahajan, and the Kapoewas-Moeroeng, the scheme sketched by SCHWANER, probably with reference to these rivers, is indeed applicable. From the high mountains their course leads first through a hilly district in which the fall of the water becomes steadily less, and where the boulders are already deposited in the lower part of the upper course in a long series of boulder-islands and karangans. There is no sharp line of demarcation between the upper and the middle course. At the termination of the upper course, which, for the Katingan, may be fixed at Toembang Samba, the height above the sea-level is already very trifling, not more than 38 metres, and consequently the middle course shows several features in common with the lower course. Thus in the entire middle course of the Katingan, and also in that of the Kahajan and the Kapoewas, we notice pintas (tĕroessans or antassans), and simultaneous with these, danaus, the remnants of deserted river-beds. The numerous great irregular windings also remind one of the lower course. No actual line of demarcation can therefore be drawn between the middle and the lower course of these rivers, and probably for this reason SCHWANER fixes it at the place where the tide counteracting the current becomes perceptible in the river. This counteraction greatly accelerates the deposition of mud, and furthers the formation of islands and deltas. In fact, in these rivers the distribution of islands is very much as SCHWANER describes, in the upper course there are many boulder-islands, in the middle course there are few, if any, in the lower course islands again appear, their number varying considerably in the different rivers. We must keep in mind, however, that, as remarked above, in none of these rivers are the middle and the lower courses clearly defined, but that many features of the lower course may be seen in the middle course. Thus, for instance, in the middle course of the Barito we find several little islands, produced through displacements of the river-bed, and the cutting off of curves.

The Kapoewas and the Mahakkam deserve special description, and I will begin with the Kapoewas which has become fairly familiar to me.

The Kapoewas may be called the great artery of West Borneo, and almost the whole residency is drained by it. There are in West Borneo only two rivers of importance, the Pawan in the south and the Sambas in the north, which do not discharge into the Kapoewas. The total river-basin of the Kapoewas covers an area of no less than ± 102000 square kilometres, more than three times the size of Holland. The area of the river-basin of the Sambas is 8300 square kilometres, that of the Pawan 10450 square kilometres in extent. The Kapoewas flows in a generally western direction, right across West Borneo, and has a total length of 1143 kilometres, about 34 kilometres longer than the Rhine. At a favorable condition of the water it is navigable for small steamers of at most 6 feet draught, up to a little above Poetoes Sibau 910 kilometres distant from its mouth; with boengs one can get as far as the pangkalan Djémoeki 164 kilometres higher still; above that place the Kapoewas must be reckoned among the mountain streams. Of its numerous tributaries, about twenty may be considered to be of some importance, but there is only one, the Mĕlawi, which is of great importance, and claims for itself no less than 22600 square kilometres, i. e., more than a fifth part of the total area of the river-basin of the Kapoewas. The width of the Kapoewas near Poetoes Sibau is 208 metres, and it increases much in its downward course. The greatest width is reached near Tajan, where it is about 1600 metres [1]). Not far below Tajan the river divides into the Mĕndawak or Djĕnoe and the Poengoer, which latter branches off into the Térentang and the Kapoewas Kĕtjil.

The delta-land comprising these branch rivers covers an area of 5400 square kilometres.

The Kapoewas (see longitudinal section on page 20 of the atlas) rises in the extreme north-west corner of the residency of West Borneo from several small mountain streams on the southern slope of the mountain range which there separates its river-basin from that of the Batang Rĕdjang in Sarawak, at an estimated height of 750 metres.

The upper part of the Kapoewas, from the sources as far as N* Gĕrogok (height 460 metres), is designated on the topographical map as Kapoewas Nikal. The fall in the Kapoewas Nikal is 1 : 83. Below N* Gĕrogok the Kapoewas winds, with considerable longitudinal development, right through the ranges of the Upper Kapoewas chain, trending on an average W.S.W.—E.N.E.; and as far as Pangkalan Djĕmoeki the fall is still moderately large (1 : 126). Below this point, and as far as N* Boengan, the river has a south-westerly course, diagonal as compared to the general trend of the mountain chains (which is E.—W. to E.N.E.— W.S.W). Here the river is navigable for boengs, but dangerous rapids and waterfalls, such as the goeroeng Aping, the goeroeng

---

1) The figures here given are partly taken from the topographical map, and partly derived from J. J. K. ENTHOVEN 9, p. 133.

For further particulars on the delta of the Kapoewas we refer to sheets Pontianak, Padang Tikar, Tajan and Soekadana of the topographical map.

Pantak, the goeroeng Matahari and the goeroeng Moenhoet, make navigation uncertain and risky. The fall from pangkalan Djémoeki (230 metres) to N⁰ Boengoen (107 metres) is 1 : 618. This latter place is 125 kilometres distant from the sources of the Kapoewas. Here the mass of water is nearly doubled by the supply from the Boengan. Below N⁰ Boengan the average direction of the course of the Kapoewas conforms more to the general trend of the mountain ranges and, with the exception of the goeroeng Dělapan, there are no further rapids of any con-

Fig. 74. VIEW IN THE LAKE DISTRICT ON DANAU SĚRYANG.

sequence. Until some distance below N⁰ Kěryau the fall of the river gradually diminishes to an average of 1 : 2114. About 19 kilometres below N⁰ Kěryau the mountains which enclose the river become rapidly lower, and before long the Kapoewas enters the great plain, which has repeatedly been mentioned in this work as the Upper Kapoewas plain. The total area of this enormous plain, which at high water is almost entirely inundated, amounts to 6855 square kilometres. The fall now decreases rapidly and soon comes down nearly to zero (see section on sheet 20 of atlas); below Loensa it is already very trifling (1 : 3600) and afterwards becomes imperceptible; between Djongkong and Sěmitau, a distance of 93 kilometres, the difference in height does not amount to more than 1.2 metres or 1 : 77666. All the important tributaries which, between Loensa and Sĕmitau, join the Kapoewas, on entering the Upper Kapoewas plain share in the sudden dwindling of the fall of the principal river, and it is interesting to note the great influence this has on the distribution of the material transported by them.

This sudden reduction in the fall of the stream, and the resulting sudden decrease of transporting power, must necessarily give rise to phenomena which are very much akin to those seen where rapid streams with great transporting power, take their course through a lake, as for instance the Rhine through the lake of Constance. When rivers thus discharge into a lake cones of débris are formed, and all the boulders, gravel, and sand, carried down by the river are deposited over a comparatively small area. Something like this takes place in the river-basin of the Upper Kapoewas. As soon as the Kapoewas and its tributaries, laden with quantities of boulders, gravel, and sand, reach the Kapoewas plain, all this material settles down within a comparatively small distance, and a succession of boulder-islands is formed, producing widenings in the stream. Very soon the components of these boulder-islands decrease in size, and at no great distance from the mountains not a single pebble is to be found in the river-beds, and only sand or mud is deposited, part of which latter remains continually suspended. This is seen with striking effect in the right side tributaries of the Kapoewas which enter the plain as mountain torrents directly from the Upper Kapoewas mountain chain, without any lower hilly ground intervening. A beautiful and typical example of this is furnished by the Embaloeh (see map V and the longitudinal section, sheet 20 of atlas). At Békatan-poort it emerges from the mountain district, and deposits all the coarse solid material it carries in a series of large boulder-islands, such as Poelau Bĕlimis, Poelau Boekang, Poelau Pĕneh, and Poelau Pandjang. This portion of the river-bed may, in a certain sense, be called the débris cone of the mountain stream. On account of this, the fall of the river does not diminish quite

30

suddenly, and it even shows a slight increase shortly after the river has emerged from the mountains and over the area which I have characterised as the débris cone. Thus in the Embaloeh, from Pĕndjawan (58 metres) up-stream as far as Loang-Goeng (79 metres), the average decline is 1 : 905. Between Pĕndjawang, just above the highest boulder-island, and Bĕnoewah Oedjoeng (48 metres), just below the lowest boulder-island, the difference in height in a distance measured along the river, of $8^1/_2$ kilometres is 10 metres which corresponds to about 1 : 800. Below this point, i. e., below the area of the débris cone the fall becomes rapidly less, so that in a distance of 92 kilometres, from Bĕnoewah Oedjoeng to N⁰ Embaloeh, it is only 9 metres, about 1 : 10200. The same thing may be seen in the Sibau and the Mĕndalam, which, after leaving the mountains, deposit almost all their boulders in a series of boulder-islands and karangans following each other in close succession. In the Mĕndalam the deposits extend over a larger area, because this river leaves the mountains diagonally, and flows in a south-westerly direction towards the Kapoewas; the diminution of the fall is therefore not so sudden or so rapid as in the Embaloeh. In the same way the Kapoewas, which leaves the mountains just above N⁰ Era and at high water carries an overwhelming mass of solid material, very soon loses the power of transporting its boulders. Most of them are found in the boulder-islands Lolong, Laap, Masoem, Loensa Ra, Pahi, Loensa, Sioet, Tambai, and Karangan Baoeng, therefore distributed over a distance of 40 kilometres. The temporary increase of the fall in the region of the débris cone, is repeated in the Kapoewas, in the tract from N⁰ Kĕryau to Loensa, with an average fall of 1 : 1448, while in the two adjoining portions of the river, both up-stream and down-stream, the fall is less, respectively 1 : 2114 and 1 : 3600. If a greater number of accurate measurements of the altitudes in this part of the Kapoewas were at our disposal, it would appear still more clearly that the temporary increase of the fall is really connected with the area where the river, after leaving the mountains, deposits its boulders on the edge of the great Kapoewas plain. Below the boulder-island Karangan Baoeng, $^1/_4$ square kilometre in size, there are only a few unimportant karangans, and a few kilometres below Poetoes-Sibau the sandbanks reveal the last traces of gravel. From this latter point down to Sĕmitau it is in vain to look for any pebbles in the bed of the Kapoewas. This striking feature of the absence of all traces of gravel though still in sight of the high mountains from which the rivers rise as mountain streams, and notwithstanding the enormous volumes of water which are transported by them, would have been changed long ago in the Kapoewas if the bed of this river had been in a moderate or cold zone, where besides the water another important transporting agent is at work, viz., floating ice, which carries boulders and gravel and is hardly if at all influenced by the variations in the fall of the water [1]).

For the transport of solid material the rivers of Borneo are now, as probably in times past, entirely dependent upon the carrying power of the water, and of the tree-trunks and drift-wood, which under favorable conditions may accumulate and transport large stones, gravel and sand. Now the transporting power of wood, like that of floating ice, is not dependent upon the fall of the stream. With floating ice however, the total mass of solid material transported is very considerable, and the chances are very great that as the ice melts or crumbles away these solid masses will sink somewhere to the bottom of the river-bed, but it is quite the exception for drift-wood to carry gravel, and where this is the case there is always a fair chance of it being carried right out to sea. In the upper course of the rivers the current is so strong on account of the many rapids, that if in some way or other stones have been caught between the roots or branches of the drifting trees, they will most likely be dropped again

---

1) I will not now consider the case of a glacier which moves down a valley so that the solid ice, without the aid of running water, acts as transporting agent.

soon; but if once they have been carried lower down, there is not then much chance of their being thrown off in the lower course where the floating trees glide majestically and almost unhindered down the wide and quiet stream.

As we have seen, the fall of the river in the Upper Kapoewas plain, is almost nil, and only fine mud is transported. These conditions, generally speaking, belong to the lower course of a river, but in this as in other respects the aspect of the Kapoewas in the Upper Kapoewas plain is exactly the same as in the lower course. With gigantic windings the river pursues its uncertain course through the plain; pintas or pintassans are of frequent occurrence, cutting off great curves of the river and forming islands (see maps IV and VIII²).

It frequently happens that a considerable volume of water collects in a pintassan, and the bend through which the water formerly flowed is closed up by mud and deserted by the stream it becomes a lake; but soon new bends are marked out there or elsewhere, new pintassans are formed, and the stream again alters its course; finally, deserted portions of the course are occupied once more, to be afterwards perhaps more or less altered in their direction, and then vacated shortly after; and thus it goes on, the river changing its course perpetually in endless variations.

These changes explain the existence in this plain (see maps I, II, IV and VIII²) of so many winding lakes arranged in one, or even sometimes in more than one series parallel with the rivers; they are old, disused portions of the river-bed. Another distinct feature of the lower course is also the depth of the stream. Over the whole of the tract Sĕmitau-Boenoet the depth is almost invariable, this portion therefore and the actual lower course below Tajan are the two best tracts for navigation on the whole of the Kapoewas. When at low-water the steamboat service between Tajan and Sintang has long been suspended, the portion Soehaid-Boenoet remains navigable for steamers of small draught because of this uniformity of depth. Only one characteristic of the lower course, the division of the stream-bed with dispersion of water, i. e., delta-formation in the true sense of the word, is not met with here, nor could it in fact exist, because near Sĕmitau the river again leaves the plain through one exit. From the same cause there is no delta-formation or dispersion of waters in the lower course of rivers which, for some reason or other, have but one opening into the sea, as, for instance, the Sambas.

As soon as the Kapoewas leaves the Upper Kapoewas plain near Sĕmitau and breaks through the deeply denuded Sĕmitau hills (see map IV) its character changes altogether. The river becomes slightly narrower and runs through a sharply defined valley, and no more pintassans and shiftings of the bed occur. Its fall moreover increases again; from being 1 : 77666 between Djongkong and Sĕmitau it increases to 1 : 17714 between Sĕmitau and Silat. In short, the river assumes the character of the middle course, and maintains this character until near Tajan, where the middle course gradually passes into the lower course. Generally speaking, the fall remains trifling in this middle course, and may even be called very small for the tract Silat-Sintang. Near and above Sintang, in fact, the large strips of land which are flooded at high water remind one somewhat of the lower course, but the characteristic development of pintassans is wanting.

As regards its fall therefore, the Kapoewas does not show the normal simple scheme of: upper course — middle course — lower course, but rather a repetition or recurrent type, for which we might propose the scheme: upper course — pseudo-lower course — middle course — lower course.

From what has been said in Chapter XII, III, this complicated type of the Kapoewas, is due to its being a true "*Durchbruchs-fluss*", which, flowing through the Sĕmitau hills first southwards and then in a westerly direction, drained what now is the Kapoewas plain at a time that this territory still occupied a higher level than the Sĕmitau hills. Afterwards the Kapoewas plain became relatively lower, and the hilly district of Semitau, with the region south of it, relatively higher, but notwithstanding this, the erosive action of the Kapoewas succeeded in keeping the bed of the river at a level low enough for the Upper Kapoewas plain to be drained by it.

The Mahakkam appears to have many points in common with the Kapoewas, and it also shows the recurrent type. It appears that its entire course might be represented by the scheme: upper course — middle course — pseudo-lower course (lake district) — middle course — lower course.

Fig. 75. VIEW ON THE LOWER COURSE OF A BORNEO RIVER.
THE SAMBAS-RIVER NEAR SAMBAS.

According to this plan we could distinguish in the Mahakkam: for the upper course, the portion of the sources as far as below the great falls above Long Dĕho; for the middle course, the portion of Long Dĕho as far as Mt. Sĕndawar; for the pseudo-lower course, the district of the great lakes to near Moeara-Kaman, for the second middle course, the portion from Moeara-Kaman to near Samarinda, where the river breaks through the coast-mountains, which are supposed to have been built up of Tertiary sediments; and for the actual lower course, the level part below Samarinda, where the river divides repeatedly and forms a typical delta projecting far out into the sea. So far no details can be given concerning the rate of fall of the Mahakkam; and fresh topographical and geological investigations will have to show how far the parallel here drawn between the Kapoewas and the Mahakkam has a "raison d'être", and also whether the resemblance in the course of the two rivers is accidental or whether there is some genetic connection underlying it [1]).

As already remarked, both the Kapoewas and the Mahakkam have a typical delta, which, particularly in the case of the Mahakkam juts far out into the sea; but the reasons why some rivers in Borneo form a delta while others under apparently similar circumstances do not do so, cannot, as yet, be determined with absolute certainty. On the south coast of Borneo there are no true deltas projecting into the sea. The Barito, the Kapoewas Moeroeng, and in a measure even the Kahajan enclose a kind of delta-land, but it does not jut out into the sea, it is merely the result of repeated bifurcations and shiftings of the bed in the lower course of these rivers, more especially is this the case with the Barito.

The conditions on the south coast of Borneo, appear to be unfavourable to the formation of deltas which must probably to some extent be attributed to the strong current which, all the year round, either in an easterly or in a westerly direction, flows parallel with the coast-line; in some measure it may also be due to very feeble positive displacements of the coast-line. Thus we see that the rivers with comparatively feeble transporting power, such as the Djĕlai, the Koemai, and the Sampit, have funnel-shaped, fairly deep and wide estuaries, while the rivers with strong transporting power possess either a few small mud-islands at the estuary and consequently a very feebly developed delta

---

1) Some information about this problem may be expected from the recent researches of NIEUWENHUIS. The name Long Dĕho is taken from his preliminary report on his travel across Borneo from Pontianak to Samarinda.

as in the case of the Katingan, or else, as in the case of the Kahajan, the Kapoewas Moeroeng and the Barito, they have their conjoint, wide, funnel-shaped estuary for the greater part filled up with silt, thus embracing a kind of delta which does not project into the sea.

On the west coast it is quite different. Here we find a well-developed delta both on the Kapoewas and the Batang Redjang, but no delta on the Sambas or on the Batang Loepar. On the east coast the Mahakkam possesses a large typical delta, but at the Bĕrau, Boeloengan, Pĕsajap, and Siboekoe, conditions are present which produce a kind of delta intermediate between that of the Barito and that of the Kapoewas. Summarising, it would appear that the conditions for delta-formation on the west and east coasts are, generally speaking, neither favourable nor un-favourable, but they are of such a nature that it is only possible for rivers of very considerable transporting power to form true deltas, and moreover the existence or non-existence of deltas depends in a great measure upon the structure of the coast area where the river discharges. The theory, sometimes advanced on other grounds, that the whole of Borneo may be passing through a period of negative displacement of the coast line, is not supported by the mode of occurrence of the deltas. The character of the estuaries of the rivers which empty into the Java Sea, points, for that portion of Borneo, rather to a positive than to a negative displacement of the coast line.

CHAPTER XIV.

About Karangans and Pintas.

I. *Karangans.*

Karangan is the name given by the natives of Borneo to any accumulation of material deposited in the bed of a river by running water. The cause of the accumulation is always a local decrease in the velocity of the stream. But this local decrease of velocity may come about in various ways. Hereon we will now give a short summary.

A. The most ordinary form is the karangan found in the convex curve of the river-bank in the upper courses of rivers (fig. 76). It originates by the fact that the velocity of the water is greater on the concave[1]) and smaller on the convex side of the curve, as compared with the average velocity which at a given level governs the transporting power of the water over a given portion of the stream. Such is the karangan κατ᾽ ἐξοχήν. If the accumulation of material is composed of sand, as is the case in the upper part of the middle course of the rivers,

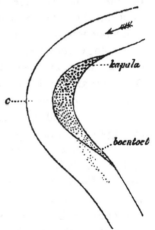

Fig. 76. NORMAL TYPE OF A
KARANGAN IN THE BEND
OF A RIVER.

---

1) In order to avoid misunderstandings we remind our readers that the concave curve of the bank is on the convex side of the river bend, and vice-versa, the convex curve of the bank is on the concave side of the river bend.

one speaks of a *Karangan passir*. If on the other hand it forms an island in the stream it is known as *poelau Karangan* (boulder-island), and if the accumulation consists of drift-wood the name of *Karangan kajoe* is given to it. Mud banks lying in the same manner in the middle- or lower-course are sometimes called *Karangan loempoet*.

On the upper side of the bend, the *kapala Karangan*, the coarsest material is first deposited, towards the middle it becomes finer, and the karangan terminates in a *boentoet Karangan* consisting of fine gravel or sand. This type remains the same everywhere, with only this difference that the higher one gets in the upper course, and consequently the greater the fall of the water, the material is coarser, and there a kapala karangan would consist of very big rounded boulders, while the boentoet would show small boulders or pebbles. In the middle course on the contrary, the karangan passir is found with coarse sand in the kapala and fine sand or mud in the boentoet. In those portions of the river which may be said to form the transition stage between the upper- and the middle course, the kapala karangan contains gravel, the boentoet exclusively sand. A karangan consisting of fine sand or mud is always overgrown with reeds on the inner edge, or the strip where at high water the shore would be. These reed borders are not found in the karangans of the upper course.

If the karangan, as is generally the case, consists of more or less partially flattened pieces of rock or boulders, the stones are usually found overlying each other in tile fashion, in such a manner that every stone dips in the direction of the current, accompanied by a more or less distinct arrangement into rows at right angles to the direction of the current. This feature is very clearly defined in the karangans of the Upper Kapoewas above Poetoes Sibau.

The boentoet karangan always extends a little way beyond the bend of the river, down-stream, and thus forms in the following rantau a distinct, gradually diminishing, shallow appendage. This projection of the boentoet, often causes the formation of a poelau karangan in the upper course of rivers, (fig. 77). For as soon as the water has cut out a second bed *b* by the side of the narrowed channel *a*, it partially cuts off the boentoet from the main body of the karangan.

For navigation with sampans the karangan passir offers no danger whatever, nor even much chance of delay, provided one steers clear of the concave side of the bend. The karangans of the upper course always cause much delay and are sometimes dangerous. Because of the narrowing of the river-bed at normal water-level in the bends, the velocity in the remaining channel *c* (fig. 76) becomes very great. This channel as a rule is not free from boulders; especially at the entrance and at the termination

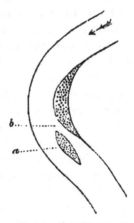

Fig. 77. KARANGAN WITH POELAU KARANGAN CUT OFF.

of the karangan where a boulder- or gravel-bank often extends diagonally across the river-bed to the other side. This causes rapids in the navigable channel, and as a rule the boat has to be poled with great force along the whole length of the karangan, carefully avoiding any rocks in the channel. The actual rapids at the top and at the boentoet are the most awkward places; and generally all the crew have to leave the vessel at these spots so as to help to drag it over the stones. At the the extreme point (bow) of the boeng a long rattan-rope is fastened which is held by some persons on the karangan to prevent the vessel from slipping (shooting) backwards. The riam near the kapala is as a rule more dangerous than the one near the boentoet, because in the former the boulders in and under the water are so much bigger. It often happens that in the narrow channel of the karangan a tree-trunk brought down by the stream, sticks fast, and causes an accumulation of drift-wood which may greatly complicate the difficulties of navigation in the rapids.

Generally, going down-stream, the boats shoot the rapids loose, one man in the front directing the course with a long pole, so as to avoid all big stones and tree-trunks. If the pilot does not know the rapids and the position of the stones therein, there is always a great chance of the boat capsizing. If the rapids are high and the current strong, rattan ropes, fastened both to the bow and to the stern of the vessel, are held by the men who run along side of the boat on the boulder-banks. Thus the risk of accidents is minimized, but the natives do not willingly resort to this method because of the delay and the extra labour which it involves.

The accumulation of transported material is often subjected to other influences than those caused by the winding of the river, which may create other kinds of karangans.

B. Every obstacle in the water may lead to the deposition and accumulation of stones in front and behind it, just as on the sea-shore any object against which the sand is blown may create a miniature sand-dune.

Every rock rising out of the water may become the nucleus of an accumulation which, if close to the shore, easily forms a karangan, if in mid-stream, a poelau karangan. Just as on the shore the sand which flies up against an object forms a sand-talus, sloping gently up to the top of the object, but rapidly descending on the other side of it, so we see here, in the river, a long stretched-out karangan attaching itself up-stream to a rock or rocky island, surrounding the island with a narrow edge of boulders, and rapidly disappearing under the water down-stream (fig. 78). Such rocks or rocky islands are of frequent occurrence in the upper courses of the rivers, and the consequence is that most karangans do not show the normal type, formed under the influence of the bends of the rivers, but rather a modified type.

C. Another cause of a local decrease of velocity in the stream and thus of the formation of deposits, may be found in the influence which the tributaries exercise on the principal stream. Various phenomena

may occur in this case which are of great importance to the pioneer geologist who often has to depend upon the examination of boulder-banks to obtain any knowledge of the geological structure of the basin of the river which has carried down these boulders. The geologist should therefore be able to decide prompt-
ly and accurately by which stream the
boulders found in a karangan have been
transported. This knowledge he can
acquire by a just estimate of the various
phenomena which may occur when
karangans are made at the point where

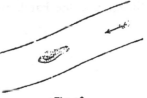

Fig. 78.

one river joins another. The manner of formation of these ka-rangans is governed principally by:

a. the relative proportion of the volume of water carried by the two rivers;

b. the proportion of the quantity of boulders transported by the two rivers, which is a complicated function of the velocity (governed by the volume of water and by the fall) and of the geological structure of the river-basin;

c. the mutual direction of the current in both rivers, at their point of junction.

In connection with these factors the following phenomena may be distinguished:

1. The quantity of gravel carried by the tributary is nil or wholly insignificant as compared with the material carried by the principal river.

a. The volume of water of the tributary is insignificant compared with that of the principal river. Result: no karangans are formed. Example: brooklets from the danaus discharging into the Kapoewas.

b. The volume of water of the tributary is not unimportant. Result: The water of the principal river is forced up a little and pushed to the other side, and the rapidity of the stream, locally, considerably diminished. In the principal river, opposite the mouth of the tributary a boulder-bank or boulder-island is then formed consisting exclusively of material transported by the principal river. The few pebbles carried by the tributary will be immediately transported by the stronger current of the principal river (see fig. 79). Example: rivulets such as the Mitau, the Riang, and the Patah Batang which discharge into the Lĕkawai.

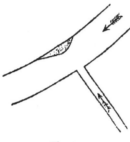

Fig. 79.

2. The volume of water in the tributary is considerable less than that in the principal river, but the fall and the quantity of débris carried by the tributary is very great. Result: At the point of junction, the fall of the water in the tributary abruptly diminishes, and a large proportion of the boulders carried by it, sinks to the bottom. A karangan is formed at and just below the mouth of the tributary, which will consist principally although not exclusively, of material down carried by it (Fig 80, at *a*). Opposite the mouth a small karangan is often formed (Fig. 80, at *b*), consisting exclusively of stones carried down by the principal river, its origin being explained in the manner described above sub 1[b]. This type is found in perfection wherever small but rapid mountain torrents flow into a great river, as for instance where the Mĕdjoewai discharges into the Kapoewas.

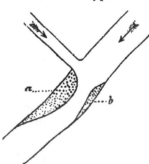

Fig. 80. KARANGANS NEAR
THE MOUTH OF THE Mĕujoewai.

3. The two rivers which meet together do not differ greatly in velocity or in the volume of water or in the quantity of débris transported by them. In such a case the result is invariably the formation of a karangan consisting of mixed material. But the mode of development of these karangans may differ very considerably

If at the point of meeting of the two rivers, the banks

happen to be high and steep, generally accompanied by a narrowing of the river-bed, the karangan is not of much importance (Fig. 81), as, for instance, at the meeting of the Kapoewas
and the Kĕryau, and of the Kapoewas and the Boengan, which
phenomenon is explained by the great velocity of the water at
the point of junction; if, on the other hand, the banks are low and

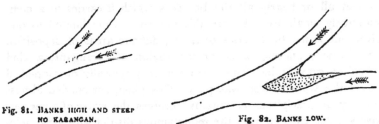

Fig. 81. BANKS HIGH AND STEEP
NO KARANGAN.

Fig. 82. BANKS LOW.

the rivers somewhat broadened, an important karangan is formed
(Fig. 82), as, for instance, at the meeting of the Boengan and
the Poenoe.

If at the point of junction the direction of the current does
not considerably differ in the two
rivers, there is but little chance for
a karangan to be formed (Fig. 81);
if, on the contrary, their currents are
somewhat opposed (Fig. 83) a karangan always occurs. A good example
of the latter is furnished by the Tĕpoewai, where it joins the
Gaäng.

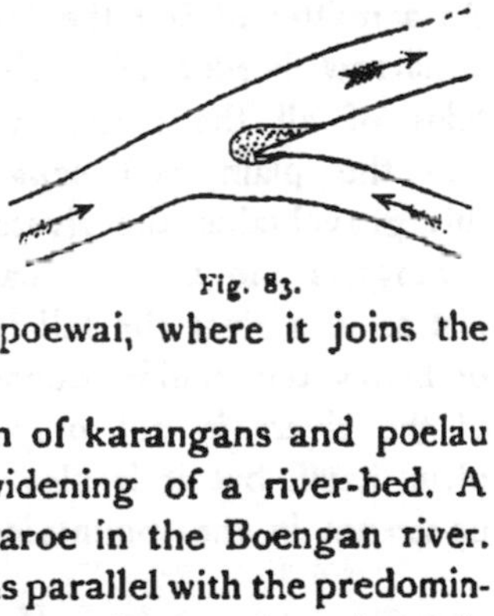

Fig. 83.

D. A fourth cause of the formation of karangans and poelau
karangans is found in the sudden widening of a river-bed. A
beautiful example of this is Poelau Daroe in the Boengan river.
There, for a short distance, this river runs parallel with the predominating trend of the mountains, while lower down, where it again
assumes its character of a transverse stream, great obstacles have
to be overcome. Lateral erosion has been hard at work here,
and created a wide plain, the island Daroe, which, at low water,
is almost entirely covered with boulders, the water of the

river flowing through a few channels between them; at high water the whole of this expanse is flooded and then forms a kind of lake.

Just as in the upper- or middle course of a river any local factor retarding the current creates a karangan, so a more general retardation, spread over a greater distance, may force a river to deposit all or nearly all the boulders which it carries in a comparatively small area. Where this influence is not limited to one river but is felt by a series of rivers, definite areas of deposition are formed, making a connected region of karangans, a kind of united gravel-talus. This feature is very typically developed all round the Upper Kapoewas plain. This plain, surrounded almost entirely by precipitous mountains, receives the water of numerous rivers which emerge from the mountainous district with great velocity, considerable fall, and corresponding by large transporting power. As soon as these rivers reach the Upper Kapoewas plain, their fall becomes reduced to almost nil, so that at the foot of the mountain range all the boulders very quickly settle down. As a matter of fact the Upper Kapoewas plain is hemmed in by a narrow border of boulders and gravel, the common gravel-talus of all the rivers which flow down from the mountains into the plain and separate it from the high land. Across this gravel-talus the rivers wind and widen, and form series of karangans and boulder-islands, until they reach the plain. And just as in a river the fall is greater near a karangan than above or below the boulder-deposit, so here, along this edge, the fall of the rivers is not only much greater than in the Kapoewas plain itself, but it is also greater than in the adjoining portion up-stream in the mountains.

II. *The Pintas.*

Not less important than the karangans are the pintas or pintassans in the rivers of Borneo. A pintas is a water-channel, by which anywhere in the river a bend is cut off and a shorter water-way is formed. The word pintas or pintassan is derived from the Malay verb *pintas*, "to cut off the way". Besides the name of pintas or pintassan which is generally used in West Borneo [1]) the names of antassan and tĕroessan [2]) are also given to these channels; literally translated both these latter words mean "the straight or shortest way".

The pintas are confined to those portions of the rivers where the fall is slight and consequently owing to the very feeble slope of the territory the river's course is undetermined and wavering. It mean-ders over the plain in wide curves which gradually become more and more pronounced, because at the bends on the concave curve the banks are continually eroded and undermined, while at the opposite side deposition of transported material takes place, until one of the shorter water-ways which at times of flood the river invariably chooses, remains also in use when the water is low, thus creating a pintas or shorter permanent communication, which cuts off the curve, (Fig. 84).

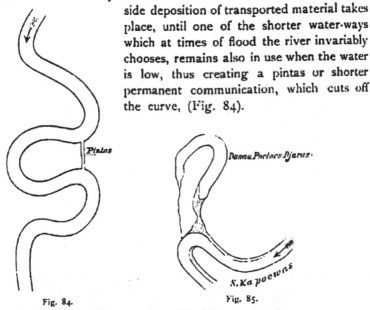

Fig. 84.   Fig. 85.

At first the river flows both through the new way of the pintas,

1) VAN LIJNDEN *31*, p. 545 uses the word *bintas*.

2) J. PIJNAPPEL. *46*, p. 149 distinguishes between antassan and tĕroessan, but I presume this to be incorrect and I find it nowhere confirmed. Both antassan and tĕroessan mean the shortest or straight way and are used indifferently.

and also through the old curve, but after a while the pintas can transport the total volume of the water, and the old curve falls into disuse and is banked up by alluvial deposits on one or both sides. A typical example may be cited from the Upper Kapoewas near Lanau (Fig. 85) where the old river-bed, the cut off curve, is called Danau Poetoes Djaras.

Later on the old deserted curve (bend) may become a curved danau (Fig. 86). But even then the process of increasing the curves continues unremittently, and a new pintas may be formed of which the above danau forms part (Fig. 87). This occurs

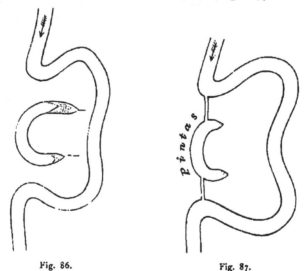

Fig. 86.   Fig. 87.

frequently in the Upper Kapoewas plain; as examples may serve the pintas Sěmoeloeng just below pintas Kaja, and above all the portion of the Kapoewas between Djongkong and Pyasa, a detailed sketch of which has been given in map VIII⁰. The new pintas in its turn may become the principal channel and the first pintas may be entirely or partially deserted and converted into a danau. Thus the process is repeated in endless variety until the effect is obtained which we see for instance in the Upper Kapoe-

was plain. On both sides of the meandering river, various portions of deserted river-beds appear as curiously shaped danaus
which, in many cases, defy one's endeavours to determine from
their positions the original course of the river.

These alterations in the river-beds however are far from rapid
in their progress. In many cases a pintas remains for many
years as merely a narrow, short channel of communication between
two bends in a river, only navigable at high water, without any
perceptible widening of the channel taking place and without
any measurable increase of the amount of water running through
it. We find this condition in the pintas Poetoes with Danau
Poetoes just above Sélimbau, the pintas Niboeng 14 kilometres
up-stream from Sélimbau, the pintas Bawak near Pyasa, and the
pintas Sarai 21 kilometres below Boenoet [1]), all on the Kapoewas,
which are known to have existed for many years. It also sometimes
happens that a pintas formed at one interval of high water is
banked up again entirely at low water, and remains in disuse
for a considerable time or perhaps for ever. An important
factor which hinders the development of the pintas is the drift-
wood. I saw a beautiful example of this on the Upper Kapoewas
in the pintas Kaja, 20 kilometres up-stream from Nª Embaloeh
(Fig. 88).

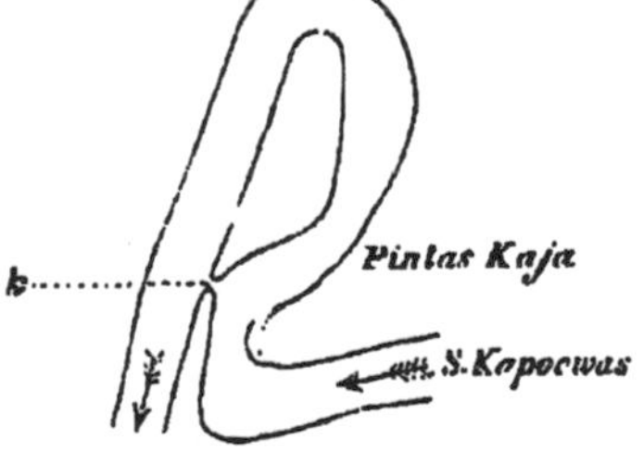

At point k the distance between
the two points of the river is not
more than 20 metres, while the
actual length along the bend is
three kilometres. At k a wide
pintas of 12 metres was formed,
which however at low water was
entirely filled up with tree-trunks,
between which great masses of
sand and mud remained suspended

Fig. 88.

and now constituted a strong obstruction. Now if this is quickly

1) These four pintas are mentioned by VAN LIJNDEN 31, p. 545 as existing in the year 1850.

followed by a very high flood it is possible that all this wood may be set adrift and that the pintas may increase considerably; but if there be a succession of floods of little importance, the chances are that so much sand and mud settle at the spot that the pintas is firmly blocked up again and for a length of time remains impracticable.

For communication with large boats these pintas are of course valueless, but they are most important for small rowing boats. I remember with pleasure the naïve boisterous joy of my rowers at sight of a pintas which proved to be still navigable, while they had expected to find it dry.

THE END.

ALPHABETICAL LIST

OF

LITERATURE QUOTED

1. O. BECCARI. Cenno di un viaggio a Borneo. Bulletino della Società geografica Italiana I, p. 193 en volg. 1868.
2. C. BOCK. Reis in Oost- en Zuid-Borneo van Koetei naar Banjermassin. 's Gravenhage 1881.
3. —— Unter den Kannibelen auf Borneo. Jena 1882.
4. M. BUYS. Twee maanden op Borneo's Westkust. Leiden 1892.
5. M. CHAPER. Notes recueillies au cours d'une exploration dans l'île de Borneo. Bull. de la Soc. géol. de France 3ième Série. Tome XIX, p. 877—882, 1891.
6. W. M. CROCKER. Notes on Sarawak and Northern Borneo. Proc. of the Royal. Geogr. Soc. New Series III, p. 193—205, 1881.
7. J. H. CROOCKEWIT. Verslag van een togt naar den Goenoeng Klam en naar het Peneing-gebergte. Natuurk. Tijdschr. voor Ned. Indië XI, p. 276—294, 1856.
8. P. VAN DIJK. Over de waarde van eenige Nederlandsch-Indische kolensoorten. Nat. Tijdschr. voor Ned. Indië. XV, p. 139—158, 1858.
9. J. J. K. ENTHOVEN. De topographische opneming der Westerafdeeling van Borneo. Album der Natuur 1892, p. 133—146 en p. 165—180.
10. R. EVERWIJN. Voorloopig onderzoek naar kolen in de landschappen Salimbouw, Djonkong en Boenoet, in de residentie Westerafdeeling van Borneo. Nat. Tijdschr. voor Ned. Indië VII, p. 379—395, 1854.
11. —— Overzicht van de mijnbouwkundige onderzoekingen, welke tot nu toe door den dienst van het Mijnwezen in de Westerafdeeling van Borneo werden verricht (met een geologische overzichtskaart). Jaarboek van het Mijnwezen 1879, I, p. 1—125.
— R. FENNEMA, see VERBEEK.
12. K. VON FRITSCH. Die Eocän-formation von Borneo und ihre Versteinerungen. Palaeontographica Suppl. bd. III. Heft 2 en 3, Cassel 1877/78 en Jaarboek van het Mijnwezen 1879, I, p. 127—258.
— H. VON GAFFRON, see PIJNAPPEL.
13. H. B. GEINITZ. Ueber Kreidepetrefacten von West-Borneo. Zeltschr. der deutsch. geol. Ges. XXXV, p. 205, 1883.
14. L. W. C. GERLACH. Reis naar het meergebied van de Kapoeas in Borneo's Westerafdeeling. Bijdr. tot de Taal-, Land- en Volkenkunde van Ned. Indië 4e Reeks, Dl. V, 1881.
J. GROLL, see VAN LIJNDEN.
15. J. HAGEMAN. Iets over den dood van Georg Müller. Tijdschr. voor Indische Taal-, Land- en Volkenkunde III, p. 487—494, 1855.

16. H. HALLIER. Rapport van de botanische tochten in Borneo's Westerafdeeling gedurende de Borneo-expeditie 1893—94. Natuurk. Tijdschr. voor Ned. Indië. LIV, p. 406—449, 1895.

17. —— Ein neues Cypripedium aus Borneo, ibidem p. 450—452.

18. —— Neue und bemerkenswerthe Pflanzen aus dem Malaiisch-Papuanischen Inselmeer. Annales du Jardin Botanique de Buitenzorg. XIII, p. 276—327, 1896.

19. —— Ueber Paphiopedilum amabile und die Hochgebirgsflora des Berges Klamm in West-Borneo. Annales du Jardin Botanique de Buitenzorg. XIV, p. 18—52, 1896.

20. A. R. HEIN. Die bildenden Künste bei den Dajaks auf Borneo. Wien 1890.

21. HENDRICH. Eine Reise nach Katingan. Berichte der Rhein: Missions Gesellschaft 1885, p. 372.

22. L. HORNER. Verslag van een geologisch onderzoek van het zuid-oostelijke gedeelte van Borneo. Verh. van het Batav. Genootschap XVII, p. 104, 1839. This report was handed in to the Netherlands Colonial Government in 1837.

23. J. A. HOOZE. Topogr., geol., miner., en mijnbouwk. beschrijving van een gedeelte der afdeeling Martapoera. Jaarboek van het Mijnwezen 1893.

24. CH. HOSE. A journey up the Baram-River to Mount Dulit and the highlands of Borneo. Geogr. Journal I, p. 193—208, 1893.

25. O. VAN KESSEL. Statistieke aanteekeningen omtrent het stroomgebied der rivier Kapoeas. Indisch Archief I, 2, p. 185—204, 1849.

26. P. G. KRAUSE. Ueber Lias von Borneo. Samml. des geol. Reichsmuseums in Leiden. 1. V, p. 154—168, 1895.

27. —— Ueber tertiäre, cretaceïsche und ältere Ablagerungen aus West-Borneo. Samml. des geol. Reichsmuseums in Leiden. 1. V, p. 169—220, 1897.

28. E. L. M. KÜHN. Schetsen uit Borneo's Westerafdeeling. Bijdr. tot de Taal-, Land- en Volkenkunde van Ned. Indië 6. II, p. 63—88 en p. 214—239, 1896 en III, p. 57—82, 1897.

29. A. LACROIX. Les transformations endomorphiques du magma granitique de la haute Ariège. C. R. de l'Acad. des Sciences. Tome CXXIII, p. 1021—1023, 1896.

30. H. LING ROTH. The natives of Sarawak and British North Borneo. 2 Vol. London 1896.

31. D. W. C. VAN LIJNDEN and J. GROLL. Aanteekeningen over de landen van het stroomgebied der Kapoeas. Natuurk. Tijdschr. voor Ned. Indië II, p. 536—636, 1851.

32. K. MARTIN. Untersuchungen über den Bau von Orbitolina von Borneo. Samml. des geol. Reichsmuseums in Leiden. 1. IV. p. 209—231, 1890.

33. —— Neue Fundpunkte von Tertiärgesteinen im indischen Archipel. Samml. des geol. Reichsmuseums in Leiden. 1. I, p. 131—179, 1883, nebst Anhang: Von Gaffron's Karte von Süd-Borneo ibid. p. 179—193.

34. —— Begeleidende woorden bij een geologische kaart van Borneo, geteekend door VON GAFFRON. Tijdschr. van het Aard. Gen. 1. VII, p. 16—22, 1883.

35. —— Versteinerungen der sogenannten alten Schieferformation von West-Borneo. Samml. des geol. Reichsmuseums in Leiden. 1. IV, p. 198—208, 1890.

36. —— Neues über das Tertiär von Java und die mesozoischen Schichten von West-Borneo. Samml. des geol. Reichsmuseums in Leiden. 1. V, p. 23—51, 1895.

37. —— Uit het jongste geologische verleden der Nederlandsche koloniën in Oost en West. Redevoering uitgesproken op de 321e verjaardag der Universiteit te Leiden op 8 Februari 1896. Leiden 1896.

38. —— Die Fauna der Mëlawigruppe, einer tertiären (eocänen?) Brackwasser Ablagerung aus dem Innern von Borneo. Samml. des geol. Reichsmuseums in Leiden. V. p. 257—315, 1899.

39. W. J. M. MICHIELSEN. Verslag eener reis door de boven-districten der Sampit- en Katinganrivieren in Maart en April 1880. Tijdschr. voor Indische Taal-, Land- en Volkenkunde XXVIII, p. 1—87, 1882.

40. G. A. F. MOLENGRAAFF. De Nederlandsche expeditie naar Centraal-Borneo in 1894. Hand. van het 5e Ned. Nat. en Geneesk. Congres p. 498—506, 1895.

41. —— Die niederländische Expedition nach Central-Borneo in den Jahren 1893 und 1894. Petermann's Mitteil. 1895, p. 201—208.

42. S. MÜLLER. Reizen en onderzoekingen in den Indischen Archipel, gedaan in de jaren 1828 tot 1836. I, p. 129—326, Amsterdam 1857; ook reeds in C. J. TEMMINCK Verh. over de natuurlijke geschiedenis der Ned. overzeesche bezittingen. Land- en Volkenkunde p. 321—346. Leiden 1839—1844.

43. A. W. NIEUWENHUIS. Die Durchquerung Borneo's durch die niederländische Expedition 1896—97. Petermann's Mitteilungen. XLIV, p. 9—13, 1898.

44. IDA PFEIFFER. Meine zweite Weltreise. Wien 1856. I.

45. TH. POSEWITZ. Borneo. Berlin 1889.

46. J. PIJNAPPEL. Beschrijving van het westelijk gedeelte van de Zuid- en Oosterafdeeling van Borneo naar 4 rapporten van VON GAFFRON. Bijdr. tot de Taal-, Land- en Volkenkunde van Ned. Indië. III, p. 143—346, 1860.

47. F. L. RANSOME. The geology of Angel island. Bull. Dep. of Geol. Univ. of California N° 7, p. 193—240, 1894.

48. A. ROTHPLETZ. Die Perm- Trias- und Jura-Formation auf Timor und Rotti im indischen Archipel. Palaeontographica. XXXIX, p. 57—166, 1892.

49. W. SALOMON. Neue Beobachtungen aus den Gebieten der Cima d'Asta und des Monte Adamello. Tscherm. miner. und petr. Mitt. XII, p. 408—415, 1891.

50. C. J. VAN SCHELLE. Verslag over het voorkomen van cinnaber bij de rivier Betoeng, zijtak der rivier Bojan aan de rivier Kapoeas, in de westerafdeeling van Borneo. Jaarb. van het Mijnwezen 1880. II, p. 15—33.

51. —— De geologische en mijnbouwkundige onderzoekingen in de westerafdeeling van Borneo. Jaarb. van het Mijnwezen 1880. II, p. 33—41.

52. —— Opmerkingen omtrent het winnen van delfstoffen in een gedeelte der residentie Westerafdeeling van Borneo. Jaarb. van het Mijnwezen 1881. I, p. 263—288.

53. —— Mededeeling omtrent eenige ingezonden ertsen en mineralen van het stroomgebied der Boven-Kapoeas en Melawirivier in de Westerafdeeling van Borneo. Jaarb. van het Mijnwezen 1883, II, p. 81—84.

54. G. POULETT SCROPE. Volcanoes, London 1872.

55. C. A. L. M. SCHWANER. Borneo. 2 Dl. Amsterdam 1853.

56. TEUSCHER. Dagboek van Teuscher's tweede reis naar de Westerafdeeling van Borneo in 1883. Tijdschr. van Land- en Tuinbouw en boschcultuur in Ned. Oost-Indië. IV, p. 146 en volg. 1888/89.

57. J. E. TEYSMANN. Verslag eener botanische reis naar de westkust van Borneo 1874/75. Natuurk. Tijdschr. voor Ned. Indië XXXV, p. 273—586, 1875.

58. H. TROMP. Zwei Missionsreisen auf Borneo. Berichte der rhein. Missionsgesellschaft 1891, p. 68—79.

59. J. C. E. TROMP. De Rambai en Sebroeang Dajaks. Tijdschr. voor Ind. Taal-, Land- en Volkenkunde. XXV, p. 108—119, 1879.

60. S. W. TROMP. Mededeelingen uit Borneo. Tijdschr. van het Kon. Ned. Aardr. Genootschap 1e Serie. VII, p. 728—763, 1890.

61. R, D. M. VERBEEK. Over het voorkomen van gesteenten der krijtformatie in de residentie Westerafdeeling van Borneo. Versl. en Med. der Kon. Akad. van Wet. 2. XVIII, p. 39—43, 1883.

62. —— Topographische en geologische beschrijving van een gedeelte van Sumatra's Westkust (met geol. atlas). Amsterdam 1883.

63. —— en R. FENNEMA. Geologische beschrijving van Java en Madoera (met geol. atlas) Amsterdam 1896.

64. P. J. VETH. Borneo's Westerafdeeling. 2 Dl. Zalt Bommel 1856.

65. F. VOGEL. Mollusken aus dem Jura von Borneo. Samml. des geol. Reichsmuseums in Leiden. 1. V, p. 127—153, 1896.

66. J. WALTHER. Die Denudation in der Wüste und ihre geologische Bedeutung. Abhandl. der Kgl. sächs. Akad. der Wiss. Math. phys. Classe XVI, p. 345—569, 1891.

67. A. WICHMANN. Die Binnenseeen von Celebes. Petermann's Mitth. 1893, Heft X, XI en XII.

68. —— Der Posso-See in Celebes. Petermann's Mitth. 1896, p. 160—165.

69. N. WING EASTON. Verslag van het Mijnwezen over het derde kwartaal van 1894. Javasche Courant van 25 Januari 1895, N° 7. Batavia 1895.

70. —— Eenige nadere opmerkingen aangaande de geologie van het Tobameer. Jaarboek van het Mijnwezen 1895.

INDEX.

1) A. W. NIEUWENHUIS, Tinea imbricata
(Manson). Archiv. für Dermatologie und Syphilis XLVI, Heft, 2, 1898.